Translations of

MATHEMATICAL MONOGRAPHS

Volume 41

Convolution Equations and Projection Methods for their Solution

I. C. Gohberg
I. A. Fel'dman

American Mathematical Society
Providence, Rhode Island

УРАВНЕНИЯ В СВЕРТКАХ
И ПРОЕКЦИОННЫЕ МЕТОДЫ ИХ РЕШЕНИЯ

И. Ц. ГОХБЕРГ, И. А. ФЕЛЬДМАН

Издательство ,,Наука"
главная редакция
физико-математической литературы
Москва 1971

Translated from the Russian by F. M. Goldware

2000 *Mathematics Subject Classification.* Primary 45-XX; Secondary 47-XX.

Library of Congress Cataloging-in-Publication Data

Gokhberg, Izrail' T͡Sudikovich.
Convolution equations and projection methods for their solution.

(Translations of mathematical monographs, v. 41)
Translation of Uravnenii͡a v svertkakh i proekt͡sionnye metody ikh reshenii͡a.
Bibliography: p.
1. Integral equations. 2. Linear operators.
I. Fel'dman, Izrail' Aronovich, joint author.
II. Title. III. Series.

QA431.G58713 515'.45 73-22275
ISBN 0-8218-1591-1 (hard cover)
ISBN 0-8218-3977-2 (soft cover)

∞ The paper used in this book is acid-free and falls within the guidelines
established to ensure permanence and durability.
Visit the AMS home page at http://www.ams.org/

10 09 08 07 06 05 04 03 02   10 09 08 07 06 05

Table of Contents

Preface to the English Translation

The present edition differs significantly in one part from the Russian one. We have in mind the end of the third chapter, where methods of inverting finite Toeplitz matrices and their continuous analogues are set forth. The last two sections of Chapter III (§§ 6 and 7) of the Russian edition are replaced by three new sections (§§ 6,7 and 8). The new presentation is more complete and is also distinguished by greater generality and simplicity. In addition, addenda are inserted in "Notes and guide to the literature" and in the bibliography. These addenda reflect publications appearing after the publication of the Russian edition.

We express thanks to the American Mathematical Society, to its Editor of Translations, Dr. S. H. Gould, and to the translator for translating the book into the English language.

We are sincerely grateful to Professor E. Hewitt for attention to the translation of our book.

The Authors

Kishinev
June 24, 1972

Dedicated to
Mark Grigor'evič
KREĬN

PREFACE

The first profound results concerning integral equations on a semiaxis, with kernels depending on the difference of the arguments, were obtained in 1931 by N. Wiener and E. Hopf. After the appearance of their classical paper [1] such equations were given the name Wiener-Hopf equations.

By the present time the theory of Wiener-Hopf equations has been quite fully developed. A great many mathematicians (mainly Soviet) participated in its elaboration. We mention only M. G. Kreĭn's fundamental paper [4]. Beginning with that paper the ideas and methods of functional analysis have had extensive application in the analysis of Wiener-Hopf equations.

The presentation adopted in this book is also based on functional analysis. It has its origins in a distinctive operational calculus.

This route leads naturally to a definite class of convolution equations which contains in particular the original Wiener-Hopf integral equations. Besides that, discrete and difference analogues, the so-called pair equations, singular integral equations on the circle, etc. belong here.

Various projection methods of solving convolution equations make up a significant part of the book. Justification of these methods is also obtained within the framework of the general scheme.

To a considerable extent the presentation is based on papers of the authors which were published in 1963–1967.

The book was conceived in 1964. At that time the authors began reading special courses at Kishinev University. Preliminary publication of the mimeographed brochure "Projection Methods of Solving Wiener-Hopf Equations" (Kishinev, 1967; MR **37** # 1915) helped us substantially in the work on the book.

We assume that the reader is acquainted with the elements of the theory of operators in Hilbert and Banach spaces and the theory of Banach algebra.

The authors sincerely thank M. G. Kreĭn for support and abiding interest in this paper, as well as N. Ja. Krupnik, A. S. Markus, A. A. Semencul and I. B. Simonenko for discussion of various questions and valuable comments.

The authors express sincere gratitude to the book's editor, F. V. Širokov. His assistance contributed significantly to the simplicity and clarity of the presentation.

Kishinev
February 18, 1970

INTRODUCTION

1. Various classes of convolution equations are considered in this book. Among these are integral and discrete Wiener-Hopf equations, integral-difference equations, equations in finite differences, various pair equations, singular integral equations, etc.

The discrete Wiener-Hopf equation

$$\sum_{k=1}^{\infty} a_{j-k}\xi_k = \eta_j \qquad (j = 1, 2, \cdots), \tag{0.1}$$

where $\xi = \{\xi_j\}_1^\infty$ is the vector sought, and $\eta = \{\eta_j\}_1^\infty$ and $\{a_j\}_{-\infty}^{\infty}$ are given vectors, is an elementary representive of such equations.

We shall consider equation (0.1) in the space l_p $(1 \leqq p < \infty)$. For the sake of simplicity let us assume that the condition

$$\sum_{-\infty}^{\infty} |a_j| < \infty \tag{0.2}$$

is fulfilled.

Let us denote by A the operator generated in the space l_p by the matrix $\|a_{j-k}\|_1^\infty$, It is easy to see that A expands into the series

$$A = \sum_{-\infty}^{\infty} a_j V_j,$$

which is convergent in the operator norm. Here $V_0 = I$, and the V_j $(j = \pm 1, \cdots)$ are shift operators defined by the equalities

$$V_j\{\xi_1, \xi_2, \cdots\} = \{0, \cdots, 0, \xi_1, \xi_2, \cdots\}, \qquad j = 1, 2, \cdots,$$

and

$$V_{-j}\{\xi_1, \xi_2, \cdots\} = \{\xi_{j+1}, \xi_{j+2}, \cdots\}, \quad j = 1, 2, \cdots.$$

The operator $V = V_1$ is isometric, while the operator $V^{-1} = V_{-1}$ is a left inverse of V. Obviously $V_j = V^j\,(j = 2, 3, \cdots)$ and $V_k = (V^{-1})^{|k|}$ $(k = -2, -3, \cdots)$.

The operator A can be regarded as the value of the function

$$a(\zeta) = \sum_{-\infty}^{\infty} a_j \zeta^j \qquad (|\zeta| = 1)$$

of the operator V, i. e. $A = a(V)$.

This definition differs in some sense from the one usually adopted in functional analysis. The fact is that a fundamental principle of constructing functions of operators is *violated* here: the spectrum of the operator V coincides with the unit disk at the same time that the function $a(\zeta)$ is defined only on the unit circle. As a result, the multiplicativity of the correspondence $a(\zeta) \to a(V)$ is violated, i. e. in general the equality

$$a_1(V)a_2(V) = a(V)$$

does not hold, where $a_1(\zeta)$ and $a_2(\zeta)$ ($|\zeta| = 1$) are functions expanding into absolutely convergent Fourier series, and $a(\zeta) = a_1(\zeta)a_2(\zeta)$. This is immediately evident in the example

$$a_1(\zeta) = \zeta, \qquad a_2(\zeta) = \zeta^{-1}, \qquad a(\zeta) = 1,$$

because $VV^{-1} \neq I$. However, if the function $a_0(\zeta)$ expands into an absolutely convergent Fourier series, if $a_+(\zeta)$ ($a_-(\zeta)$) expands into an absolutely convergent Fourier series in nonnegative (nonpositive) powers of ζ, and if $b(\zeta) = a_-(\zeta)a_0(\zeta)a_+(\zeta)$, then

$$b(V) = a_-(V)a_0(V)a_+(V). \tag{0.3}$$

(The order of multiplying is vital!) Relation (0.3), the *partial multiplicativity* of the correspondence, permits a distinctive operational calculus to be constructed not only for this particular operator V but also for the unilaterally invertible operators of quite a broad class.

The question of the invertibility of functions of unilaterally invertible operators of the form

$$a(V) = \sum_{-\infty}^{\infty} a_j V^j$$

is central here.

Let V be an *arbitrary* isometric operator operating in the Banach space $\mathfrak{B}$, and let it have a left inverse V^{-1} such that $|V^{-1}| = 1$. Then the following theorem is valid.

THEOREM 0.1. *Let the function $a(\zeta)$ ($|\zeta| = 1$) expand into an absolutely convergent Fourier series. For the operator $A = a(V)$ to be invertible, even if only on one side, it is necessary and sufficient that the function $a(\zeta)$ vanish nowhere on the unit circle ($a(\zeta) \neq 0$, $|\zeta| = 1$).*

If this condition is fulfilled, the operator A is invertible on the left, on the right, or on both sides, depending on whether the number

$$\kappa = \text{ind}\, a(\zeta) = (1/2\pi)\,[\arg a(e^{i\varphi})]_{\varphi=0}^{2\pi}$$

is positive, negative, or equal to zero.[1)]

[1)] The symbol $[\ \ ]_0^{2\pi}$ denotes the increment of the function on the segment $[0, 2\pi]$.

Let us explain how the appropriate inverse operator of $a(V)$ whose existence is guaranteed by Theorem 0.1 is *constructed.*

If the function $a(\zeta)$ does not vanish on the circle $|\zeta| = 1$, then it admits *factorization,* i. e. is representable in the form

$$a(\zeta) = a_-(\zeta)\zeta^\kappa a_+(\zeta) \qquad (|\zeta| = 1), \tag{0.4}$$

where the functions $a_-(\zeta)$ and $1/a_-(\zeta)$ expand into absolutely convergent series in nonpositive powers of ζ; analogously $a_+(\zeta)$ and $1/a_+(\zeta)$ expand into series in nonnegative powers of ζ.

Using the partial multiplicativity (0.3), we obtain

$$a(V) = a_-(V)V^\kappa a_+(V).$$

Then the operator $[a(V)]^{-1}$, which is inverse to $a(V)$ on the appropriate side, is given by the equality

$$[a(V)]^{-1} = a_+^{-1}(V)V^{-\kappa}a_-^{-1}(V).$$

(Let us recall that V^{-1} is a left inverse of V.)

This is an existence theorem, and this inversion method is pivotal to the entire general theory.

2. Making practical use of the above described method of inverting the operator $a(V)$ is sometimes difficult. Therefore various projection methods for solving convolution equations play an important part in the book. We examine in particular the reduction method, the Galerkin method and some generalizations of the latter.

The most prevalent among the projection methods of solving the *general* infinite system of linear equations

$$\sum_{k=-\infty}^{\infty} a_{jk}\xi_k = \eta_j \qquad (j = 0, \pm 1, \cdots) \tag{0.5}$$

is the *reduction method.* Here an approximate solution is sought in the form

$$\xi^{(n)} = \{\cdots, 0, \xi_{-n}^{(n)}, \cdots, \xi_n^{(n)}, 0, \cdots\},$$

where the vector $\{\xi_j^{(n)}\}_{j=-n}^{n}$ is the solution of the finite *truncated* system

$$\sum_{k=-n}^{n} a_{jk}\xi_k = \eta_j \qquad (j = 0, \pm 1, \cdots, \pm n). \tag{0.6}$$

Sometimes a convenient procedure leads to the objective, i. e. system (0.6), beginning with some n, turns out to be uniquely solvable, and the vectors $\xi^{(n)}$ converge to a solution of system (0.5) in the norm of the appropriate space.

This is automatically true if system (0.5), when considered in the Hilbert[2)]

---

2) $\tilde{l}_2$ is the Hilbert space of sequences $\xi = \{\xi_j\}_{-\infty}^{\infty}$ of complex numbers indexed from $-\infty$ to ∞. The symbol $\tilde{l}_p$ is used analogously.

space $\tilde{l}_2$, is solvable for any right side in $\tilde{l}^2$ and if

$$\sum_{j,k=-\infty}^{\infty} |a_{jk} - \delta_{jk}|^2 < \infty.$$

This and other well-known cases of applying the reduction method in $\tilde{l}_2$ are embraced on the whole by the following. The operator A defined by system (0.5) must be bilaterally invertible and representable as a sum of positive definite and completely continuous operators.[3]

Unfortunately this classical method is inapplicable even to the simple system

$$\xi_{j+1} = \eta_j \qquad (j = 0, \pm 1, \cdots), \tag{0.7}$$

because here the matrix of the truncated system (0.6) is degenerate for any n. At the same time the operator A is unitary here. A similar situation also arises in the case of the more general equation

$$\sum_{k=-\infty}^{\infty} a_{j-k}\xi_k = \eta_j \qquad (j = 0, \pm 1, \cdots), \tag{0.8}$$

where $\sum_{-\infty}^{\infty} |a_j| < \infty$.

If the corresponding function

$$a(\zeta) = \sum_{-\infty}^{\infty} a_j \zeta^j \tag{0.9}$$

does not vanish on the unit circle, then by virtue of Wiener's well-known theorem on division by an absolutely convergent Fourier series the operator $A = \|a_{j-k}\|_{j,k=-\infty}^{\infty}$ will be invertible in the space $\tilde{l}_p$ $(p \geqq 1)$, and a solution of equation (0.8) will be obtained by the formula

$$\xi_j = \sum_{k=-\infty}^{\infty} b_{j-k}\eta_k \qquad (j = 0, \pm 1, \cdots),$$

where the b_j are the Fourier coefficients of the function $1/a(\zeta)$.

The question of the applicability of the classical reduction method to equation (0.8) is completely solved by the following proposition.

THEOREM 0.2. *Let the function $a(\zeta)$ expand into an absolutely convergent Fourier series. For the truncated equation*

$$\sum_{k=-n}^{n} a_{j-k}\xi_k = \eta_j \qquad (j = 0, \pm 1, \cdots, \pm n)$$

to have, beginning with some n, the unique solution $\{\xi_j^{(n)}\}_{j=-n}^{n}$ and for the vectors

[3] Under these conditions the operator A admits reduction relatively to *any* orthonormal basis (cf. 1°, §2, and Theorem 3.1, Chapter II).

$$\xi^{(n)} = \{\cdots, 0, \xi^{(n)}_{-n}, \cdots, \xi^{(n)}_n, 0, \cdots\}$$

to converge (*as* $n \to \infty$) *in the norm of the space* $\bar{l}_p$ *to a solution of the full equation* (0.8) *for any right side* $\{\eta_j\}^{\infty}_{-\infty} \in \bar{l}_p$, *it is necessary and sufficient that*

a) $a(\zeta) \neq 0$ $(|\zeta| = 1)$,

b) $\kappa = \operatorname{ind} a(\zeta) = 0$.

Condition a) in Theorem 0.2 is natural; it is equivalent to the invertibility of operator A. If *only* condition a) is fulfilled, it turns out that after shifting the "center" of equation (0.8) by the *index* κ, i.e. after writing it in the form

$$\sum_{k=-\infty}^{\infty} a_{j-k+\kappa}\xi_k = \eta_{j+\kappa} \qquad (j = 0, \pm 1, \cdots),$$

we obtain the equation to which we now apply the reduction method. The equation, truncated relative to the new "center"

$$\sum_{k=-n}^{n} a_{j-k+\kappa}\xi_k = \eta_{j+\kappa} \qquad (j = 0, \pm 1, \cdots, \pm n),$$

will have, beginning with some n, a unique solution, and the vectors $\xi^{(n)}$ will converge to a solution of the *full* equation (0.8).

Thus considerations involving applicability of the reduction method distinguish the indicial diagonal $j - k = \kappa$ in the matrix $\|a_{j-k}\|^{\infty}_{j,k=-\infty}$; for the method to be applicable, exactly this diagonal must become principal in the matrices of the truncated systems.

An analogue of Theorem 0.2 holds for the discrete Wiener-Hopf equation

$$\sum_{k=1}^{\infty} a_{j-k}\xi_k = \eta_j \qquad (j = 1, 2, \cdots), \tag{0.10}$$

considered in the space l_p $(p \geqq 1)$.

Now, in contrast with the foregoing, conditions a) and b) together serve as the condition for the invertibility of the corresponding operator A. It is amusing that although equation (0.8) is essentially simpler than equation (0.10), proof of the convergence of the reduction method for (0.8) is derived from the corresponding theorem for equation (0.10).

An analogous situation arises in the continuous case. For example, let us cite the theorem for the Wiener-Hopf equation

$$\varphi(t) - \int_0^{\infty} k(t-s)\varphi(s)\, ds = f(t) \qquad (0 < t < \infty). \tag{0.11}$$

Let us assume that the kernel $k(t)$ belongs to $L_1(-\infty, \infty)$. Here the analogue of the function $a(\zeta)$ is the Fourier transform

$$K(\lambda) = \int_{-\infty}^{\infty} k(t)e^{i\lambda t}\, dt.$$

THEOREM 0.3 *For the truncated equation*

$$\varphi(t) - \int_0^\tau k(t-s)\varphi(s)\,ds = f(t) \qquad (0<t<\tau)$$

to have, beginning with some τ, *the unique solution* $\varphi_\tau(t) \in L_p(0, \tau)$, *and for the functions*

$$\tilde{\varphi}_\tau(t) = \begin{cases} \varphi_\tau(t), & 0<t\leqq\tau, \\ 0, & \tau<t<\infty, \end{cases}$$

which are continued as zero, to converge, as $\tau \to \infty$, *in the norm of* $L_p(0, \infty)$ *to a solution of equation* (0.11) *for any right side* $f(t) \in L_p(0, \infty)$, *it is necessary and sufficient that*

a) $1 - K(\lambda) \neq 0 \; (-\infty < \lambda < \infty)$,

b) $\kappa = (1/2\pi)[\arg(1 - K(\lambda))]_{\lambda=-\infty}^{\infty} = 0$.

All of these theorems on the reduction method are comprised in the general propositions concerning the applicability of projection methods to functions of unilaterally invertible operators.

Besides the case of the continuous function $a(\zeta)$ we consider the case of *discontinuous* $a(\zeta)$. For the time being let us confine ourselves to formulating just a few corollaries of the general theorem, passing in the process from the circle to the real axis.

Let A be a bounded linear operator of multiplication by the continuous function $\omega(t)$, let it operate in $L_2(0, 1)$ by the rule

$$(A\varphi)(t) = \omega(t)\varphi(t),$$

and let $\|a_{jk}\|_{-\infty}^{\infty}$ be the matrix of this operator in the orthogonal basis

$$\varphi_0(t) = 1, \quad \varphi_j(t) = \sin 2\pi jt, \quad \varphi_{-j}(t) = \cos 2\pi jt \qquad (j = 1, 2, \cdots).$$

THEOREM 0.4. *For the reduction method to be applicable to the system of equations with matrix* $\|a_{jk}\|_{-\infty}^{\infty}$ *it is necessary and sufficient that*

1) $\omega(t) \neq 0 \; (0 \leqq t \leqq 1)$,

2) $|[\arg \omega(t)]_{t=0}^{t=1}| < \pi$.

The reduction method described for the equation

$$\omega(t)\varphi(t) = f(t)$$

appearing in the theorem is actually the Galerkin method. Thus the theorem just formulated gives the necessary and sufficient condition for the applicability of the Galerkin method to this elementary equation on a basis of the simplest orthogonal system. It is curious that the "necessary" condition 2) appears even in this, one would think, "trivial" case.

With this we complete our consideration of the "scalar" case.

The next step is considering the "matricial" case, i. e. the case when the a_j ($j = 0, \pm 1, \dots$) are matrices and $k(t)$ is an nth order matrix-function. For such equations the convergence of one or another approximating method is determined by the *left* and *right indices* of the respective matrix-functions. It appears that now the vanishing of all left and right indices is necessary and sufficient for the *convergence* of the reduction method, whereas only the vanishing of the right indices is necessary and sufficient for the *invertibility* of the corresponding operator.

3. The reader already has some idea of the book's content. Let us pass to a review of it by chapters.

The theory of elementary classes of convolution equations is set forth in the first chapter.

The third chapter is devoted to justifying the reduction method for discrete Wiener-Hopf equations, the Galerkin method for integral Wiener-Hopf equations, etc. The essential role here, as in the subsequent chapters, is played by the general theorems concerning the projection methods for solving linear equations which are set forth in the second chapter.

The applicability of one or another projection method is determined by the index of the corresponding function. An iterative method for calculating the index of a rational function is set forth in § 5 of the third chapter. In the same chapter, in review order, theorems are cited and questions formulated concerning projection methods for solving Wiener-Hopf equations on a "multi-dimensional" half-space. In addition a method for inverting a finite Toeplitz matrix and its continuous analogue is set forth in the same chapter.

In the fourth chapter several results of the first and third chapters are generalized to the case of piecewise continuous functions of isometric operators in Hilbert space.

In the following two chapters the theory and projection methods are set forth for various classes of pair convolution equations. In particular, singular integral equations on the circle are put here.

The seventh chapter is devoted to integral-difference Wiener-Hopf equations.

The eighth contains generalizations to the matricial case of several results of the preceding chapters.

Asymptotic methods of solving homogeneous convolution equations are set forth in the Appendix.

CHAPTER I

GENERAL THEOREMS CONCERNING WIENER-HOPF EQUATIONS

The basic aspects of Wiener-Hopf integral equations and their discrete analogues are set forth in this chapter. A certain operational calculus is taken as the basis of the theory of such equations. Theorems concerning various classes of Wiener-Hopf equations are obtained as corollaries of abstract propositions concerning functions of unilaterally invertible operators.

§ 1. Polynomials in unilaterally invertible operators

1. *Auxiliary propositions.* Let $\mathfrak{B}$ be a Banach space, $\mathfrak{L}_1$ and $\mathfrak{L}_2$ two of its subspaces intersecting only at zero. The set of all vectors of the form $x_1 + x_2$, where x_1 and x_2 respectively range over the subspaces $\mathfrak{L}_1$ and $\mathfrak{L}_2$, is called the *direct* sum $\mathfrak{L}_1 \dotplus \mathfrak{L}_2$ of the subspaces $\mathfrak{L}_1$ and $\mathfrak{L}_2$. The direct sum $\mathfrak{L}_1 \dotplus \mathfrak{L}_2$ is obviously a linear set. However it is not in general a subspace (i. e. it may be nonclosed). An appropriate example is cited below.

Let $\mathfrak{L}$ be a subspace of the Banach space $\mathfrak{B}$. A subspace $\mathfrak{N}$ is called a *direct* complement of $\mathfrak{L}$ in $\mathfrak{B}$ if the direct sum of the subspaces $\mathfrak{L}$ and $\mathfrak{N}$ equals the whole space $\mathfrak{B}$. A direct complement, of course, is not unique.

If $\mathfrak{B} = \mathfrak{H}$ is Hilbert space, each of its subspaces $\mathfrak{L}$ has a direct complement. *For example, $\mathfrak{L}$'s orthogonal complement is one.* In general not every subspace has a direct complement in a Banach space.

We refer the reader to K. Hoffman's book **[1]** (Russian p. 220) for details relating to the following example. Let C be the Banach space of all continuous functions on the unit circle, and let A be the subspace of C which consists of all functions holomorphic in the open disk $|\zeta| < 1$ and continuous in the closed disk $|\zeta| \leqq 1$. *The subspace A does not have a direct complement in the space C.* This example permits in turn an example of a nonclosed direct sum to be given. Let A_- be the subspace of C which consists of all functions holomorphic in the region $|\zeta| > 1$ which are continuous on the set $|\zeta| \geqq 1$ and vanish at infinity. The intersection $A \cap A_-$ obviously consists only of zero. From the example cited it follows that the direct sum $A \dotplus A_-$ does not coincide with the whole space C. Since this direct sum is obviously dense in C because it contains all polynomials in positive and negative powers, it is not closed.

If $\mathfrak{L}$ is a subspace of the space $\mathfrak{B}$, the dimension of the quotient space $\mathfrak{B}/\mathfrak{L}$ (for the definition, see, for example, Kantorovič and Akilov **[1]**, Chapter XII, § 1.3) is called the *codimension* of the subspace $\mathfrak{L}$.

If the dimension or codimension of the subspace $\mathfrak{L}$ *is finite*, $\mathfrak{L}$ *has a direct complement in* $\mathfrak{B}$. As a matter of fact, this direct complement can be constructed in the first case as follows. Let $x_1,\cdots, x_n$ be some basis of the subspace $\mathfrak{L}$ and $f_1,\cdots,f_n$ a biorthogonal system of functionals of the dual space $\mathfrak{B}^*$ (i. e. $f_j(x_k) = \delta_{jk}$, $j, k = 1,\cdots, n$). Let us denote by $\mathfrak{N}$ the set of all common zeros of the functionals f_j. It is easy to see that the subspace $\mathfrak{N}$ is a direct complement of $\mathfrak{L}$.

Now let the codimension of the subspace $\mathfrak{L}$ be finite, and let $\xi_1,\cdots, \xi_n$ be a basis of the quotient space $\mathfrak{B}/\mathfrak{L}$. Let us choose an arbitrary vector x_k from each class ξ_k. It is easy to see that the subspace $\mathfrak{N}$ with basis $x_1,\cdots, x_n$ is a direct complement of $\mathfrak{L}$.

Let $\mathfrak{B}_1$ and $\mathfrak{B}_2$ be two Banach spaces. Let us denote by $\mathfrak{A}(\mathfrak{B}_1, \mathfrak{B}_2)$ the space of all bounded linear operators acting from $\mathfrak{B}_1$ into $\mathfrak{B}_2$. The set $\mathfrak{A}(\mathfrak{B}, \mathfrak{B})$ ($= \mathfrak{A}(\mathfrak{B})$) is a Banach algebra in the uniform topology.

With each operator $A \in \mathfrak{A}(\mathfrak{B}_1, \mathfrak{B}_2)$ are associated two linear sets: Ker A ($\subset \mathfrak{B}_1$), the *kernel* of A, i. e. the set of all solutions of the homogeneous equation $Ax = 0$, and Im A ($\subset \mathfrak{B}_2$) the *image* of operator A, i. e. the set of values of A. Along with the symbol Im A we shall also use the notation $A\mathfrak{B}_1$. Let us denote by $A\mathfrak{L}$ the image of A on the subset $\mathfrak{L} \subset \mathfrak{B}_1$.

We shall call the operator $A \in \mathfrak{A}(\mathfrak{B}_1,\mathfrak{B}_2)$ *invertible* if an operator $A^{-1} \in \mathfrak{A}(\mathfrak{B}_2,\mathfrak{B}_1)$ exists such that $A^{-1}Ax = x$ ($x \in \mathfrak{B}_1$) and $AA^{-1}x = x$ ($x \in \mathfrak{B}_2$).

By virtue of a well-known theorem of Banach, the operator $A \in \mathfrak{A}(\mathfrak{B}_1, \mathfrak{B}_2)$ is invertible if and only if Ker $A = \{0\}$ and Im $A = \mathfrak{B}_2$. It is easy to see that for any operator $A \in \mathfrak{A}(\mathfrak{B}_1, \mathfrak{B}_2)$ the set Ker A is *closed* and consequently is a *subspace*. The set Im A can be *nonclosed*. For example, *if A is a completely continuous infinite-dimensional operator*, Im A *is nonclosed*. Indeed, A in a natural way induces a completely continuous, one-to-one mapping $\tilde{A}$ from the quotient space $\mathfrak{B}_1/\text{Ker } A$ onto Im A. If Im A is assumed closed, then by virtue of Banach's theorem $\tilde{A}$ is invertible. This contradicts the fact that $\tilde{A}$ is a completely continuous infinite-dimensional operator.

An operator $P \in \mathfrak{A}(\mathfrak{B})$ is called a *projection* if $P^2 = P$. Together with P the operator $I-P$ is also a projection, where I is the identity operator in $\mathfrak{B}$. The set of values of the projection P is closed, since Im P = Ker $(I - P)$. If P is a projection, the space $\mathfrak{B}$ decomposes into the direct sum of the subspaces Im P and Ker P. Indeed, the equality $x = Px + (I-P)x$ gives the required decomposition of any vector $x \in \mathfrak{B}$.

Conversely, let $\mathfrak{B}$ decompose into the direct sum of the subspaces $\mathfrak{L}$ and $\mathfrak{N}$. Then every vector $x \in \mathfrak{B}$ is uniquely representable as $x = x_1 + x_2$ ($x_1 \in \mathfrak{L}$, $x_2 \in \mathfrak{N}$). It is easy to see that an operator P defined by the equality $Px - x_1$ is a projection in $\mathfrak{B}$, where Im $P = \mathfrak{L}$ and Ker $P = \mathfrak{N}$. In this case we say that P *projects* $\mathfrak{B}$ *onto* $\mathfrak{L}$ *parallel to* $\mathfrak{N}$.

2. *Unilaterally invertible operators.* Again let $\mathfrak{A} = \mathfrak{A}(\mathfrak{B})$ be the algebra of all bounded linear operators acting in the Banach space $\mathfrak{B}$. An operator $A \in \mathfrak{A}$ is called *invertible on the left* (*on the right*) if an operator $A^{(-1)} \in \mathfrak{A}$ exists such that

$$A^{(-1)}A = I \qquad (AA^{(-1)} = I),$$

where I is the identity operator in $\mathfrak{B}$. The operator $A^{(-1)}$ is called a *left* (*right*) *inverse* of A.

If the set of values Im X of an operator $X \in \mathfrak{A}$ is a subspace having a direct complement in $\mathfrak{B}$, any of these direct complements is denoted by Coker X.

1°. *For the operator $A \in \mathfrak{A}$ to be left invertible it is necessary and sufficient that:*

1) *the equation $A\varphi = 0$ have a unique zero solution,*

2) *the set of values* Im A *of A be a subspace having a direct complement in $\mathfrak{B}$.*

If these conditions are fulfilled, then the general form of an operator $A^{(-1)}$ inverse to A on the left is given by the equality

$$A^{(-1)} = A_0^{-1} P, \tag{1.1}$$

where A_0^{-1} is an operator inverse to A if the latter is regarded as an operator from $\mathfrak{B}$ onto Im A *and P ($\in \mathfrak{A}$) is an arbitrary projection onto* Im A. *In this connection*[1)]

$$\dim \operatorname{Ker} A^{(-1)} = \dim \operatorname{Coker} A. \tag{1.2}$$

Indeed, let conditions 1) and 2) be fulfilled, and let P ($\in \mathfrak{A}$) be an arbitrary projection onto Im A. Equality (1.1) defines an operator $A^{(-1)} \in \mathfrak{A}$ which is a left inverse of A. If $A^{(-1)}$ is any left inverse of A, it is representable in the form (1.1), where the projection P is defined by the equality $P = AA^{(-1)}$.

The sufficiency of conditions 1) and 2) is thus established.

The necessity of condition 1) is obvious. It is also obvious that the set Im A must be a subspace. The operator $P = AA^{(-1)}$ will in this connection be a projection whose set of values coincides with Im A. Consequently the subspace Im $(I - P)$ is a direct complement of Im A.

Equality (1.2) follows immediately from (1.1) and the relation

$$\dim \operatorname{Ker} A^{(-1)} = \dim \operatorname{Ker} P.$$

The following proposition is established just as simply.

2°. *For an operator $A \in \mathfrak{A}$ to be right invertible, it is necessary and sufficient that:*

1) *the set of values* Im A *coincide with $\mathfrak{B}$,*

2) *the subspace* Ker A *have a direct complement in $\mathfrak{B}$.*

If these conditions are fulfilled and $\mathfrak{N}$ is a direct complement of Ker A, *the general form of an operator $A^{(-1)}$ inverse to A on the right is given by the equality*

$$A^{(-1)} = A_0^{-1} + B,$$

[1)] Equality (1.2) is taken in the sense that either the numbers dim Ker $A^{(-1)}$ and dim Coker A are finite and equal or they are both infinite.

where A_0^{-1} is an operator inverse to A considered as an operator operating from $\mathfrak{N}$ onto $\mathfrak{B}$, and B $(\in \mathfrak{A})$ is an arbitrary operator annihilated on the left by A (i. e. $\operatorname{Im} B \subset \operatorname{Ker} A$). *In this connection*

$$\dim \operatorname{Coker} A^{(-1)} = \dim \operatorname{Ker} A.$$

Let A $(\in \mathfrak{A})$ be an operator invertible on some side and $A^{(-1)}$ a corresponding inverse of A. The validity of the following proposition is verified without difficulty.

3°. *If A is left invertible, the equation*

$$Ax = y \tag{1.3}$$

is solvable if and only if

$$(I - AA^{(-1)})_y = 0.$$

Under fulfillment of this condition the vector $x = A^{(-1)} y$ is the unique solution of equation (1.3).

If A is right invertible, the vector $x = A^{(-1)} y$ is for any $y \in \mathfrak{B}$ a solution of equation (1.3), *and the subspace* $\operatorname{Ker} A$ *coincides with the set of values of the projection $I - A^{(-1)}A$.*

We shall say that the operator A $(\in \mathfrak{A})$ is invertible *only* on the left if it is invertible on the left and not invertible on the right. The expression "invertible only the right" has an analogous meaning.

It follows immediately from propositions 1° and 2° that if A is invertible only on one side a corresponding inverse operator $A^{(-1)}$ is *not unique.*

4°. *Let operators A and C $(\in \mathfrak{A})$ be invertible (i. e. bilaterally invertible) and let $B \in \mathfrak{A}$ be invertible on one side. Then the operator $D = ABC$ is invertible on the same side as B, where*

$$\dim \operatorname{Ker} D = \dim \operatorname{Ker} B, \qquad \dim \operatorname{Coker} D = \dim \operatorname{Coker} B.$$

The proof of this proposition is obvious.

Let a left invertible operator A be represented in the form

$$A = B\Lambda, \tag{1.4}$$

where B is invertible and Λ is left invertible.

By using proposition 3° it is easy to obtain the condition for the solvability of equation (1.3) for this case. Let us form the projection

$$\begin{aligned} P_A &= I - AA^{(-1)} = I - B\Lambda(B\Lambda)^{(-1)} \\ &= I - B\Lambda\Lambda^{(-1)} B^{-1} = B(I - \Lambda\Lambda^{(-1)}) B^{-1} = BP_\Lambda B^{-1}. \end{aligned}$$

A necessary and sufficient condition for the solvability of equation (1.3) is given by the equality

$$BP_\Lambda B^{-1} F = 0$$

or the equivalent equality

$$P_A B^{-1} F = 0. \tag{1.5}$$

Introducing the auxiliary equation

$$B\psi = f, \tag{1.6}$$

with the operator B invertible, we can now formulate the following proposition.

5°. *If a left invertible operator A admits representation* (1.4), *equation* (1.3) *is solvable if and only if the condition*

$$P_A\psi = (I - AA^{(-1)})\,\psi = 0 \tag{1.7}$$

is fulfilled for the solution ψ of the auxiliary equation (1.6).

Under fulfillment of this condition the solution φ of equation (1.3) *is given by the formula*

$$\varphi = A^{(-1)}\,\psi. \tag{1.8}$$

The second case which occurs to us for a *left* invertible operator A consists of the following.

Let us suppose that the operator

$$B = \Pi A \tag{1.9}$$

is bilaterally invertible for the *right* invertible operator Π.

Let us introduce the auxiliary equation

$$B\,\psi = \Pi f \tag{1.10}$$

with the operator B invertible.

6°. *For equation* (1.3) *to be solvable, it is necessary and sufficient that the condition*

$$P_\Pi A\psi = (I - \Pi^{(-1)}\Pi)\,A\psi = P_\Pi f. \tag{1.11}$$

be fulfilled for the solution ψ of the auxiliary equation (1.10).

If this condition is fulfilled, the vector ψ is the solution of the original equation (1.3).

Indeed, from (1.9) we obtain the fact that the operator $B^{-1}\,\Pi$ is a *left* inverse of A. By virtue of proposition 3° a necessary and sufficient condition for the solvability of equation (1.3) may be written in the form

$$P_A f = [I - A(B^{-1}\Pi)]\,f = 0. \tag{1.12}$$

The projection P_A is now representable in the form $P_A = P_\Pi P_A + (I - P_\Pi)\,P_A$, But

$$\begin{aligned}(I - P_\Pi)P_A &= \Pi^{(-1)}\,\Pi\,(I - AB^{-1}\Pi)\\ &= \Pi^{(-1)}\,\Pi - \Pi^{(-1)}\,(\Pi A)\,B^{-1}\,\Pi = \Pi^{(-1)}\,\Pi - \Pi^{(-1)}\,\Pi = 0.\end{aligned}$$

Consequently condition (1.12) is equivalent to the condition

$$P_\Pi P_A f = 0, \tag{1.13}$$

which coincides with (1.11).

Finally, if condition (1.13) is fulfilled, then (1.12) is also fulfilled, from which it is evident that the solution $\psi = B^{-1}\Pi f$ of the auxiliary equation satisfies the original equation.

Now let us establish the dual proposition for operators A which are right invertible. We consider two cases here, also.

First let *right* invertible A be represented in the form $A = \Pi B$, where the operator B is invertible and Π is right invertible.

Let us take the equation

$$B\psi = \Pi^{(-1)} f \tag{1.14}$$

as the auxiliary.

7°. *The solution* ψ *of the auxiliary equation* (1.14) *is always one of the solutions of the original equation* (1.3). *The general solution of the homogeneous equation*

$$A\chi = 0 \tag{1.15}$$

is given by the formula

$$\chi = B^{-1} P_{\Pi} h, \tag{1.16}$$

where h is an arbitrary vector.

Indeed, the first assertion is verified immediately.

To establish the second let us calculate in accordance with 3° the corresponding projection[2)]

$$P_A = I - A^{(-1)} A = I - (\Pi B)^{(-1)} \Pi B = I - B^{-1} \Pi^{(-1)} \Pi B = B^{-1} P_{\Pi} B.$$

The general solution of homogeneous equation (1.15) is given by the formula

$$\chi = P_A h = B^{-1} P_{\Pi} Bh,$$

which is equivalent to (1.16)

Let us now formulate the proposition dual to 6°. Let operator A be right invertible, and let the *left* invertible operator Λ be such that

$$A\Lambda = B, \tag{1.17}$$

where B is bilaterally invertible.

Here the equation

$$B\psi = f \tag{1.18}$$

will be the auxiliary.

8°. *The vector*

$$\varphi = \Lambda\psi \tag{1.19}$$

where ψ *is the solution of* (1.18), *is always one of the solutions of the original equation* (1.3).

[2)] We denote it by the same symbol P_A, although now this is another projection.

The general solution of the homogeneous equation (1.15) *is given by the formula*

$$\chi = (I - \Lambda B^{-1} A)P_\Lambda h, \tag{1.20}$$

where h is arbitrary vector.

The first assertion is obvious. Condition (1.17) indicates that the operator ΛB^{-1} is a right inverse of A. Let us form the projection

$$P_A = I - \Lambda B^{-1}A.$$

According to 3°, the general solution of equation (1.15) is given by the formula

$$\chi = P_A h = (I - \Lambda B^{-1} A)h. \tag{1.21}$$

The projection P_A is now representable in the form

$$P_A = P_A P_\Lambda + P_A (I - P_\Lambda).$$

But $P_A (I - P_\Lambda) = (I - \Lambda B^{-1} A)\, \Lambda\Lambda^{(-1)} = \Lambda\Lambda^{(-1)} - \Lambda B^{-1} (A\Lambda)\, \Lambda^{(-1)} = 0$. Therefore formula (1.21) is equivalent to the formula

$$\chi = P_A P_\Lambda h.$$

9°. *If the operator* $A\,(\in \mathfrak{A})$ *is invertible on one side, then all operators* $X\,(\in \mathfrak{A})$ *satisfying the condition*[3)]

$$|X - A| < 1/|A^{(-1)}|$$

are invertible on the same side, with

$$\dim \operatorname{Ker} X = \dim \operatorname{Ker} A, \qquad \dim \operatorname{Coker} X = \dim \operatorname{Coker} A.$$

This proposition follows from 4° if account is taken of the fact that when A is left invertible, X is representable in the form

$$X = [I - (A - X)A^{(-1)}]A,$$

while, when A is right invertible,

$$X = A[I - A^{(-1)} (A - X)].$$

3. *Inversion of polynomials in unilaterally invertible operators.* In this subsection $V\,(\in \mathfrak{A})$ will denote an operator invertible only on the left, and $V^{(-1)}$, any of its left inverses. Let us suppose that the following condition is fulfilled.[4)]

(Σ I). *The spectra of the operators V and $V^{(-1)}$ are contained in the unit disk* $|\zeta| \leqq 1$.

LEMMA 1.1. *Let the operators V and $V^{(-1)}$ satisfy condition* (Σ I). *Then the spec-*

[3)] Wherever it does not lead to ambiguity the norm of an operator or vector is denoted by the simple strokes | |.

[4)] In what follows we shall need one more important condition (ΣII), and, in Chapter VIII, the conditions (ΣIII) and (ΣIV).

trum of each coincides with the unit disk. If $|\lambda| < 1$. *the operator* $V-\lambda I$ *is invertible only on the left, while the operator* $V^{(-1)} - \lambda I$ *is invertible only on the right; if, however,* $|\lambda| = 1$, *neither of the operators* $V - \lambda I$ *nor* $V^{(-1)} - \lambda I$ *has an inverse on either side.*

PROOF. Let $|\lambda| < 1$. Then by virtue of condition (Σ I) the operator $I-\lambda V^{(-1)}$ is invertible, while by virtue of proposition 4° the operator $V - \lambda I = (I - \lambda V^{(-1)})V$ is invertible only on the left. Now let $|\lambda_0| = 1$. Let us assume that the operator $V - \lambda_0 I$ is invertible only on the left, for example, and $(V - \lambda_0 I)^{(-1)}$ is its left inverse. Then by virtue of proposition 9° the operator

$$V - \lambda I = V - \lambda_0 I - (\lambda - \lambda_0)I$$

will be invertible only on the left for $|\lambda - \lambda_0| < |(V - \lambda_0 I)^{(-1)}|^{-1}$. This contradicts the invertibility of the operator $V - \lambda I$ for $|\lambda| > 1$.

Analogously we establish the fact that for $|\lambda_0| = 1$ the operator $V - \lambda_0 I$ cannot be invertible only on the right or invertible, as well as all assertions concerning the spectrum of the operator $V^{(-1)}$. The lemma is proved.

Let us introduce the following notation: $\mathfrak{R}(V)$ is the linear span (nonclosed!) of all operators $V^{(j)}$, $j = 0, \pm 1, \cdots$, where $V^{(j)} = V^j, j = 0, 1, \cdots$, and $V^{(j)} = (V^{(-1)})^j$, $j = -1, -2, \cdots$, and $\mathfrak{R}^+(V)$ and $\mathfrak{R}^-(V)$ are the linear spans (nonclosed) of the operators V^j, $j = 0, 1, \cdots$, and $V^{(j)}$, $j = 0, -1, \cdots$, respectively. It is not difficult to verify that each operator $R \in \mathfrak{R}(V)$ is uniquely representable as the finite sum $R = \sum \alpha_j V^{(j)}$, where the α_j are complex numbers. A one-to-one correspondence is established between the set $\mathfrak{R}(V)$ and the set of all polynomials (with complex coefficients) in integral powers of ζ under which the polynomial $R(\zeta) = \sum \alpha_j \zeta^j$ corresponds to the operator $R = \sum \alpha_j V^{(j)} \in \mathfrak{R}$. Let us call this polynomial the *symbol* of the operator R.

We need the following easily established proposition.

10°. *Let* $R_\pm \in \mathfrak{R}^\pm(V)$, $R_0 \in \mathfrak{R}(V)$. *Define the polynomial* $R(\zeta)$ *by the equality* $R(\zeta) = R_-(\zeta)R_0(\zeta)R_+(\zeta)$. *Then the equality*

$$R = R_- R_0 R_+$$

holds for the operator R ($\in \mathfrak{R}(V)$) *corresponding to the symbol* $R(\zeta)$.

The following theorem plays an essential role in what follows.

THEOREM 1.1. *Let operators* V *and* $V^{(-1)}$ *satisfy condition* (ΣI). *For an operator* $R \in \mathfrak{R}$ *to be invertible in* $\mathfrak{A}$ *on at least one side, it is necessary and sufficient that its symbol not vanish*:

$$R(\zeta) \neq 0 \qquad (|\zeta| = 1). \tag{1.22}$$

If condition (1.22) *is fulfilled,* R *will be respectively invertible, only left invertible, or only right invertible, depending on whether the number*

$$\kappa = \operatorname{ind} R(\zeta) \tag{1.23}$$

is equal to zero, positive, or negative.

Let us introduce the following definition. Let A be an operator of some class, and let $\nu = \nu(A)$ be some real functional. We shall say that the invertibility of A is *consistent* with the functional ν if A is invertible only on the left, invertible only on the right, or simply invertible depending on whether the number ν is positive, negative, or equal to zero.

Then the second assertion of Theorem 1.1 can be reformulated as follows.

If condition (1.22) *is fulfilled, the invertibility of R is consistent with the index κ of the function $R(\zeta)$.*

PROOF. Under fulfillment of condition (1.22) the polynomial $R(\zeta)$, as is easily seen, is representable in the form

$$R(\zeta) = c \prod_{j=1}^{k} (1 - \zeta_j^+ \zeta^{-1}) \zeta^\kappa \prod_{j=1}^{s} (\zeta - \zeta_j^-), \tag{1.24}$$

where $|\zeta_j^+| < 1$ for $j = 1,\cdots,k$ and $|\zeta_j^-| > 1$ for $j = 1,\cdots,s$; and κ is defined by equality (1.23).

By virtue of proposition 10°, ζ can be replaced by V and ζ^{-1} by $V^{(-1)}$ in (1.24). Making this substitution, we obtain

$$R = c \prod_{j=1}^{k} (I - \zeta_j^+ V^{(-1)}) V^{(\kappa)} \prod_{j=1}^{s} (V - \zeta_j^- I).$$

The operators

$$R_- = c \prod_{j=1}^{k} (I - \zeta_j^+ V^{(-1)}) \quad \text{and} \quad R_+ = \prod_{j=1}^{s} (V - \zeta_j^- I)$$

are invertible by virtue of condition (Σ I). Therefore the operator R will be invertible from the same side as the operator $V^{(\kappa)}$. Moreover, the operator

$$R^{(-1)} = R_+^{-1} V^{(-\kappa)} R_-^{-1}$$

will be the inverse, only the left inverse, or only the right inverse of R depending on whether the number κ is equal to zero, positive, or negative.

To complete the proof of the theorem it is sufficient to show that if the symbol $R(\zeta)$ vanishes at even one point of the unit circle, the operator R does not have an inverse in $\mathfrak{A}$ on either side.

Let $R(\zeta_0) = 0$ ($|\zeta_0| = 1$). By virtue of proposition 10° the equality $R(\zeta) = R_1(\zeta)(\zeta - \zeta_0)$ implies that

$$R = R_1 (V - \zeta_0 I) \qquad (R_1 \in \mathfrak{R}(V)).$$

By virtue of Lemma 1.1 R does not have a left inverse. Since, on the other hand,

$R(\zeta) = (\zeta^{-1} - \zeta_0^{-1}) R_2(\zeta)$ and $R = (V^{(-1)} - \zeta_0^{-1}I)R_2$ $(R_2 \in \mathfrak{R}(V))$, it follows that R does not have a right inverse by virtue of the same Lemma 1.1. The theorem is proved.

§ 2. Continuous functions of unilaterally invertible operators

A generalization of the results of § 1.3 to the case of broad classes of *continuous* functions of unilaterally invertible operators will be obtained in subsequent sections of this chapter. In the present section functions of unilaterally invertible operators are introduced, and their elementary properties are established.

As in § 1, we shall assume that a left invertible bounded linear operator V operating in the space $\mathfrak{B}$ satisfies condition (Σ I).

1. Let us denote by $\mathfrak{N}(V)$, $\mathfrak{N}^+(V)$ and $\mathfrak{N}^-(V)$ the closure (in the norm of $\mathfrak{A}$) of the linear manifolds $\mathfrak{R}(V)$, $\mathfrak{R}^+(V)$ and $\mathfrak{R}^-(V)$, respectively. Obviously $\mathfrak{N}^+(V)$ and $\mathfrak{N}^-(V)$ are commutative subalgebras of the algebra $\mathfrak{A}$, while $\mathfrak{N}(V)$ is a subspace (but not a subalgebra) of $\mathfrak{A}$.

LEMMA 2.1. *Let R be an arbitrary operator of $\mathfrak{R}(V)$, and $R(\zeta)$, its symbol. Then the spectral radius $r(R)$ of the operator R is given by the formula*

$$r(R) = \max_{|\zeta|=1} |R(\zeta)|. \tag{2.1}$$

PROOF. Let ζ_0 be an arbitrary point of the unit circle. By virtue of Lemma 1.1 the operator $R - R(\zeta_0)I$ is noninvertible. Therefore $r(R) \geqq |R(\zeta_0)|$, and hence

$$r(R) \geqq \max_{|\zeta|=1} |R(\zeta)|.$$

On the other hand, for any number λ satisfying the condition

$$|\lambda| > \max_{|\zeta|=1} |R(\zeta)|$$

the operator $R - \lambda I$ is invertible. This follows from the relations

$$R(\zeta) - \lambda \neq 0 \quad (|\zeta| = 1), \qquad \arg\, [(R(e^{i\varphi}) - \lambda)]_0^{2\pi} = 0$$

and Theorem 1.1. Thus

$$r(R) \leqq \max_{|\zeta|=1} |R(\zeta)|.$$

The lemma is proved.

Equality (2.1) implies the relation

$$\max_{|\zeta|=1} |R(\zeta)| \leqq |R| \qquad (R \in \mathfrak{R}(V)). \tag{2.2}$$

This relation permits each operator $A \in \mathfrak{N}(V)$ to be put into correspondence with some function continuous on the unit circle. Indeed, let $A \in \mathfrak{N}(V)$, and let R_n, $n = 1, 2, \cdots$, be some sequence of operators of $\mathfrak{R}(V)$ which converges to

A. By virtue of (2.2) the sequence of polynomials $R_n(\zeta)$ converges uniformly on the unit circle; consequently its limit $A(\zeta)$ ($|\zeta| = 1$) is a continuous function. It is easy to see that this function $A(\zeta)$ does not depend on the choice of the sequence $R_n \in \mathfrak{R}(V)$ converging to A. The function $A(\zeta)$ thus constructed for the operator $A \in \mathfrak{R}(V)$ satisfies the inequality

$$\max_{|\zeta|=1} |A(\zeta)| \leqq |A| \tag{2.3}$$

by virtue of (2.2).

The function $A(\zeta)$ is called the *symbol* of the operator A.

Let us denote by $\mathfrak{R}(\zeta)$ the set of symbols $A(\zeta)$ ($|\zeta| = 1$) corresponding to all of the operators $A \in \mathfrak{R}(V)$.

2. Some information concerning the class of symbols $\mathfrak{R}(\zeta)$ is given by the following two propositions.

1°. *The set* $\mathfrak{R}(\zeta)$ *contains the algebra of all functions whose Fourier coefficients* $\{a_j\}_{-\infty}^{\infty}$ *satisfy the condition*

$$\sum_{-\infty}^{\infty} \alpha_j |a_j| < \infty, \qquad \textit{where } \alpha_j = |V^{(j)}|. \tag{2.4}$$

As a matter of fact, if condition (2.4) is fulfilled, then the series $\sum_{-\infty}^{\infty} a_j V^{(j)}$ converges in the operator norm, and its sum A belongs to $\mathfrak{R}(V)$. Obviously $A(\zeta) = \sum_{-\infty}^{\infty} a_j \zeta^j$.

The operator $V \in \mathfrak{A}$ is called *strictly isometric* if 1) it is isometric, 2) Im $V \neq \mathfrak{B}$, and 3) a left inverse operator $V^{(-1)} \in \mathfrak{A}$ ($V^{(-1)}V = I$) exists with norm equal to 1.

Each isometric (nonunitary) operator in Hilbert space is strictly isometric.

Every strictly isometric operator V, together with its left inverse opperator $V^{(-1)}$, obviously satisfies condition (Σ I), § 1.

Proposition 1° implies in particular that the set $\mathfrak{R}(\zeta)$ constructed relative to a strictly isometric operator V contains all functions $\varphi(\zeta)$ ($|\zeta| = 1$) with absolutely convergent Fourier series.

Let us note the following proposition concerning two *extreme* cases of the structure of $\mathfrak{R}(\zeta)$.

2°. *If V is an isometric operator operating in Hilbert space $\mathfrak{H}$, then $\mathfrak{R}(\zeta)$ coincides precisely with the set of all functions continuous on the unit circle, the strict equality*

$$|A| = \max_{|\zeta|=1} |A(\zeta)|$$

holding for every operator $A \in \mathfrak{R}(V)$.

If, however, V is a strictly isometric operator operating in the space l_1 in accordance with the rule

$$V\{\xi_1, \xi_2, \cdots\} = \{0, \xi_1, \xi_2, \cdots\}, \qquad V^{(-1)}\{\xi_j\}_1^{\infty} = \{\xi_{j+1}\}_1^{\infty},$$

then $\mathfrak{R}(\zeta)$ coincides precisely with the set of all functions expanding into absolutely *convergent Fourier series. Moreover, the strict equality* $|A| = \sum_{-\infty}^{\infty} |a_j|$ *holds, where the* a_j *are the Fourier coefficients of* $A(\zeta)$.

Let $\mathfrak{H}$ be Hilbert space while U is a *unitary* extension of the operator V which operates in some *extended* Hilbert space $\tilde{\mathfrak{H}} \supset \mathfrak{H}$, and P is an orthogonal projection projecting $\tilde{\mathfrak{H}}$ onto $\mathfrak{H}$. Let us note that in the case of Hilbert space the operator $V^{(-1)}$ is uniquely defined for a strictly isometric operator V: $V^{(-1)} = V^*$, the operators V and V^* being respectively the restrictions of the operators PUP $(= UP)$ and PU^*P $(= PU^*)$ to the subspace $\mathfrak{H} = \operatorname{Im} P$.

Let R be an arbitrary operator of $\mathfrak{K}(V)$ and $R(\zeta)$ the polynomial corresponding to it, i. e. its symbol. As a function of the unitary operator U, the operator $R(U)$ satisfies the inequality

$$|R(U)| \leqq \max_{|\zeta|=1} |R(\zeta)|.$$

It is easy to see that R is the restriction of $PR(U)P$ to $\operatorname{Im} P$. Therefore

$$|R| \leqq |PR(U)P| \leqq \max_{|\zeta|=1} |R(\zeta)|.$$

Comparing this relation with (2.2) leads to the equality

$$|R| = \max_{|\zeta|=1} |R(\zeta)|,$$

from which (2.5) and the first assertion are easily derived.

The second assertion follows from the easily verified equality $R = \sum |a_j|$, where $R \in \mathfrak{K}(V)$, and the a_j are the coefficients of the polynomial $R(\zeta)$.

3. The general estimate (2.3) of the function $A(\zeta)$ implies that if the operator A belongs to $\mathfrak{R}^+(V)$ $(\mathfrak{R}^-(V))$, its symbol $A(\zeta)$ $(|\zeta| = 1)$ admits holomorphic continuation into the interior (the exterior) of the unit disk.

As already noted, the class $\mathfrak{R}^+(V)$ $(\mathfrak{R}^-(V))$ is a commutative Banach algebra. The following lemma affords a description of the set of maximal ideals of this algebra.

Let us recall that every element x of a commutative Banach algebra $\mathfrak{R}$ can be placed into correspondence with a function $x(M)$ on the set of maximal ideals of that algebra (cf. Gel'fand, Raĭkov and Šilov [1]).

LEMMA 2.2. *Let* ζ *be an arbitrary fixed point of the unit disk* $|\zeta| \leqq 1$. *The set* M_ζ *of all elements* A *of the algebra* $\mathfrak{R}^+(V)$ *for which* $A(\zeta) = 0$ *is a maximal ideal of that algebra. There are no other maximal ideals in* $\mathfrak{R}^+(V)$. *The function* $A(M_\zeta)$ *on the set of maximal ideals which corresponds to the element* $A \in \mathfrak{R}^+(V)$ *coincides with the symbol* $A(M_\zeta) = A(\zeta)$.

An analogous lemma holds for the algebra $\mathfrak{R}^-(V)$ (with the substitution of the inequality $|\zeta| \geqq 1$ for $|\zeta| \leqq 1$).

PROOF. Let $|\zeta| \leqq 1$. Let us define the functional $f_\zeta(A) = A(\zeta)$ $(A \in \mathfrak{R}^+(V))$ on the algebra $\mathfrak{R}^+(V)$.

The functional f_ζ is obviously linear, multiplicative, and, by virtue of estimate (2.3), bounded. The set of all elements on which f_ζ vanishes is a maximal ideal of the algebra $\mathfrak{R}^+(V)$. Let us denote it by M_ζ.

Now let M_0 be some maximal ideal of $\mathfrak{R}^+(V)$. As is well known, the spectrum of the element V coincides with the set of values of the function $V(M)$, which corresponds to the element V, on the set of maximal ideals. On the other hand, it is easy to obtain by Lemma 1.1 the fact that the spectrum of V in $\mathfrak{R}^+(V)$ is the unit disk. Therefore $V(M_0) = \zeta_0$, where $|\zeta_0| \leqq 1$. This implies that $R(M_0) = R(\zeta_0)$ for any $R \in \mathfrak{K}^+(V)$.

Extending the latter equality through continuity, we obtain $A(M_0) = A(\zeta_0)$ for any $A \in \mathfrak{R}^+(V)$. As is well known, the ideal M_0 coincides with the set of elements $A \in \mathfrak{R}^+(V)$ for which $A(M_0) = 0$. Thus M coincides with M_{ζ_0}, and $A(M_{\zeta_0}) = A(\zeta_0)$. The lemma is proved.

§ 3. Inversion of continuous functions of unilaterally invertible operators

For the time being we have imposed the single condition (Σ I) on the operators V and $V^{(-1)}$. It has permitted us to obtain the theorem concerning inversion of a polynomial in these operators, as well as to introduce continuous functions of V.

To invert these functions of V it is necessary to introduce additional restrictions on the operators V and $V^{(-1)}$. Two variants of such restrictions are cited below. Let us emphasize that, as before, the operators V and $V^{(-1)}$ satisfy condition (Σ I), § 1.

1. *First variant.* In this subsection we shall assume that the operators V and $V^{(-1)}$ satisfy in addition the condition

(Σ II). *If $R_1, R_2 \in \mathfrak{K}(V)$ and the symbol of the operator R is the product of the symbols of R_1 and R_2: $R(\zeta) = R_1(\zeta)R_2(\zeta)$, then*

$$|R| \leqq |R_1|\,|R_2|. \tag{3.1}$$

Every *strictly isometric* operator V, together with its "pair" $V^{(-1)}$, satisfies condition (Σ II). Indeed, let $R_1, R_2 \in \mathfrak{K}(V)$ and $R(\zeta) = R_1(\zeta)R_2(\zeta)$. It is easy to see that for sufficiently large natural n the equality

$$R = V^{(-n)} R_1 R_2 V^n$$

is valid, which implies (3.1).

Let us define a new (commutative!) multiplication on the linear manifold $\mathfrak{K}(V)$ by setting

$$R_1 \circ R_2 = R \qquad (R_1, R_2 \in \mathfrak{K}(V)),$$

where the operator R ($\in \mathfrak{K}(V)$) is determined by the polynomial $R(\zeta) = R_1(\zeta)R_2(\zeta)$. This multiplication will also be distributive and associative. In addition, by virtue of condition (ΣII),

$$|R_1 \circ R_2| \leqq |R_1|\,|R_2| \qquad (R_1, R_2 \in \mathfrak{K}(V)).$$

This relation permits the new multiplication to be extended through continuity to all pairs of elements of $\mathfrak{R}(V)$. After the introduction of the new multiplication, the class $\mathfrak{R}(V)$, which previously was not even an algebra, becomes a *commutative* Banach algebra with generators V and $V^{(-1)}$. We shall denote this algebra by $\mathfrak{R}_0(V)$. The element $V^{(-1)}$ is the inverse of V in $\mathfrak{R}_0(V)$.

The following proposition is derived without difficulty from the definition of the new multiplication and proposition 10° of § 1.

1°. *Let $B_\pm \in \mathfrak{R}^\pm(V)$ and $X \in \mathfrak{R}(V)$ Then the operator B_-XB_+ belongs to $\mathfrak{R}(V)$, with $(B_-B_+)\circ X = B_-XB_+$.*

Let us establish one more auxiliary proposition.

2°. *The spectrum of each of the operators V and $V^{(-1)}$ as elements of the algebra $\mathfrak{R}_0(V)$ coincides with the unit circle.*

Indeed, Lemma 1.1 and proposition 1° imply that the spectral radii of the elements V, $V^{(-1)} \in \mathfrak{R}_0(V)$ equal unity, and that all of the points of the unit circle are points of the spectrum of the elements V and $V^{(-1)}$. Since for $|\lambda| < 1$ the element $V - \lambda I$ is representable in the form

$$V - \lambda I = (I - \lambda V^{(-1)})\circ V,$$

where both factors are invertible in $\mathfrak{R}_0(V)$, the spectrum of the element V coincides with the unit circle. This is also valid for the element $V^{(-1)}$, since it is V's inverse in $\mathfrak{R}_0(V)$.

Let us now cite a lemma concerning the structure of the set of maximal ideals of the algebra $\mathfrak{R}_0(V)$.

LEMMA 3.1. *Let ζ be an arbitrary fixed point of the unit circle. The set M_ζ of all elements A of the algebra $\mathfrak{R}_0(V)$ for which $A(\zeta) = 0$ is a maximal ideal of $\mathfrak{R}_0(V)$. There are no other maximal ideals in $\mathfrak{R}_0(V)$. The function $A(M_\zeta)$ on the set of maximal ideals which corresponds to the element A coincides with its symbol $A(M_\zeta) = A(\zeta)$.*

After the spectrum of the element V is found (proposition 2°), the proof of Lemma 3.1 is developed essentially like the proof of the similar Lemma 2.2.

THEOREM 3.1. *Let the operators V and $V^{(-1)}$ satisfy conditions (Σ I) and (Σ II). For an operator $A \in \mathfrak{R}(V)$ to be invertible in $\mathfrak{A}$ on at least one side, it is necessary and sufficient that its symbol not vanish:*

$$A(\zeta) \neq 0 \qquad (|\zeta| = 1). \tag{3.2}$$

If condition (3.2) is fulfilled, A will be invertible, invertible only on the left, or invertible only on the right, depending on whether the number

$$\kappa = \kappa(A) = \operatorname{ind} A(\zeta)$$

is equal to zero, positive, or negative.

In other words, *the invertibility of A is consistent with the index of its symbol $A(\zeta)$.*

For the sake of abbreviating the formulations, let us stipulate the following:

We shall say that *the invertibility of some operator $A \in \mathfrak{A}(\mathfrak{B})$ is determined by a function $f(\zeta)$ corresponding to it and continuous on the unit circle* if A is invertible on any side if and only if $f(\zeta)$ never vanishes: $f(\zeta) \neq 0$ $(|\zeta| = 1)$. Moreover, when this condition is fulfilled, the invertibility of A is consistent with the index of the function $f(\zeta)$.

Now we can reformulate Theorem 3.1 as follows.

Let the operators V and $V^{(-1)}$ satisfy conditions (Σ I) and (Σ II). Then the invertibility of an arbitrary operator A of $\mathfrak{R}(V)$ is determined by its symbol $A(\zeta)$.

PROOF. By virtue of Lemma 3.1 condition (3.2) is necessary and sufficient for the element A to be invertible in the algebra $\mathfrak{R}_0(V)$.

Let us suppose that condition (3.2) is fulfilled, and let $R \in \mathfrak{K}(V)$ be an element sufficiently close to A. Then R is also invertible in $\mathfrak{R}_0(V)$, with

$$|R^{-1} \circ A - I| < 1, \tag{3.3}$$

where R^{-1} is the inverse of R in $\mathfrak{R}_0(V)$. From (3.3) and

$$\max_{|\zeta|=1} |A(\zeta)| \leqq |A| \tag{3.4}$$

we get the inequality

$$\max_{|\zeta|=1} |R^{-1}(\zeta)\, A(\zeta) - 1| < 1,$$

which implies that

$$R(\zeta) \neq 0 \quad (|\zeta| = 1) \qquad \text{and} \qquad \kappa(R) = \kappa(A).$$

It was established in the proof of Theorem 1.1 that under these conditions operator R can be represented in the form $R = R_- V^{(\kappa)} R_+$, where $R_\pm$ are invertible operators of $\mathfrak{K}^\pm(V)$.

The equality $A = R \circ (I + C)$, where $C = R^{-1} \circ A - I$, and proposition 1° imply that for $\kappa \geqq 0$

$$A = R_-(I + C) V^\kappa R_+, \tag{3.5}$$

and for $\kappa < 0$

$$A = R_- V^{(\kappa)} (I + C) R_+. \tag{3.6}$$

Since by virtue of (3.3) the operator $I + C$ is invertible, for $\kappa > 0$ the operator

$$A^{(-1)} = R_+^{-1} V^{(-\kappa)} (I + C)^{-1} R_-^{-1}$$

will be only a left inverse of A, for $\kappa < 0$ the operator

$$A^{(-1)} = R_+^{-1} (I + C)^{-1} V^{-\kappa} R_-^{-1}$$

will be only a right inverse of A, and for $\kappa = 0$ the operator

$$A^{-1} = R_+^{-1}(I + C)^{-1}R_-^{-1}$$

will be the inverse of A.

To complete the proof of the theorem it is sufficient to show that if the symbol $A(\zeta)$ vanishes at even one point of the unit circle, A cannot be invertible on either side.

Let us assume that $A(\zeta_0) = 0$ ($|\zeta_0| = 1$) but A is invertible on at least one side. Then, by virtue of proposition 9° of § 1, all operators $B \in \mathfrak{A}$ for which

$$|B - A| < |A^{(-1)}|^{-1} \; (= \delta)$$

are also invertible on the same side.

Let $R_1 \in \mathfrak{R}(V)$ be some operator such that $|R_1 - A| < \delta/2$. Then, by virtue of (3.4),

$$\max_{|\zeta|=1} |R_1(\zeta) - A(\zeta)| < \delta/2.$$

Theorem 1 will be established under fulfillment of condition (Σ I) for a certain set of operators of $\mathfrak{R}(V)$. The question of its validity for any operator of $\mathfrak{R}(V)$ remains open.

Let U be a bounded *invertible* extension of the operator V which operates in some Banach space $\tilde{\mathfrak{B}} \supset \mathfrak{B}$ and satisfies the following conditions:

1) *The spectral radii of the operators* U, $U^{-1}(\in \mathfrak{A}(\tilde{\mathfrak{B}}))$ *equal unity.*

2) *A projection* P $(\in \mathfrak{A}(\tilde{\mathfrak{B}}))$ *exists, projecting* $\tilde{\mathfrak{B}}$ *onto* $\mathfrak{B}$ *such that* $PU^{-1}P = PU^{-1}$ (*invariance of the subspace* Im $(I - P)$ *relative to* U^{-1}) *and* $PU^{-1}x = V^{-1}x$ $(x \in \mathfrak{B})$.

Let us note that a space $\tilde{\mathfrak{B}}$ and an invertible operator $U \in \mathfrak{A}(\tilde{\mathfrak{B}})$ which satisfy conditions 1) and 2) always exist. For example, the direct sum of two copies of the Banach space $\mathfrak{B}$ may be taken as $\tilde{\mathfrak{B}}$, and the operators P, U and U^{-1} $(\in \mathfrak{A}(\tilde{\mathfrak{B}}))$ may be defined by the equalities

$$P\{f_1, f_2\} = \{f_1, 0\},$$

$$U = \begin{Vmatrix} V & I - VV^{(-1)} \\ 0 & V^{(-1)} \end{Vmatrix}, \quad U^{-1} = \begin{Vmatrix} V^{(-1)} & 0 \\ I - VV^{(-1)} & V \end{Vmatrix}.$$

Conditions 1) and 2) imply that the operators U, U^{-1} and P satisfy the following conditions:

$UP = PUP$ (*invariance of* Im P *relative to* U),

$UP \neq PU$ (this means that U *maps* Im P *into its right side*), (3.7)

$PU^{-1} = PU^{-1}P$ (*invariance of the subspace* Im $(I - P)$ *relative to* U^{-1}),

and in particular $|R_1(\zeta_0)| < \delta/2$.

The operator $R = R_1 - R_1(\zeta_0)I$ $(\in \mathfrak{R}(V))$ lies in a δ-neighborhood of A and is therefore invertible on at least one side. But $R(\zeta_0) = 0$, which contradicts Theorem 1.1. The theorem is proved.

REMARK 1. In proving the necessity of condition (3.2) the fact that the operators V and $V^{(-1)}$ satisfy condition (Σ II) was *not utilized.*

REMARK 2. Let us recall that the totality of all elements x of a commutative Banach algebra $\mathfrak{C}$ such that

$$\lim_{n\to\infty} \sqrt[n]{|x^n|} = 0$$

is called the *radical* of $\mathfrak{C}$.

As is well known (cf. Gel'fand, Raĭkov and Šilov **[1]**, Chapter I, § 4), an element x belongs to the radical if and only if its corresponding function $x(M)$ on the set of maximal ideals is indentically equal to zero. The correspondence between the functions $x(M)$ and the elements x of an algebra $\mathfrak{C}$ is one-to-one for algebras without a radical.

The question of whether the algebra $\mathfrak{R}_0(V)$ in general has a nonnull radical remains open. Let us note that all of the specific algebras $\mathfrak{R}_0(V)$ cited in the present chapter have a null radical.

2. *Second variant.* It is not assumed here that the operators V and $V^{(-1)}$ satisfy condition (Σ II).

Let us show that the spectrum of each of the operators U and U^{-1} coincides with the unit circle. It is obvious that both of these spectra are on the circle. Assuming that the spectrum of U does not coincide with the circle, we obtain the result that the resolvent set of U consists of one connected component. For all the points λ of this component the operator $V - \lambda I$ will obviously be left invertible. This leads to a contradiction, since V is invertible only on the left and for sufficiently large $|\lambda|$ the operator $V - \lambda I$ is invertible.

Let us introduce the sets $\mathfrak{K}(U)$, $\mathfrak{K}^{\pm}(U)$, $\mathfrak{R}(U)$ and $\mathfrak{R}^{\pm}(U)$ analogously to the way it was done in preceding sections. In contrast to $\mathfrak{R}(V)$, the set $\mathfrak{R}(U)$ is a commutative Banach algebra. To each operator $R = \sum \alpha_j U^j \in \mathfrak{K}(U)$ there corresponds a polynomial, its *symbol* $R(\zeta) = \sum \alpha_j \zeta^j$; in this connection

$$\max_{|\zeta|=1} |R(\zeta)| \leq |R|.$$

This relation permits each operator $A \in \mathfrak{R}(U)$ to be put into correspondence with a function $A(\zeta)$, the *symbol* of A, which is continuous on the unit circle, with

$$\max_{|\zeta|=1} |A(\zeta)| \leq |A|.$$

It can be proved without difficulty that the unit circle is the set of maximal ideals of the algebra $\mathfrak{R}(U)$ and the function $A(\zeta)$ is the function on the maximal ideals which corresponds to the element $A \in \mathfrak{R}(U)$. Hence the following proposition is easily derived.

3°. *For an operator $A \in \mathfrak{R}(U)$ to be invertible in $\mathfrak{A}(\mathfrak{B})$ it is necessary and sufficient that its symbol not vanish*:

$$A(\zeta) \neq 0 \qquad (|\zeta| = 1). \tag{3.8}$$

Under fulfillment of this condition the inverse operator A^{-1} belongs to $\mathfrak{R}(U)$.

Let us denote by $\hat{\mathfrak{R}}(U)$ the set of all restrictions to $\mathfrak{B} = \operatorname{Im} P$ of operators of the form $P\tilde{A}P$, where $\tilde{A}$ ranges over $\mathfrak{R}(U)$. Obviously the set $\hat{\mathfrak{R}}(U)$ is contained in $\mathfrak{R}(V)$, but it does not in general coincide with $\mathfrak{R}(V)$.[5)]

It is easy to see that if the operator A belongs to $\hat{\mathfrak{R}}(U)$ and is the restriction to $\mathfrak{B}$ of the operator $P\tilde{A}P$, where $\tilde{A} \in \mathfrak{R}(U)$, the continuous functions associated with the operators $A \in \mathfrak{R}(V)$ and $\tilde{A} \in \mathfrak{R}(U)$ coincide, i. e. the symbols of these operators relative to the algebras $\mathfrak{R}(V)$ and $\mathfrak{R}(U)$ coincide.

THEOREM 3.2. *The invertibility of any operator A of $\hat{\mathfrak{R}}(U)$ is determined by its symbol $A(\zeta)$.*

PROOF. We shall denote the restriction of an operator to a subspace $\mathfrak{B}$ by the symbol $|\mathfrak{B}$ to the right of the operator. Let condition (3.8) be fulfilled; let $A = P\tilde{A}P \,|\, \mathfrak{B}$ $(A \in \mathfrak{R}(U))$ and let $R \in \hat{\mathfrak{R}}(U)$ be an operator such that

$$R^{-1}\tilde{A} = I + C, \qquad |PCP| < 1 \qquad (C \in \mathfrak{R}(U)). \tag{3.9}$$

The following representation of R emerges, as in the proof of Theorem 3.1:

$$R = R_- U^\kappa R_+,$$

where $R_\pm$ are invertible operators of $\hat{\mathfrak{R}}^\pm(U)$, with $R_\pm^{-1} \in \mathfrak{R}^\pm(U)$.

The equalities

$$A_+P = PA_+P, \; PA_- = PA_-P \qquad (A_\pm \in \mathfrak{R}^\pm (U)) \tag{3.10}$$

follow immediately from invariance relations (3.7). Therefore

$$P\tilde{A}P = PR_-(I + C)U^\kappa R_+P = PR_-P[P(I + C)U^\kappa P]\, PR_+P.$$

The operators $PR_\pm^{-1} P$ are obviously the inverses of the operators $PR_\pm P$ operating in $\operatorname{Im} P$.

If $\kappa = 0$, then A is invertible, since the operator $P + PCP$ is invertible in $\mathfrak{A}(\mathfrak{B})$ by virtue of (3.9). We have

$$P(I + C)U^\kappa P = (P + PCP)PU^\kappa P \qquad (\kappa > 0),$$
$$P(I + C)U^\kappa P = PU^\kappa P(P + PCP) \qquad (\kappa < 0).$$

by virtue of relations (3.10).

5) One can verify this on the basis of the following example. Let $\mathfrak{B}$ be the Banach space of sequences $\xi = \{\xi_j\}_{-\infty}^{\infty}$ with norm $|\xi| = \Sigma_{j=-\infty}^{-1} \; |\xi_j| + (\Sigma_{j=0}^{\infty} \; |\xi_j|^2)^{1/2}$ (the direct sum of l_1 and l_2). Let operators U and P be defined by the equalities

$$U \{\xi_j\}_{-\infty}^{\infty} = \{\xi_{j-1}\}_{-\infty}^{\infty}, \qquad P \{\xi_j\}_{-\infty}^{\infty} = \{\eta_j\}_{-\infty}^{\infty},$$

$$\eta_j = \begin{cases} \xi_j, \; j = 0, \, 1, \cdots, \\ 0, \; j = -1, \, -2, \cdots, \end{cases}$$

and let the operators V and $V^{(-1)}$ be restrictions to the space $\mathfrak{B} = \operatorname{Im} P$ of the operators PUP and $PU^{-1} P$, respectively.

Relations (3.7) and (3.10) imply that the operator $PU^{\kappa}P$ is invertible only on the left in $\mathfrak{A}(\mathfrak{B})$ for $\kappa > 0$ and invertible only on the right for $\kappa < 0$.

Thus the sufficiency of the theorem's conditions has been established. If the function $A(\zeta)$ vanishes on the unit circle, A is not invertible from either side. This follows from the remark to Theorem 3.1.

§ 4. General propositions concerning the invertibility of functions of unilaterally invertible operators

1. Two variants of the inversion of continuous functions of unilaterally invertible operators were presented in the preceding section. Both of these may be comprised in one general scheme consisting of the following.

Let the linear manifold $\mathfrak{R}_N$ be contained in $\mathfrak{R}(V)$, and with respect to some special norm $|\ |_N$ let it be a Banach space with the following properties:[6]

1) The linear manifold $\mathfrak{K}(V)$ is contained in $\mathfrak{R}_N$ and is dense in $\mathfrak{R}_N$ (in the norm $|\ |_N$).

2) $|X| \leqq C|X|_N$ $(C>0;\ X \in \mathfrak{R}_N)$.

3) The inverse operators $(V^{(\pm1)} - \lambda I)^{-1}$, which always exist for $|\lambda| > 1$, belong to $\mathfrak{R}_N$.

4) If $R_1, R_2 \in \mathfrak{K}(V)$ and $R(\zeta) = R_1(\zeta)R_2(\zeta)$, then

$$|R|_N \leqq |R_1|_N |R_2|_N.$$

The meaning of all of these conditions consists of the following. It is important for us to achieve fulfillment of a condition of the type of 4). We failed to achieve it in the old norm, and therefore some new (larger) norm $|\ |_N$ has been taken in which it does hold. However, it is necessary in this connection to narrow the class $\mathfrak{R}(V)$ under consideration to $\mathfrak{R}_N$.

As in § 3.1, a new multiplication

$$R_1 \circ R_2 = R \qquad (R_1, R_2 \in \mathfrak{K}(V))$$

is introduced on the linear manifold $\mathfrak{K}(V)$, where the operator R $(\in \mathfrak{K}(V))$ is determined by the polynomial $R(\zeta) = R_1(\zeta)R_2(\zeta)$. By virtue of property 4) this multiplication is continuous in the norm $|\ |_N$, and consequently it admits extension to all pairs of elements of $\mathfrak{R}_N$. Thus $\mathfrak{R}_N$ becomes a *commutative* Banach algebra (in the sense of the new multiplication).

Then by repeating the argument of § 3.1 we ascertain that Theorem 3.1 remains valid for any operator $A \in \mathfrak{R}_N$ (fulfillment of condition (Σ II) is not required in this connection).

If the operators V and $V^{(-1)}$ satisfy condition (Σ II), then obviously $\mathfrak{R}(V)$ satisfies conditions 1) – 4) in the original norm.

2. Let us now ascertain that the results of § 3.2 are also comprised in the scheme set forth. Let us show that conditions 1) – 4) are fulfilled for the linear manifold $\mathfrak{R}_N = \tilde{\mathfrak{R}}(U)$ in the norm

$$|X|_N = \inf |\tilde{X}|, \tag{4.1}$$

where the greatest lower bound is taken with respect to all operators $\tilde{X} \in \mathfrak{R}(U)$ for which the restriction of $P\tilde{X}P$ to $\mathfrak{B} = \operatorname{Im} P$ equals X. Let us denote by $\mathfrak{N}$ the set of all operators $\tilde{A}$ of $\mathfrak{R}(U)$ such that $P\tilde{A}P = 0$. The set $\mathfrak{N}$ is a closed ideal of $\mathfrak{R}(U)$.[7] As a matter of fact, it is sufficient to ascertain that if $\tilde{A} \in \mathfrak{N}$ and $R \in \mathfrak{K}(U)$, then $\tilde{A}R \in \mathfrak{N}$. This follows from the equality

$$P\tilde{A}RP = P\tilde{A}R_+P + P\tilde{A}R_-P = P\tilde{A}PR_+P + PR_-P\tilde{A}P = 0,$$

[6] The subscript N serves merely to denote a *new* linear manifold and a *new* norm.

[7] It can be ascertained without difficulty that $\mathfrak{N}$ belongs to the radical of the algebra $\mathfrak{R}(U)$.

where

$$R = \sum_{-n}^{n} \alpha_j U^j, \qquad R_+ = \sum_{0}^{n} \alpha_j U^j, \qquad R_- = \sum_{-n}^{-1} \alpha_j U^j.$$

Equality (4.1), which can be written in the form

$$|X|_N = \inf_{T \in \mathfrak{N}} |\tilde{X} + T|, \qquad P\tilde{X}P|\mathfrak{B} = X,$$

means that $\mathfrak{R}_N$ is isomorphic and isometric to the factor-algebra $\mathfrak{R}(U)/\mathfrak{N}$.

Now let us verify the fulfillment of conditions 1) — 4). The first two are obviously fulfilled. The third condition follows from the inequality

$$|V^{(n)}|_N \leqq |U^n| \qquad (n = 0, \pm 1, \cdots),$$

and the fourth, from the relations

$$|R|_N = \inf_{T \in \mathfrak{N}} |R(U) + T| = \inf_{T \in \mathfrak{N}} |R_1(U)R_2(U) + T|$$
$$\leqq \inf_{T_1, T_2 \in \mathfrak{N}} |(R_1(U) + T_1)(R_2(U) + T_2)| \leqq |R_1|_N |R_2|_N.$$

3. Now let us show that two special, in general distinct algebras $\mathfrak{R}_N$ satisfying conditions 1) – 4) are always contained in $\mathfrak{R}(V)$.

The first of these, the algebra $\tilde{\mathfrak{R}}_N$ (in a certain sense *minimal*), consists of all operators A each of which is representable as the sum of an absolutely convergent series $A = \sum_{-\infty}^{\infty} a_j V^{(j)}$, the new norm being defined by the equality

$$|A|_N = \sum_{-\infty}^{\infty} |a_j| \, |V^{(j)}|.$$

Only condition 4) needs verification. Let

$$R_1 = \sum \alpha_j V^{(j)}, \qquad R_2 = \sum \beta_j V^{(j)} \quad \text{and} \quad R(\zeta) = R_1(\zeta) R_2(\zeta).$$

Then

$$|R_1|_N |R_2|_N = \sum_{j,k} |\alpha_j| \cdot |\beta_k| \cdot |V^{(j)}| \cdot |V^{(k)}| \geqq \sum_{j,k} |\alpha_j| \cdot |\beta_k| \cdot |V^{(j+k)}| \geqq |R|_N.$$

Let us now construct the second algebra, the algebra $\bar{\mathfrak{R}}_N$ (in a certain sense *maximal*). Let R be an arbitrary operator of $\mathfrak{K}(V)$. Obviously it can be represented (in many ways!) in the form

$$R = R_- R_+ \qquad (R_\pm \in \mathfrak{K}^\pm(V)). \tag{4.2}$$

Let us introduce a new (commutative) multiplication $R \circ X$ for any $X \in \mathfrak{R}(V)$ and $R \in \mathfrak{K}(V)$ by setting

$$R \circ X = R_- X R_+, \tag{4.3}$$

where $R_\pm$ is defined by (4.2). It is easy to see that this multiplication does not depend on the representation method (4.2) and is defined uniquely by the operator R. If $X \in \mathfrak{K}(V)$, multiplication (4.3) coincides with the one introduced earlier.

For each $R \in \mathfrak{K}(V)$ let us introduce the linear operator T_R operating in $\mathfrak{R}(V)$ by the rule

$$T_R X = R \circ X. \tag{4.4}$$

Since the operator T_R is obviously closed and defined on all of $\mathfrak{R}(V)$, it is bounded. It is easy to see that for any $R_1, R_2 \in \mathfrak{K}(V)$ we have

$$T_{R_1} T_{R_2} = T_{R_1 \circ R_2}. \tag{4.5}$$

Let us introduce the norm $| \ |_N$ on the linear manifold $\mathfrak{R}(V)$ by setting $|R|_N = |T_R|$. That it satisfies the relations

$$|R|_N \geqq |R|, \qquad |R_1 \circ R_2|_N \leqq |R_1|_N \cdot |R_2|_N \tag{4.6}$$

follows from (4.4) and (4.5).

Let us denote by $\bar{\mathfrak{R}}_N$ the closure of $\mathfrak{R}(V)$ in the norm $| \ |_N$. Obviously $\bar{\mathfrak{R}}_N \in \mathfrak{A}(\mathfrak{R}(V))$. Let us show that $\bar{\mathfrak{R}}_N$ can be identified with some part of $\mathfrak{R}(V)$. Since $| \ |_N$ is by virtue of (4.6) stronger than the original norm $| \ |$, it is sufficient to ascertain that these norms are *consistent*, i. e. it is sufficient to ascertain that the relations

$$\lim_{n\to\infty} |R_n - S|_N = 0, \qquad \lim_{n\to\infty} |R_n| = 0,$$

where $S \in \mathfrak{A}(\mathfrak{R}(V))$, imply $S = 0$.

By virtue of the equalities

$$R_n \circ V^{(m)} = \begin{cases} R_n V^{(m)}, & m \geqq 0, \\ V^{(m)} R_n, & m < 0 \end{cases}$$

we have for any $R \in \mathfrak{R}(V)$

$$SR = \lim_{n\to\infty} T_{R_n} R = \lim_{n\to\infty} R_n \circ R = \lim_{n\to\infty} (R_- R_n + R_n R_+) = 0,$$

where

$$R = \sum_j \alpha_j V^{(j)}, \qquad R_+ = \sum_{j \geqq 0} \alpha_j V^{(j)}, \qquad R_- = \sum_{j<0} \alpha_j V^{(j)}.$$

Thus $S = 0$.

Relations (4.6) imply that the algebra $\bar{\mathfrak{R}}_N$ satisfies conditions 1) – 4).

§ 5. Factorization of functions and its application to the inversion of operators

In the proof of Theorem 1.1 (the theorem on invertibility of polynomials in unilaterally invertible operators) the representation of the polynomial $R(\zeta)$ in the form (1.24):

$$R(\zeta) = c \prod_{j=1}^{k} (1 - \zeta_j^+ \zeta^{-1}) \zeta^\kappa \prod_{j=1}^{s} (\zeta - \zeta_j^-),$$

where $|\zeta_j^+| < 1$ and $|\zeta_j^-| > 1$, was essential. This *factorization of the polynomial* permitted an effective formula for inverting the operator R on the appropriate side to be obtained easily.

Effective formulas for inverting operators of $\mathfrak{R}(V)$ can also be obtained for a significantly more general case.

Let the operators V and $V^{(-1)}$ satisfy condition (Σ I), and let the correspondence between the sets $\mathfrak{R}(V)$ and $\mathfrak{R}(\zeta)$ be one-to-one. Let us denote by $\mathfrak{R}^+(\zeta)$ ($\mathfrak{R}^-(\zeta)$) the set of symbols $A(\zeta)$ corresponding to all operators $A \in \mathfrak{R}^+(V)$ ($\mathfrak{R}^-(V)$). If the symbol $A(\zeta) \in \mathfrak{R}(\zeta)$ admits *factorization* of the form

$$A(\zeta) = A_-(\zeta)\zeta^\kappa A_+(\zeta), \tag{5.1}$$

where $A_\pm(\zeta) \in \mathfrak{R}^\pm(\zeta)$, $A_\pm^{-1}(\zeta) \in \mathfrak{R}^\pm(\zeta)$, then A can be represented in the form

$$A = A_- V^{(\kappa)} A_+, \tag{5.2}$$

and the operator $A^{(-1)}$, the inverse of A on the appropriate side, is obtained by means of the equality

$$A^{(-1)} = A_+^{-1} V^{(-\kappa)} A_-^{-1}, \tag{5.3}$$

where the operators $A_\pm^{-1}$ correspond to the symbols $A_\pm^{-1}(\zeta)$.

Similar reasoning will be applied rather frequently. Its basic aim is to obtain an equality of the type (5.2), the *factorization* of the operator A, which permits that operator to be easily inverted.

As a rule, factorization of an operator is induced by *preliminary* factorization of the corresponding function. However, passage from functions to operators can be effected easily only if the algebra $\mathfrak{R}_0(V)$ does not have a radical.

Sometimes one manages to obtain factorization of the type (5.2) *immediately*, without the preliminary factorization of functions. This is important if the question of the radical is open.

After these general remarks let us pass to the question of the possibility of factoring the continuous functions of one or another algebra. The expansion of a continuous function $a(\zeta)$ ($|\zeta| = 1$) into the three factors

$$a(\zeta) = a_-(\zeta)\zeta^{(\kappa)} a_+(\zeta) \tag{5.4}$$

is called its *factorization*. Here κ is an integer; $a_+(\zeta)$ ($|\zeta| = 1$) is a function admitting continuation which is holomorphic inside the unit disk and continuous in the region $|\zeta| \leqq 1$, where $a_+(\zeta) \neq 0$ ($|\zeta| \leqq 1$).

Analogous properties (with the change of the region $|\zeta| \leqq 1$ to $|\zeta| \geqq 1$) must be possessed by $a_-(\zeta)$.

Let us agree to call some Banach algebra $\mathscr{C}$ consisting of functions continuous on the unit circle Γ an *R-algebra* if it contains the set of all rational functions with poles *not belonging to* Γ and if this set of rational functions is dense in $\mathscr{C}$.

It is easy to see that the linear span of the functions ζ^j ($j = 0, \pm 1, \cdots$) is a dense set in every R-algebra.

Let us denote by $\mathscr{C}^+$ ($\mathscr{C}^-$) the closure (in the norm of the R-algebra $\mathscr{C}$) of the linear span of the functions $1, \zeta, \cdots$ ($1, \zeta^{-1}, \cdots$).

It is not difficult to verify that if $\mathscr{C}$ is an R-algebra, then all functions of $\mathscr{C}^+$ ($\mathscr{C}^-$) admit holomorphic continuation into the region $|\zeta| < 1$ ($|\zeta| > 1$) and the sets of maximal ideals of the algebras $\mathscr{C}$, $\mathscr{C}^+$ and $\mathscr{C}^-$ are homeomorphic respectively to the sets $|\zeta| = 1$, $|\zeta| \leqq 1$ and $|\zeta| \geqq 1$.

Let us introduce one more subalgebra, the subalgebra $\mathscr{C}_0^-$, which is the closure of the linear span of the functions $\zeta^{-1}, \zeta^{-2}, \cdots$. Obviously $\mathscr{C}_0^-$ is the subalgebra of $\mathscr{C}^-$ which consists of all functions which vanish at infinity.

Let us call the R-algebra $\mathscr{C}$ *decomposing* if it is the direct sum of its subalgebras $\mathscr{C}^+$ and $\mathscr{C}_0^-$: $\mathscr{C} = \mathscr{C}^+ \dotplus \mathscr{C}_0^-$.[8)]

[8)] Instead of this decomposition, it is of course possible to take the analogous one $\mathscr{C} = \mathscr{C}_0^+ \dotplus \mathscr{C}^-$.

Let $f(\zeta)$ belong to some R-algebra $\mathscr{C}$ and let f_j ($j = 0, \pm 1, \cdots$) be its Fourier coefficients: $f(\zeta) = \sum_{-\infty}^{\infty} f^j \zeta^j$. Let us set

$$f_+(\zeta) = \sum_{0}^{\infty} f_j \zeta^j \qquad (|\zeta| = 1).$$

Obviously $f_+(\zeta) \in L_2$. It is easy to see that decomposition of the algebra $\mathscr{C}$ is equivalent to the fact that for any function $f(\zeta)$ of $\mathscr{C}$ the function $f_+(\zeta)$ also belongs to $\mathscr{C}$:

The Wiener algebra W of all functions $a(\zeta)$ ($|\zeta| = 1$) expanding into the absolutely convergent Fourier series

$$a(\zeta) = \sum_{-\infty}^{\infty} a_j \zeta^j \qquad \left(\sum_{-\infty}^{\infty} |a_j| < \infty\right)$$

can serve as an example of a decomposing algebra.

The norm of the function $a(\zeta)$ in W is defined by the equality

$$\|a(\zeta)\| = \sum_{-\infty}^{\infty} |a_j|.$$

The algebra C of all continuous functions on the unit circle is nondecomposing. Indeed, let us consider the function

$$f(\zeta) = \frac{1}{2i} \sum_{n=2}^{\infty} \frac{1}{n \log n} (\zeta^n - \zeta^{-n}) \qquad (|\zeta| = 1).$$

The series on the right side of the equality converges uniformly (cf. N. K. Bari [1], Chapter VIII, § 13); consequently the function $f(\zeta) \in C$. The function

$$f_+(\zeta) = \frac{1}{2i} \sum_{n=2}^{\infty} \frac{1}{n \log n} \zeta^n \qquad (|\zeta| = 1)$$

is continuous everywhere on the unit circle except the point $\zeta = 1$, at which it has a discontinuity of the second kind (cf. Bari [1], loc. cit.).

THEOREM 5.1. *For every nonvanishing function $a(\zeta)$ of the R-algebra $\mathscr{C}$ to admit factorization* (5.4) *with factors $a_\pm(\zeta) \in \mathscr{C}^\pm$, it is necessary and sufficient that the algebra $\mathscr{C}$ be decomposing. If this condition is fulfilled, the number κ in factorization* (5.4) *is defined by the equality* $\kappa = \operatorname{ind} a(\zeta)$.

The proof of this theorem is based on the following general lemma concerning the factorization of elements near the identity element.

LEMMA 5.1. *Let $\mathfrak{A}$ be a Banach algebra with identity element e, and let $\mathfrak{A}_\pm$ be subalgebras of $\mathfrak{A}$ whose direct sum equals $\mathfrak{A}$, P a projection projecting $\mathfrak{A}$ onto $\mathfrak{A}_+$ parallel to $\mathfrak{A}_-$, and $Q = I - P$. If the element $a \in \mathfrak{A}$ satisfies the condition*

$$|a| < \min(|P|^{-1}, |Q|^{-1}), \tag{5.5}$$

the element $e - a$ admits the following factorization:

$$e - a = (e + b_-)(e + b_+), \tag{5.6}$$

where $b_\pm \in \mathfrak{A}_\pm$ and $(e + b_\pm)^{-1} - e \in \mathfrak{A}_\pm$.

PROOF. Let us consider the equation

$$x - Pax = e \tag{5.7}$$

in the algebra $\mathfrak{A}$.

By virtue of condition (5.5) this equation has a unique solution which obviously has the form $x = e + x_+$, where $x_+ \in \mathfrak{A}_+$. The equality

$$(e - a)(e + x_+) = e + b_- \qquad (b_- \in \mathfrak{A}_-) \tag{5.8}$$

follows from (5.7)

Analogously, considering the equation $x - Qxa = e$, we obtain the equality

$$(e + x_-)(e - a) = e + b_+ \qquad (x_- \in \mathfrak{A}_-, \, b_+ \in \mathfrak{A}_+)$$

or

$$(e + b_+)(e - a)^{-1} = e + x_-. \tag{5.9}$$

Multiplying equalities (5.8) and (5.9) termwise, we obtain

$$(e + b_+)(e + x_+) = (e + x_-)(e + b_-)$$

or

$$b_+ + x_+ + b_+x_+ = b_- + x_- + x_-b_-.$$

Since the algebras $\mathfrak{A}_+$ and $\mathfrak{A}_-$ intersect only at zero, both sides of the latter equality are zero; consequently

$$(e + b_+)(e + x_+) = (e + x_-)(e + b_-) = e. \tag{5.10}$$

If the algebra $\mathfrak{A}$ is commutative, the proof of the lemma is completed with this, since (5.10) implies that the elements $e + b_\pm$ are invertible and their inverses have the same form, while (5.8) implies factorization (5.6).

In the general case (5.10) implies only unilateral invertibility of the elements $e + b_\pm$.

To prove that the elements $e + x_\pm$ are bilateral inverses of $e + b_\pm$, let us observe that the entire argument just developed remains valid if the element a is replaced by λa, where $0 \leqq \lambda \leqq 1$. Under this substitution the elements $b_\pm$ are replaced by analytic functions $b_\pm(\lambda)$, $0 \leqq \lambda \leqq 1$, the element $e + b_+(\lambda)$ being right invertible at each point λ and the element $e + b_+(0) = e$ being bilaterally invertible. The bilateral invertibility of the element $e + b_+ = e + b_+(1)$ follows from the following proposition (cf. the book [5] of Gohberg and Kreĭn).

Let $\mathfrak{A}$ be a Banach algebra with identity element, and let $a(\lambda)$ be a continuous function mapping the segment $\Delta = \{\lambda : 0 \leqq \lambda \leqq 1\}$ into $\mathfrak{A}$. If at each point $\lambda \in \Delta$

the element $a(\lambda)$ *is invertible on one side and if* $a(\lambda_0)$ *is bilaterally invertible at the one point* $\lambda_0 \in \Delta$, *then* $a(\lambda)$ *is bilaterally invertible at all points* $\lambda \in \Delta$.

This proposition implies that $e + x_\pm$ is the inverse of $e + b_\pm$. The lemma is proved.

PROOF OF THEOREM 5.1. Let $\mathscr{C}$ be a decomposing algebra, let P be a projection projecting $\mathscr{C}$ onto $\mathscr{C}^+$ parallel to $\mathscr{C}_0^-$, $Q = I - P$, and let the function $a(\zeta) \in \mathscr{C}$ not vanish on the unit circle. Let us denote by $r(\zeta)$ a polynomial in integral powers of ζ such that the condition

$$|b| < \min(|P|^{-1}, |Q|^{-1}) \tag{5.11}$$

is fulfilled for the function $b(\zeta) = a(\zeta)r^{-1}(\zeta) - 1$.

When applied to the algebras $\mathfrak{A} = \mathscr{C}$, $\mathfrak{A}_+ = \mathscr{C}^+$ and $\mathfrak{A}_- = \mathscr{C}_0^-$, Lemma 5.1 reduces to the factorization of the function $1 + b(\zeta)$, i.e. to the equality

$$a(\zeta)r^{-1}(\zeta) = b_-(\zeta)b_+(\zeta), \tag{5.12}$$

where

$$b_\pm(\zeta) \in \mathscr{C}_\pm, \qquad b_+(\zeta) \neq 0 \quad (|\zeta| \leqq 1), \qquad b_-(\zeta) \neq 0 \quad (|\zeta| \geqq 1).$$

Condition (5.11) and the easily verifiable relation

$$\max_{|\zeta|=1} |a(\zeta)r^{-1}(\zeta) - 1| \leqq |b|$$

imply that

$$r(\zeta) \neq 0 \quad (|\zeta| = 1) \qquad \text{and} \qquad \text{ind } r(\zeta) = \text{ind } a(\zeta) = \kappa.$$

As already noted in the proof of Theorem 1.1, under these conditions the polynomial $r(\zeta)$ admits the factorization

$$r(\zeta) = r_-(\zeta)\zeta^\kappa r_+(\zeta), \tag{5.13}$$

where $r_\pm(\zeta)$ $(\in \mathscr{C}^\pm)$ are polynomials, with

$$r_+(\zeta) \neq 0 \quad (|\zeta| \leqq 1), \qquad r_-(\zeta) \neq 0 \quad (|\zeta| \geqq 1).$$

The factorization of the function $a(\zeta)$ follows from equalities (5.12) and (5.13).

Let us pass to the proof of the necessity of the theorem's conditions. Let $a(\zeta)$ be an arbitrary function of $\mathscr{C}$ and $b(\zeta) = \exp a(\zeta)$. Then

$$b(\zeta) \neq 0 \quad (|\zeta| = 1), \qquad \text{ind } b(\zeta) = 0;$$

consequently $b(\zeta)$ admits the factorization

$$b(\zeta) = b_-(\zeta)b_+(\zeta) \qquad (b_\pm(\zeta) \in \mathscr{C}^\pm,\ b_-(\infty) = 1).$$

Taking into account that ind $b(\zeta) = 0$, we obtain by virtue of a theorem of G. E. Šilov [1] the fact that the functions $a_\pm(\zeta) = \log b_\pm(\zeta) \in \mathscr{C}^\pm$, while $a_-(\zeta) \in \mathscr{C}_0^-$. Thus $a(\zeta) = a_+(\zeta) + a_-(\zeta)$, i.e. the algebra $\mathscr{C}$ is decomposing.

Let us note that the sufficiency of the theorem's conditions can be obtained immediately from the aforementioned theorem of G. E. Šilov. As a matter of fact, if $a(\zeta) \neq 0$ ($|\zeta| = 1$) and $\kappa = \operatorname{ind} a(\zeta)$, then by Šilov's theorem the function $c(\zeta) = \log \zeta^{-\kappa} a(\zeta) \in \mathscr{C}$. Since the algebra $\mathscr{C}$ is decomposing, $c(\zeta)$ can be represented in the form

$$c(\zeta) = c_+(\zeta) + c_-(\zeta) \qquad (c_\pm(\zeta) \in \mathscr{C}^\pm),$$

which implies that

$$a(\zeta) = e^{c_-(\zeta)}\, \zeta^\kappa e^{c_+(\zeta)}.$$

It is easy to verify that this equality is a factorization of the function $a(\zeta)$.

Now let a unilaterally invertible operator V satisfy conditions (Σ I) and (Σ II). Then $\mathfrak{R}(\zeta)$ is an R-algebra. If this algebra is decomposing and the correspondence between $\mathfrak{R}(V)$ and $\mathfrak{R}(\zeta)$ is one-to-one[9] (i.e. the algebra $\mathfrak{R}_0(V)$ does not have a radical), effective formulas for inverting the operators of $\mathfrak{R}(V)$ can be obtained.

Indeed, let the condition $A(\zeta) \neq 0$ ($|\zeta| = 1$) be fulfilled for an operator $A \in \mathfrak{R}(V)$. Then by virtue of Theorem 5.1 the function $A(\zeta)$ admits factorization (5.1), from which formula (5.3) follows for an operator inverse to A on the appropriate side.

Let us note that if the space $\mathfrak{R}(\zeta)$ is decomposing in the sense that every function $a(\zeta) \in \mathfrak{R}(\zeta)$ is uniquely representable in the form

$$a(\zeta) = a_+(\zeta) + a_-(\zeta) \qquad (a_\pm(\zeta) \in \mathfrak{R}^\pm(\zeta),\ a_-(\infty) = 0),$$

then $\mathfrak{R}(\zeta)$ automatically turns out to be a Banach R-algebra.

§ 6. Solution of equations with unilaterally invertible operators of $\mathfrak{R}(V)$

1. It is assumed in this section that the operators V and $V^{(-1)}$ satisfy conditions (Σ I) and (Σ II). Let the symbol of an operator A of $\mathfrak{R}(V)$ not vanish:

$$A(\zeta) \neq 0 \qquad (|\zeta| = 1). \tag{6.1}$$

The solution of the equation

$$A\varphi = f \tag{6.2}$$

can be reduced to an equation with an invertible operator.

This can be done by the schemes described by propositions 5° – 8° of § 1. Namely, let $\kappa = \operatorname{ind} A(\zeta) > 0$. Then the operator A can be represented in the form

$$A = BV^\kappa, \tag{6.3}$$

where $B \in \mathfrak{R}(V)$ is an operator with symbol $B(\zeta) = \zeta^{-\kappa} A(\zeta)$. Obviously $\operatorname{ind} B(\zeta) = 0$; consequently B is invertible. Representation (6.3) permits proposition 5°, § 1, to be applied.

[9] Let us note that the results cited below also retain their validity without this condition. The fact is that a theorem concerning factorization can be proved for the algebra $\mathfrak{R}_0(V)$.

Multiplying equality (6.3) on the left by $V^{(-\kappa)}$ and taking into account that $V^{(-\kappa)}BV^{\kappa} = B$, we obtain

$$V^{(-\kappa)}A = B. \tag{6.3'}$$

This equality in turn permits proposition 6° to be applied.

In case $\kappa < 0$ the equalities

$$A = V^{(\kappa)}B \tag{6.4}$$

and

$$AV^{-\kappa} = B \tag{6.4'}$$

are established analogously, where B is an operator with symbol $B(\zeta) = \zeta^{-\kappa}A(\zeta)$. These equalities permit propositions 7° and 8° to be applied.

Let us call the operator B appearing here an operator with a *canceled* index.

2. The following theorem makes the structure of the kernel of an operator A more precise.

THEOREM 6.1. *Let condition* (6.1) *be fulfilled for an operator* $A \in \mathfrak{R}(V)$, *and let* $\kappa < 0$. *Then*

$$\dim \operatorname{Ker} A = |\kappa| \dim \operatorname{Ker} V^{(-1)}.$$

If in addition A is representable in the form

$$A = A_- V^{(\kappa)} A_+ \quad (A_\pm \in \mathfrak{R}^\pm(V),\ A_\pm^{-1} \in \mathfrak{R}^\pm(V)), \tag{6.5}$$

then there exists a subspace $\mathfrak{L} \subset \mathfrak{B}$ *of dimension* $\dim \operatorname{Ker} V^{(-1)}$ *such that*

$$\operatorname{Ker} A = \mathfrak{L} \dotplus V\mathfrak{L} \dotplus \cdots \dotplus V^{|\kappa|-1}\mathfrak{L}. \tag{6.6}$$

PROOF. Let us first consider the elementary case when $A = V^{(-n)}$, $n = 1,2,\cdots$. Let $\mathfrak{L}_n = \operatorname{Ker} V^{(-n)}$, and let us show that $\mathfrak{L}_n$ can be represented as the direct sum

$$\mathfrak{L}_n = \mathfrak{L}_1 \dotplus V\mathfrak{L}_1 \dotplus \cdots \dotplus V^{n-1}\mathfrak{L}_1. \tag{6.7}$$

It is sufficient for this purpose to establish the equality

$$\mathfrak{L}_{k+1} = \mathfrak{L}_k \dotplus V^k\mathfrak{L}_1 \qquad (k = 1, 2, \cdots). \tag{6.8}$$

The subspaces $\mathfrak{L}_k$ and $V^k\mathfrak{L}_1$ intersect only at zero, since the equality $V^k g = f$ ($f \in \mathfrak{L}_k$, $g \in \mathfrak{L}_1$) implies $g = V^{(-k)}f = 0$. The inclusion $\mathfrak{L}_k \dotplus V^k\mathfrak{L}_1 \subset \mathfrak{L}_{k+1}$ obviously holds.

Let x be an arbitrary vector of $\mathfrak{L}_{k+1}$. Let us set $z = V^{(-k)}x$ and $y = x - V^k z$. Then $z \in \mathfrak{L}_1$, $y \in \mathfrak{L}_k$, and thus equality (6.8) and consequently also (6.7) are established.

Now let us pass to the general case. As we established in the proof of Theorem 3.1 (cf. equality (3.6)), the operator A can be represented in the form

$$A = R_- V^{(\kappa)}(I + C)R_+,$$

where the operators $R_\pm$ ($\in \mathfrak{R}(V)$) and $I + C$ are invertible.

Equality (6.7) implies that the subspace Ker A can be represented as the direct sum

$$\text{Ker } A = B\mathfrak{L}_1 \dotplus BV\mathfrak{L}_1 \dotplus \cdots \dotplus BV^{|\kappa|-1}\mathfrak{L}_1, \tag{6.9}$$

where $B = R_+^{-1}(I + C)^{-1}$. This implies the first assertion of the theorem.

If equality (6.5) is assumed fulfilled, we obtain (6.9) from (6.7), where the operator $B = A_+^{-1}$ commutes with V^k ($k = 1, 2, \cdots$). Thus equality (6.9) takes the form

$$\text{Ker } A = \mathfrak{L} \dotplus V\mathfrak{L} \dotplus \cdots \dotplus V^{|\kappa|-1}\mathfrak{L},$$

where $\mathfrak{L} = B\mathfrak{L}_1$.

REMARK. The last assertion of the theorem remains valid if instead of representation (6.5) it is required that dim Ker $V^{(-1)} < \infty$ and for any vector $f \in \mathfrak{B}$ ($f \neq 0$) the vectors $f, Vf, \cdots, V^n f$ ($n = 1, 2, \cdots$) be linearly independent.

Indeed, let us form the subspace

$$\mathfrak{N} = \mathfrak{L} \dotplus V\mathfrak{L} \dotplus \cdots \dotplus V^{|\kappa|-1}\mathfrak{L},$$

where $\mathfrak{L} = B\mathfrak{L}_1$ and as before B equals $R_+^{-1}(I + C)^{-1}$. Let us show that $\mathfrak{N} =$ Ker A. Since the dimensions of these subspaces coincide (dim $\mathfrak{N} =$ dim Ker A), it is sufficient to establish the inclusion $\mathfrak{N} \subset$ Ker A, or, what is the same thing, the equality

$$V^{(\kappa)}(I + C)R_+ f = 0 \qquad (f \in \mathfrak{N}).$$

Let us observe that for any operator F of $\mathfrak{N}(V)$ the equality $V^{(-n)}FV^n = F$ is valid for any $n \geqq 0$. This is obvious if $F \in \mathfrak{R}(V)$, and in the general case it is established by passage to the limit. Therefore for any natural n

$$V^{(-n)}(I + C)\, R_+ V^n = (I + C)\, R_+.$$

Let f be an arbitrary vector of $\mathfrak{N}$. Since it can be represented in the form

$$f = \sum_{j=0}^{|\kappa|-n} V^j B f_j \qquad (f_j \in \mathfrak{L}_1, j = 1, 2, \cdots, |\kappa| - 1),$$

it follows that

$$V^{(\kappa)}(I + C)R_+ f = \sum_{j=0}^{|\kappa|-1} V^{(\kappa+j)} f_j = 0.$$

Now let us pass to applications of the general theory.

§ 7. Discrete Wiener-Hopf equations

1. Let l_2 be the Hilbert space of sequences of complex numbers $\{\xi_j\}_1^\infty$, and V an isometric operator defined in l_2 by the equality

$$V\{\xi_j\}_1^\infty = \{0, \xi_1, \xi_2, \cdots\}.$$

The operator $V^{(-1)}$ defined by the equality

$$V^{(-1)}\{\xi_j\}_1^\infty = \{\xi_{j+1}\}_1^\infty$$

is obviously a left inverse of V, and $|V| = |V^{(-1)}| = 1$.

Let $a(\zeta)$ be an arbitrary function continuous on the unit circle, and a_j ($j = 0, \pm 1, \cdots$) its Fourier coefficients. By virtue of proposition 2° of § 2 there exists a unique operator $A \in \mathfrak{R}(V)$ whose symbol coincides with $a(\zeta)$: $A(\zeta) = a(\zeta)$. (Here the algebra $\mathfrak{R}_0(V)$ is without radical!) Let us show that the operator A is defined in l_2 by the matrix

$$\|a_{j-k}\|_{j,k=1}^\infty. \tag{7.1}$$

Let $R_n(\zeta) = \sum_{-n}^{n} \alpha_j^{(n)}\zeta^j$ ($n = 1, 2, \cdots$) be a sequence of polynomials converging uniformly to $a(\zeta)$. The sequence of operators $R_n \in \mathfrak{R}(V)$ defined by the matrices

$$\|a_{j-k}^{(n)}\|_{j,k=1}^\infty \qquad (\alpha_j^{(n)} = 0 \text{ for } |j| > n)$$

converges to some operator $B \in \mathfrak{R}(V)$, because

$$|R_{n+p} - R_n| = \max_{|\zeta|=1} |R_{n+p}(\zeta) - R_n(\zeta)| \to 0$$

by virtue of the strict equality (2.5).

Since the respective Fourier coefficients of the functions $R_n(\zeta)$ tend to the Fourier coefficients of the function $a(\zeta)$, we have $B = A$ and $A(\zeta) = a(\zeta)$.

By virtue of what has been proved the function $a(\zeta)$ can be called the *symbol* of the operator defined in l_2 by matrix (7.1).

Theorem 3.1 now yields the following result.

THEOREM 7.1. *Let $a(\zeta)$ be an arbitrary function continuous on the unit circle, and a_j ($j = 0, \pm 1, \cdots$) its Fourier coefficients. For an operator A defined in l_2 by the matrix* (7.1) *to be invertible on at least one side, it is necessary and sufficient that its symbol not vanish.*

If the latter condition is fulfilled, the invertibility of A will be consistent with the index of the function $a(\zeta)$.

In other words, *the invertiblity of A is determined by its symbol.*

We obtain as a realization of propositions 5°–8° the following proposition concerning a discrete Wiener-Hopf equation, i.e. concerning the equation

$$\sum_{k=1}^\infty a_{j-k}\xi_k = \eta_j \qquad (j = 1, 2, \cdots), \tag{7.2}$$

considered in the space l_2.

Let us first explain how equations with canceled index (cf. equalities (6.3), (6.3′)

and (6.4)) look in (7.2). If $\kappa > 0$, an equation with canceled index has the form

$$\sum_{k=1}^{\infty} a_{j-k}\xi_k = \eta_j \qquad (j = \kappa + 1, \kappa + 2, \cdots), \tag{7.3}$$

i.e. it is obtained from the original by discarding the first κ equations (see (1.10) and (6.3')).

If, however, $\kappa < 0$, the "canceled" equation has the form

$$\sum_{k=-\kappa+1}^{\infty} a_{j-k}\xi_k = \eta_j \qquad (j = 1,2,\cdots), \tag{7.4}$$

i.e. it is obtained from the original if we set $\xi_1 = \cdots = \xi_{-\kappa} = 0$ (see (1.18) and (6.4')). Let us observe also that the condition of solvability in case $\kappa > 0$ takes the form

$$\sum_{k=1}^{\infty} a_{j-k}\xi_k = \eta_j \qquad (j = 1,2,\cdots,\kappa). \tag{7.5}$$

These are the first κ equations discarded when obtaining (7.3) (see (1.11)).

All assertions of the general propositions 6° and 8° of § 1 are retained here and become very clear. The canceled equation is always uniquely solvable. The original equation, not always solvable for $\kappa > 0$, is so only under fulfillment of condition (7.5), in which $\xi = \{\xi_j\}$ is the solution of the canceled equation. A solution of the original equation reduces in an obvious way to the solution of the "partial" system (7.3) and its "complement" (7.5).

If, however, $\kappa < 0$, then by forming a vector ξ from the solution of equation (7.4):

$$\xi = \{\underbrace{0, 0, \cdots, 0}_{-\kappa}, \xi_{-\kappa+1}, \cdots, \xi_n, \cdots\},$$

we obtain one of the solutions of the original equation (7.2).

The general solution of the homogeneous system

$$\sum_{k=1}^{\infty} a_{j-k}\xi_k = 0 \qquad (j = 1, 2, \cdots) \tag{7.6}$$

is given by the equality

$$\xi = \{x_1, x_2, \cdots, x_{-\kappa}, \xi_{-\kappa+1}, \xi_{-\kappa+2}, \cdots\},$$

where the x_j $(j = 1, 2, \cdots, -\kappa)$ are arbitrary numbers and the vector $\{\xi_j\}_{-\kappa+1}^{\infty}$ is the unique solution of the system

$$\sum_{k=-\kappa+1}^{\infty} a_{j-k}\xi_k = -\sum_{k=1}^{-\kappa} a_{j-k}x_k \qquad (j = 1,2,\cdots)$$

(see proposition 8°, § 1). We leave it to the reader to formulate realizations of propositions 5° and 7° of § 1.

Let us note that Theorem 7.1 and the subsequent assertions are valid in each of the spaces l_p ($p \geqq 1$), m, c and c_0 only if the function $a(\zeta)$ belongs to the corresponding algebra $\mathfrak{R}(\zeta)$. (For example, if $a(\zeta)$ expands into an absolutely convergent Fourier series, it belongs to all of the corresponding algebras.) For all of these spaces the algebra $\mathfrak{R}_0(V)$ does not have a radical.

2. The general solution of the homogeneous system (7.6) admits yet another description. This system can be considered in one of the above-enumerated spaces l_p ($p \geqq 1$), m, c or c_0 on the assumption that the function $a(\zeta)$ belongs to the corresponding algebra $\mathfrak{R}(\zeta)$. By virtue of the remark to Theorem 6.1, there is a vector

$$\xi = \{\xi_j\}_1^\infty \tag{7.7}$$

in the space under consideration such that the vectors

$$\xi^{(r)} = \{\underbrace{0,\cdots,0}_{r-1}, \xi_1, \xi_2, \cdots\} \qquad (r = 1,2,\cdots,|\kappa|)$$

form a basis for all solutions of system (7.6).

3. Through the example of a discrete Wiener-Hopf equation let us now illustrate the rule for constructing an inverse operator of A involving factorization. Let the operator $A \in \mathfrak{R}(V)$ again be defined by matrix (7.1). If its symbol $A(\zeta)$ ($= a(\zeta)$) admits factorization (5.1):

$$A(\zeta) = A_-(\zeta)\zeta^\kappa A_+(\zeta), \tag{7.8}$$

then an operator $A^{(-1)}$, the appropriate inverse of A, is defined by equality (5.3). Direct verification shows that the matrix of $A^{(-1)}$ is determined here as follows. Let us represent the function $1/A(\zeta)$ in the form

$$A^{-1}(\zeta) = B_-(\zeta)B_+(\zeta),$$

where

$$B_+(\zeta) = \begin{cases} A_+^{-1}(\zeta), & \kappa \geqq 0, \\ A_+^{-1}(\zeta)\zeta^{-\kappa}, & \kappa < 0, \end{cases} \qquad B_-(\zeta) = \begin{cases} A_-^{-1}(\zeta)\zeta^{-\kappa}, & \kappa \geqq 0, \\ A_-^{-1}(\zeta), & \kappa < 0. \end{cases}$$

Let us denote by $\gamma_j^{(1)}$ and $\gamma_j^{(2)}$ ($j = 0,1,\cdots$) the Fourier coefficients of the functions $B_+(\zeta)$ and $B_-(\zeta)$, respectively. Then

$$A^{(-1)} = \|\gamma_{jk}\|_{j,k=1}^\infty, \quad \text{where } \lambda_{jk} = \sum_{r=1}^{\min(j,k)} \gamma_{j-r}^{(1)}\gamma_{k-r}^{(2)}.$$

Let us note, moreover, that if the symbol $A(\zeta)$ admits the factorization (7.8), then as the vector (7.7) we can take the vector $\xi = \{c_0, c_1, \cdots\}$, where the c_j ($j = 0,1,\cdots$) are the Fourier coefficients of the function $A_+^{-1}(\zeta)$.

§ 8. Wiener-Hopf integral equations

Now let us illustrate the general theory through an example of Wiener-Hopf integral equations.

Let us denote by E one of the Banach[10] spaces L_p $(1 \leqq p < \infty)$, M, M_c, M_u, C or C_0 of complex-valued functions defined on the positive semi-axis $[0, \infty]$. Criteria are established in this section for the invertibility and unilateral invertibility of operators of the form

$$(A\varphi)(t) = \varphi - \int_0^\infty k(t-s)\varphi(s)\,ds \qquad (0 \leqq t < \infty), \tag{8.1}$$

operating in E, where $k(t) \in \tilde{L}_1$ $(= L_1(-\infty, \infty))$. The proof is carried out in accordance with the two schemes set forth in §§ 3.1 and 3.2.

Let us consider the operators

$$\begin{aligned} (V\varphi)(t) &= \varphi(t) - 2\int_{-\infty}^t e^{s-t}\varphi(s)\,ds, \\ (V^{(-1)}\varphi)(t) &= \varphi(t) - 2\int_t^\infty e^{t-s}\varphi(s)\,ds \end{aligned} \qquad (0 < t < \infty)$$

in the space E.

The operator $V^{(-1)}$ is a left inverse of the operator V. It is not difficult to verify that V and $V^{(-1)}$ satisfy condition $(\Sigma\,\mathrm{I})$.

1. Let $\tilde{E}$ be one of the Banach spaces $\tilde{L}_p$ $(1 \leqq p < \infty)$, $\tilde{M}_c$, $\tilde{M}_u$, $\tilde{C}$ or $\tilde{C}_0$ of complex-valued functions defined on the entire real axis. The space $\tilde{E}$ can be regarded as an extension of the corresponding space E.[11] First let us apply the second scheme of § 3.

It is not difficult to verify[12] that the operators U, U^{-1} and P defined in E by the equalities

$$\begin{aligned} (U\varphi)(t) &= \varphi(t) - 2\int_{-\infty}^t e^{s-t}\varphi(s)\,ds. \\ (U^{-1}\varphi)(t) &= \varphi(t) - 2\int_t^\infty e^{t-s}\varphi(s)\,ds, \end{aligned} \qquad (-\infty < t < \infty)$$

$$(P\varphi)(t) = \begin{cases} \varphi(t), & 0 < t < \infty, \\ 0, & -\infty < t < 0, \end{cases}$$

satisfy conditions 1) and 2) of § 3.2, and that the set $\mathfrak{R}(U)$ consists of all operators

[10] The spaces M_c and M_u are subspaces of the space M of all measurable bounded functions. The first consists of all continuous functions, and the second, of all uniformly continuous functions. C denotes the space of all continuous functions $f(t)$ $(0 \leq t < \infty)$ for which the limit $f(\infty) = \lim_{t\to\infty} f(t)$ exists. The space C_0 is distinguished from C by the condition $f(\infty) = 0$.

[11] The space $\tilde{C}$ consists of continuous functions on $(-\infty, +\infty)$ having limits $f(+\infty)$ and $f(-\infty)$, where $f(+\infty) = f(-\infty)$. $\tilde{C}_0$ is distinguished as before by the condition $f(+\infty) = f(-\infty) = 0$.

[12] It is simplest to carry out the verification of these conditions by means of the Fourier transform.

of the form

$$(R\varphi)(t) = c\varphi(t) - \int_{-\infty}^{\infty} r(t-s)\varphi(s)\,ds, \tag{8.2}$$

where

$$r(t) = \begin{cases} e^{-t}p_1(t), & 0 < t < \infty, \\ e^{t}p_2(t), & -\infty < t < 0, \end{cases} \tag{8.3}$$

c is a constant, and $p_1(t)$ and $p_2(t)$ are arbitrary polynomials in nonnegative powers of t.

If the function $k(t) \in \tilde{L}_1$, then, as is well known, the operator

$$(\mathscr{K}\varphi)(t) = \int_{-\infty}^{\infty} k(t-s)\varphi(s)\,ds$$

is bounded in any of the spaces $\tilde{E}$, where

$$|\mathscr{K}|_{\tilde{E}} \leqq |k(t)|_{\tilde{L}_1}. \tag{8.4}$$

Since the set of all functions of the form (8.3) is dense in $\tilde{L}_1$, by virtue of estimate (8.4) the algebra $\mathfrak{R}(U)$ automatically contains all operators $\tilde{A}$ of the form

$$(\tilde{A}\varphi)(t) = \varphi(t) - \int_{-\infty}^{\infty} k(t-s)\varphi(s)\,ds, \tag{8.5}$$

where $k(t) \in \tilde{L}_1$. There is no precise description of this algebra for an arbitrary space $\tilde{E}$; therefore we are here restricted to operators only of the form (8.5) with $k(t) \in \tilde{L}_1$.

We shall place into correspondence with each operator of the form (8.5) the function

$$\mathscr{A}(\lambda) = 1 - \int_{-\infty}^{\infty} k(t)e^{i\lambda t}\,dt \qquad (-\infty < \lambda < \infty).$$

The properties of the Fourier transform imply that the functions $((\lambda - i)/(\lambda + i))^n$ correspond in this connection to the operators U^n, $n = 0, \pm 1, \cdots$.

The operators U and U^{-1} were specially selected to obtain such Fourier transforms.

Thus if an operator R of $\mathfrak{R}(U)$ has the form (8.2), its symbol $R(\zeta)$ ($|\zeta| = 1$) (a polynomial in positive and negative powers of ζ) is defined by the equality

$$R\left(\frac{\lambda - i}{\lambda + i}\right) = c - \int_{-\infty}^{\infty} r(t)e^{i\lambda t}\,dt.$$

It is easy to obtain from this the fact that if an operator $\tilde{A}$ of $\mathfrak{R}(U)$ has form (8.5), its symbol $A(\zeta)$ ($|\zeta| = 1$) is defined by the equalities

$$A(\zeta) = \mathscr{A}\left(i\frac{1+\zeta}{1-\zeta}\right), \qquad \mathscr{A}(\lambda) = 1 - \int_{-\infty}^{\infty} k(t)e^{i\lambda t}\, dt \tag{8.6}$$
$$(-\infty \leqq \lambda \leqq \infty).$$

We shall also call the *symbol* of the operator A the function $\mathscr{A}(\lambda)$ on the real axis $-\infty < \lambda < \infty$ which is related to the symbol $A(\zeta)$ by a bilinear transformation of the argument.

Let us now apply Theorem 3.2, the basic theorem concerning invertibility in the second scheme, to the operator $A = P\tilde{A}P \mid E \in \hat{\mathfrak{R}}(U)$, where the operator $\tilde{A}$ is defined by (8.5). As a preliminary, let us note that the restriction of $P\tilde{A}P$ to E can be identified with the operator (8.1) operating in E. Theorem 3.2 is realized for this operator as follows.

THEOREM 8.1. *Let $k(t) \in \tilde{L}_1$. For an operator A defined by equality* (8.1) *to be invertible in E on at least one side, it is necessary and sufficient that its symbol not vanish*:

$$\mathscr{A}(\lambda) = 1 - \int_{-\infty}^{\infty} k(t)e^{i\lambda t}\, dt \neq 0 \qquad (-\infty < \lambda < \infty). \tag{8.7}$$

If this condition is fulfilled, the invertibility of A will be consistent with the index of the symbol $\mathscr{A}(\lambda)$:

$$\operatorname{ind} \mathscr{A}(\lambda) = (1/2\pi)\,[\arg \mathscr{A}(\lambda)]_{\lambda=-\infty}^{\infty}.$$

In other words, the invertibility of A is determined by its symbol.

2. Now let us turn to the first scheme of § 3. Since the set $\hat{\mathfrak{R}}(U)$ is contained in $\mathfrak{R}(V)$, all operators of the form (8.1) get into $\mathfrak{R}(V)$. Moreover, the symbol $A(\zeta)$ ($|\zeta| = 1$) of an operator A of the form (8.1) is defined as before by equalities (8.6).

Theorem 8.1 can now be obtained from Theorem 3.1 if a lemma concerning fulfillment of condition (Σ II) is established.

LEMMA 8.1. *The operators V and $V^{(-1)}$ satisfy condition* (Σ II), *i. e., if $R_1, R_2 \in \mathfrak{R}(V)$ and $R(\zeta) = R_1(\zeta)R_2(\zeta)$, then $|R|_E \leq |R_1|_E |R_2|_E$.*

PROOF. Relation (8.2) implies that the operator $R_j \in \mathfrak{R}(V)$ ($j = 1,2$) admits the representation

$$(R_j\varphi)(t) = \sum_k \alpha_k^{(j)} (V^{(k)}\varphi)(t) = c_j\varphi(t) - \int_0^{\infty} r_j(t-s)\varphi(s)ds \qquad (j = 1,2),$$

where c_j is a constant and the function $r_j(t)$ has form (8.3). Moreover, the polynomial $R_j(\zeta)$ is defined by the equality

$$R_j\left(\frac{\lambda - i}{\lambda + i}\right) = c_j - \int_{-\infty}^{\infty} r_j(t)e^{i\lambda t}\, dt \qquad (-\infty \leqq \lambda \leqq \infty).$$

Properties of the Fourier transform imply that the operator R with symbol $R(\zeta) = R_1(\zeta)\, R_2(\zeta)$ operates in accordance with the formula

$$(R\varphi)(t) = c_1c_2\varphi(t) - c_1 \int_0^\infty r_2(t-s)\varphi(s)\, ds$$
$$- c_2 \int_0^\infty r_1(t-s)\varphi(s)\, ds + \int_0^\infty r(t-s)\varphi(s)\, ds, \tag{8.8}$$

where $r(t) = \int_{-\infty}^{\infty} r_1(t-s)r_2(s)\, ds$.

On the other hand, it is easy to find the product of R_1 and R_2:

$$(R_1R_2\varphi)(t) = c_1c_2\varphi(t) - c_1 \int_0^\infty r_2(t-s)\varphi(s)\, ds$$
$$- c_2 \int_0^\infty r_1(t-s)\varphi(s)\, ds + \int_0^\infty \varphi(s)\, ds \int_s^\infty r_1(t-s-u)r_2(u)\, du$$

or, in comparison with (8.8),

$$(R_1R_2\varphi)(t) = (R\varphi)(t) - \int_0^\infty \varphi(s)\, ds \int_{-\infty}^{-s} r_1(t-s-u)r_2(u)\, du. \tag{8.9}$$

Let us consider first the case when $E = L_p$ $(1 \leqq p < \infty)$. We introduce the operators

$$(U_\tau\varphi)(t) = \begin{cases} \varphi(t-\tau), & t \geqq \tau, \\ 0, & t < \tau, \end{cases} \qquad (U_{-\tau}\varphi)(t) = \varphi(t+\tau) \qquad (0 < \tau < \infty)$$

in the space $E = L_p$. The operators U_τ and $U_{-\tau}$ $(\tau > 0)$ are bounded in L_p, and $|U_\tau| = |U_{-\tau}| = 1$.

If an operator A operating in L_p has the form

$$(A\varphi)(t) = \int_0^\infty k(t,s)\varphi(s)\, ds,$$

then

$$(U_{-\tau} AU_\tau\varphi)(t) = \int_0^\infty k(t+\tau, s+\tau)\varphi(s)\, ds.$$

In particular, if the kernel of A is a difference, i.e. $k(t,s) = h(t-s)$, then $U_{-\tau} AU_\tau = A$.

By virtue of the remarks just made, (8.9) implies the equality

$$(U_{-\tau} R_1R_2U_\tau\varphi)(t) = (R\varphi)(t) - \int_0^\infty \varphi(s)\, ds \int_{-\infty}^{-(s+\tau)} r_1(t-u-s)r_2(u)\, du. \tag{8.10}$$

For each $\tau > 0$ the function

$$\gamma_\tau(t,s) = \int_{-\infty}^{-(s+\tau)} r_1(t-u-s)r_2(u)\, du$$

satisfies the relation

$$|\gamma_\tau(t,s)| \leqq h_\tau(t-s), \quad \text{where } h_\tau(t) = \int_{-\infty}^{-\tau} |r_1(t-u)r_2(u)|\, du \ (\in \tilde{L}_1),$$

with

$$|h_\tau|_{\tilde{L}_1} \leq |r_1|_{\tilde{L}_1} \int_{-\infty}^{-\tau} |r_2(u)|\, du. \tag{8.11}$$

It is easy to see that the estimate

$$\left| \int_0^\infty \gamma_\tau(t,s)f(s)\, ds \right|_{L_p} \leqq |h_\tau|_{\tilde{L}_1} \cdot |f|_{L_p}$$

holds for any function $f(t)$ of L_p with compact support. This implies that the operator

$$(\Gamma_\tau\varphi)(t) = \int_0^\infty \gamma_\tau(t, s)\varphi(s)\, ds$$

is bounded in L_p, the estimate $|\Gamma_\tau|_{L_p} \leqq |h_\tau|_{\tilde{L}_1}$ being valid for the L_p norm.

From (8.11) it follows that $|\Gamma_\tau|_{L_p} \to 0$ as $\tau \to \infty$.

Let ε be an arbitrary positive number. Let us choose τ so large that $|\Gamma_\tau|_{L_p} < \varepsilon$. Then, by virtue of (8.10),

$$|R|_{L_p} \leq |U_{-\tau}R_1R_2U_\tau|_{L_p} + \varepsilon \leqq |R_1|_{L_p}|R_2|_{L_p} + \varepsilon,$$

which implies the relation

$$|R|_{L_p} \leq |R_1|_{L_p}|R_2|_{L_p}.$$

Thus the lemma is proved for the case $E = L_p$. It is proved in particular for $E = L_1$; this implies its validity for the space M conjugate to L_1, and therefore for all the remaining spaces.

3. Before formulating a corollary of Theorem 6.1, let us note three simple facts. The operator $V^{(n)}$, $n = \pm 1, \pm 2, \cdots$, is defined by the equalities

$$(V^{(n)}\varphi)(t) = \varphi(t) - \int_0^\infty l_n(t - s)\varphi(s)\, ds,$$

where

$$l_n(t) = \begin{cases} 0, & 0 < t < \infty, \\ \sum_{j=1}^{|n|} C_{|n|}^j \dfrac{2^j}{(j-1)!}\, t_{j-1}e^t, & -\infty < t < 0 \end{cases} \qquad (n = -1, -2, \cdots);$$

$$l_n(t) = l_{-n}(t) \qquad (n = 1, 2, \cdots).$$

(The function $l_n(t)$ is determined from the equality

$$1 - \int_{-\infty}^\infty l_n(t)e^{i\lambda t}\, dt = \left(\frac{\lambda - i}{\lambda + i}\right)^n.)$$

It is easy to see that the projection $P_n = I - V^nV^{(-n)}$ $(n = 1, 2, \cdots)$ operates in accordance with the formula

$$(P_n\varphi)(t) = \sum_{j=0}^{n-1} \sqrt{2}\, \Lambda_j(2t)\, e^{-t} \int_0^\infty \varphi(s)\sqrt{2}\, \Lambda_j(2s)\, e^{-s} ds,$$

where the $\Lambda_j(t)$ $(j = 0, 1, \cdots)$ are Laguerre polynomials, normed in $L_2(0, \infty)$.

The general solution of the homogeneous equation $V^{(-n)}\varphi = 0$ $(n = 1, 2, \cdots)$ has the form

$$\varphi(t) = \sum_{j=1}^n c_j t^{j-1} e^{-t} \qquad (0 < t < \infty),$$

where the c_j are arbitrary numbers.

Let us now explain how the cancellation of the index of the Wiener-Hopf integral equation

$$\varphi(t) - \int_0^\infty k(t-s)\varphi(s)\,ds = f(t) \qquad (0 < t < \infty) \tag{8.12}$$

takes place.

First of all, let us note that the equation with canceled index has the form

$$\varphi(t) - \int_0^\infty k_1(t-s)\varphi(s)\,ds = g(t), \tag{8.13}$$

where for any κ the kernel k_1 is given by the equality

$$k_1(t) = k(t) + l_{-\kappa}(t) - \int_{-\infty}^{\infty} k(t-s)\, l_{-\kappa}(s)\,ds.$$

The right side $g(t)$ is given by the formula

$$g(t) = \begin{cases} (V^{(-\kappa)}f)(t) & \text{for } \kappa > 0, \\ f(t) & \text{for } \kappa < 0. \end{cases}$$

We are here following cancellation formulas (6.3′) and (6.4′), although, of course, the index could also be canceled by formulas (6.3) and (6.4). Now let us obtain a realization of propositions 6° and 8° of § 1.

A canceled equation is always uniquely solvable under the standard condition $\mathscr{A}(\lambda) = 1 - K(\lambda) \neq 0, \ -\infty < \lambda < \infty$.

For $\kappa > 0$ the original equation (8.12) is solvable if and only if

$$\int_0^\infty (A\varphi - f)(t)\, t^{j-1}e^{-t}\,dt = 0 \qquad (j = 1,2,\cdots,\kappa),$$

where φ is the solution of the canceled equation. Under fulfillment of this condition, in fact, φ is also a solution of the original equation. If, however, $\kappa < 0$, the function

$$\psi(t) = \varphi(t) - \int_0^\infty l_{-\kappa}(t-s)\varphi(s)\,ds$$

is one of the solutions of the original equation (8.12). The general solution of the homogeneous equation

$$A\psi = 0 \tag{8.14}$$

is given by the equality

$$\psi(t) = \chi(t) - \int_0^\infty l_{-\kappa}(t-s)\chi(s)ds + \sum_{j=1}^{|\kappa|} c_j t^{j-1}e^{-t}$$

where the c_j are arbitrary numbers and $\chi(t)$ is a solution of the equation

$$\chi(t) - \int_0^\infty k_1(t-s)\chi(s)\,ds = -\sum_{j=1}^{|\kappa|} c_j t^{j-1}e^{-t} + \int_0^\infty k(t-s)\sum_{j=1}^{|\kappa|} c_j s^{j-1}e^{-s}\,ds.$$

Let us cite in addition a result concerning the structure of the kernel of A which in a certain sense is the continuous analogue of Theorem 6.1. For the sake of brevity let us here call an ordered system of functions $\{\varphi_0(t),\cdots,\varphi_n(t)\}$ a *D-chain* when the following conditions are satisfied:

1) All functions $\varphi_j(t)\ (j = 0,1,\cdots,n)$ are absolutely continuous in any finite interval.

2) $\varphi_{k+1}(t) = d\varphi_k/dt$ and $\varphi_k(0) = 0\ (k = 0,\cdots,n-1)$; and $\varphi_n(0) \neq 0$.

THEOREM 8.2. *If $\kappa < 0$, the kernel of A will be one and the same in all spaces E. This kernel has a basis which is a D-chain.*

This theorem can be obtained as a corollary of a more general proposition (cf. Theorem 9.1).

4. Under fulfillment of the standard condition (8.7) the results of § 5 permit a formula to be found for the inverse operator $A^{(-1)}$ on the appropriate side.

Let us note in this regard that all definitions and results of § 5 concerning the factorization of continuous functions on the circle are carried over without difficulty to functions defined on the real axis.

Let us cite some of them. The representation of a function $G(\lambda) \in \tilde{C}$ in the form[13]

$$G(\lambda) = G_-(\lambda)\left(\frac{\lambda - i}{\lambda + i}\right)^{\kappa} G_+(\lambda) \tag{8.15}$$

is called a *factorization* of it, where κ is an integer and $G_+(\lambda)$ and $G_-(\lambda)$ are functions holomorphic inside and continuous up to the boundary in the half-planes $\operatorname{Im}\lambda \geqq 0$ and $\operatorname{Im}\lambda \leqq 0$, respectively, where

$$G_+(\lambda) \neq 0 \quad (\operatorname{Im}\lambda \geqq 0), \qquad G_-(\lambda) \neq 0 \quad (\operatorname{Im}\lambda \leqq 0).$$

The Banach algebra $\mathscr{C} \subset \tilde{C}$ is called an *R-algebra* if it contains the set of all rational functions with poles outside the real axis, this set being dense in $\mathscr{C}$.

Let us denote by $\mathscr{C}^+$ ($\mathscr{C}^-$) the closure (in the norm of $\mathscr{C}$) of the linear span of the functions

$$\left\{\left(\frac{\lambda - i}{\lambda + i}\right)^j\right\}_{j=0}^{\infty} \quad \left(\left\{\left(\frac{\lambda - i}{\lambda + i}\right)^j\right\}_{j=-\infty}^{0}\right)$$

and by $\mathscr{C}_0^-$ the closure of the linear span of the functions

$$\left\{\left(\frac{\lambda - i}{\lambda + i}\right)^j\right\}_{j=-\infty}^{-1}.$$

Let us call an R-algebra $\mathscr{C}$ *decomposing* if it is the direct sum of the subalgebras $\mathscr{C}^+$ and $\mathscr{C}_0^-$. The following theorem is an analogue of Theorem 5.1.

[13] Let us recall that the space $\tilde{C}$ consists of all functions f continuous on $(-\infty, \infty)$, with the additional condition $f(\infty) = f(-\infty)$.

THEOREM 8.3. *In order for every everywhere (including infinity) nonvanishing function $G(\lambda)$ of an R-algebra $\mathscr{C}$ to admit factorization* (8.15) *with factors $G_\pm(\lambda) \in \mathscr{C}^\pm$, it is necessary and sufficient that the algebra $\mathscr{C}$ be decomposing.*

If this condition is fulfilled, the number κ is defined by the equality

$$\kappa = \operatorname{ind} G(\lambda) = (1/2\pi)\ [\arg G(\lambda)]_{\lambda=-\infty}^{\infty}.$$

We need this theorem for the case when the set $\mathfrak{L}$ of all functions of the form

$$\mathscr{F}(\lambda) = c + \int_{-\infty}^{\infty} f(t) e^{i\lambda t}\, dt,$$

where c is a constant and $f(t)$ is an arbitrary function of $\tilde{L}_1$, serves as the algebra $\mathscr{C}$. The norm

$$|\mathscr{F}| = |c| + \int_{-\infty}^{\infty} |f(t)|\, dt$$

is introduced on $\mathfrak{L}$.

Moreover, the algebras $\mathscr{C}^+, \mathscr{C}^-$ and $\mathscr{C}_0^-$ consist respectively of all functions of the form

$$\mathscr{F}(\lambda) = c + \int_0^{\infty} f(t) e^{i\lambda t}\, dt,$$

$$\mathscr{F}(\lambda) = c + \int_{-\infty}^{0} f(t) e^{i\lambda t}\, dt,$$

$$\mathscr{F}(\lambda) = \int_{-\infty}^{0} f(t) e^{i\lambda t}\, dt,$$

where $f(t)$ ranges over $\tilde{L}_1$.

Let us denote these sets by $\mathfrak{L}^+$, $\mathfrak{L}^-$ and $\mathfrak{L}_0^-$. It is obvious that $\mathfrak{L} = \mathfrak{L}^+ \dotplus \mathfrak{L}_0^-$, i. e. that the algebra $\mathfrak{L}$ decomposes.

Let condition (8.7) be fulfilled, and let the equality

$$\mathscr{A}(\lambda) = \mathscr{A}_-(\lambda) \left(\frac{\lambda - i}{\lambda + i}\right)^{\kappa} \mathscr{A}_+(\lambda)$$

afford a factorization of the symbol $\mathscr{A}(\lambda)$ in the algebra $\mathfrak{L}$. Let us represent the function $\mathscr{A}^{-1}(\lambda)$ in the form

$$\mathscr{A}^{-1}(\lambda) = G_-(\lambda)\, G_+(\lambda),$$

where

$$G_+(\lambda) = \begin{cases} \mathscr{A}_+^{-1}(\lambda), & \kappa \geqq 0, \\ \mathscr{A}_+^{-1}(\lambda) \left(\dfrac{\lambda - i}{\lambda + i}\right)^{-\kappa}, & \kappa < 0, \end{cases}$$

$$G_-(\lambda) = \begin{cases} \mathscr{A}_-^{-1}(\lambda)\left(\dfrac{\lambda - i}{\lambda + i}\right)^{-\kappa}, & \kappa \geqq 0, \\ \mathscr{A}_-^{-1}(\lambda), & \kappa < 0. \end{cases}$$

We obtain the following formula for the operator A^{-1}:

$$(A^{-1}f)(t) = f(t) + \int_0^\infty \gamma(t, s) f(s)\, ds,$$

where

$$\gamma(t, s) = \gamma_1(t - s) + \gamma_2(s - t) + \int_0^{\min(t,s)} \gamma_1(t - r)\gamma_2(s - r)\, dr,$$

and the functions $\gamma_j(t)$ $(0 < t < \infty; j = 1,2)$ are defined by the relations

$$G_+(\lambda) = 1 + \int_0^\infty \gamma_1(t) e^{i\lambda t}\, dt, \qquad G_-(\lambda) = 1 + \int_0^\infty \gamma_2(t) e^{-i\lambda t}\, dt.$$

§ 9. Functions of generating operators

The results of the previous section can be generalized. The fact is that an operator A defined in one of the spaces $L_p(0, \infty)$ $(1 \leqq p < \infty)$ by equality (8.1) can be represented in the form

$$A = I - \int_{-\infty}^{\infty} k(t) U_t\, dt, \tag{9.1}$$

where the operator U_t is defined for $t \geqq 0$ by the equality

$$(U_t f)(s) = \begin{cases} f(s - t), & s \geqq t, \\ 0, & s < t, \end{cases}$$

and for $t < 0$ by the equality $U_t f(s) = f(s - t)$. The generalization consists of the fact that a general class of operators of the form (9.1) will be considered, where the U_t satisfy certain natural conditions.

1. Let U_t $(0 \leqq t < \infty;\ U_0 = I)$ be an arbitrary strongly continuous semigroup of linear isometric operators acting in a separable Banach space $\mathfrak{B}$. In what follows we shall suppose that the semigroup U_t $(0 \leqq t < \infty)$ possesses the following properties:

1) $U_t \mathfrak{B} \neq \mathfrak{B}$ $(0 < t < \infty)$.

2) *Each operator* U_t *has a bounded left inverse* $U_t^{(-1)}$ *which can be so chosen that the totality* $U_t = U_{-t}^{(-1)}$ $(-\infty < t < 0)$ *forms a strongly continuous semigroup and* $|U_t| = 1$ $(-\infty < t \leqq 0)$.

In other words, we are assuming that for each $t > 0$ the operator U_t is strictly isometric and, in addition, that the left inverse operators $U_t^{(-1)}$ are "consistent."

In the case when $\mathfrak{B}$ is Hilbert space we can always manage to fulfill condition 2) by setting $U_{-t} = U_t^*$ $(0 \leqq t < \infty)$.

Let us denote by T the generating operator of the semigroup U_t $(0 \leqq t < \infty)$. As is well known, T is a closed operator with dense domain of definition, and all points λ with $\operatorname{Re} \lambda > 0$ are regular points of this operator. (See Hille and Phillips [1].)

Let us prove the following assertion concerning the operator T.

1°. *For all points* λ *in the left half-plane* $\operatorname{Re} \lambda < 0$

$$|(T - \lambda I) f| \geqq |\operatorname{Re} \lambda|\, |f| \qquad (f \in D_T) \tag{9.2}$$

and the magnitude $\mathfrak{m} = \dim(\mathfrak{B}/\mathrm{Im}\,(T - \lambda I))$ *takes one and the same positive value.*

Indeed, let us set $F(t) = e^{-\lambda t} U_t f$. Then, since

$$dU_t f/dt = U_t Tf \qquad (f \in D_T),$$

it follows that $dF/dt = e^{-\lambda t} U_t(Tf - \lambda f)$, whence

$$|dF/dt| = e^{\mu t} \, |Tf - \lambda f| \qquad (\mu = -\mathrm{Re} - \lambda).$$

Since

$$F(t) = f + \int_0^t \frac{dF(\tau)}{d\tau} d\tau \qquad (F(0) = f),$$

we have

$$|F(t)| \leqq |f| + \frac{e^{\mu t} - 1}{\mu} |Tf - \lambda f|.$$

On the other hand, $|F(t)| = e^{\mu t}|f|$. These last two relations imply (9.2).

The constancy of the number $\mathfrak{m}$ follows from estimate (9.2) and from the following theorem (see Gohberg and Kreĭn [**1**], Theorem 9.2) which we cite without proof.

Let A be a closed operator, and let the estimate

$$|(A - \lambda I)x| \geqq \rho(\lambda)|x|,$$

where x ranges over D_A and $\rho(\lambda) > 0$, be fulfilled for each point λ of some connected open set.

Then the quantity $\dim \{\mathfrak{B}/\mathrm{Im}\,(A - \lambda I)\}$ *is constant within the limits of this region.*

Let us show that $\mathfrak{m}$ is different from zero. Let us assume the contrary. Then the spectrum of the operator T is not within the left half-plane, and

$$|(T - \lambda I)^{-1}| \leqq 1/|\mathrm{Re}\,\lambda|.$$

However, by virtue of a theorem of Hille and Yosida (see Hille and Phillips [**1**], Theorem 11.7.1) the operator T also does not have its spectrum in the right half-plane, and the relation

$$|(T - \lambda I)^{-1}| \leqq 1/\mathrm{Re}\,\lambda \qquad (\mathrm{Re}\,\lambda > 0),$$

is also fulfilled.

Thus the spectrum of operator T is purely imaginary, and

$$|(T - \lambda I)^{-1}| \leqq 1/|\mathrm{Re}\,\lambda| \qquad (\mathrm{Re}\,\lambda \neq 0).$$

But then, according to Theorem 12.3.2 in Hille and Phillips [**1**], T is the generating operator of a strongly continuous *group* of contractions V_t $(-\infty < t < \infty)$. Obviously $V_t = U_t$ $(0 \leqq t < \infty)$; consequently the operators U_t will have inverses V_{-t}, which contradicts the condition $U_t\mathfrak{B} \neq \mathfrak{B}$ $(0 < t < \infty)$. The proposition is proved.

As is well known (cf. Hille and Phillips [**1**]),

$$(T - \lambda I)^{-1} f = -\int_0^\infty e^{-\lambda t} U_t f\, dt \qquad (\mathrm{Re}\,\lambda > 0, f \in \mathfrak{B})$$

and therefore an operator V $(\in \mathfrak{A})$ defined by the equality

$$V = (T + I)(T - I)^{-1}$$

can be represented in the form

$$Vf = f - 2\int_0^\infty e^{-t} U_t f\, dt \qquad (f \in \mathfrak{B}).$$

Let T_1 be the generating operator of the semigroup U_{-t} $(0 \leqq t < \infty)$. Let us form the operator

V_1 ($\in \mathfrak{A}$) by setting

$$V_1 f = (T_1 + I)(T_1 - I)^{-1} f = f - 2\int_0^\infty e^{-t} U_{-t} f\, dt \qquad (f \in \mathfrak{B}).$$

We easily obtain $V_1 V = I$ by taking into consideration that, for any two functions $k_1(t), k_2(t) \in \bar{L}_1$ satisfying the conditions $k_1(t) = 0$ for $t < 0$ and $k_2(t) = 0$ for $t > 0$, the equality

$$\int_{-\infty}^{\infty} k_2(t) U_t \int_{-\infty}^{\infty} k_1(s) U_s f\, ds\, dt = \int_{-\infty}^{\infty} k(t) U_t f\, dt \qquad (f \in \mathfrak{B})$$

holds, where $k(t) = \int_{-\infty}^{\infty} k_2(t - s)\, k_1(s)\, ds$.

Thus the operator V_1 is a left inverse of V: $V_1 = V^{(-1)}$.

2°. *The operator T_1 is an extension of the operator $-T$.*

Indeed, as just proved,

$$(T_1 + I)(T_1 - I)^{-1}(T + I)(T - I)^{-1} = I$$

or

$$(I + 2(T_1 - I)^{-1})(I + 2(T - I)^{-1}) = I.$$

Thus

$$(T_1 - I)^{-1} + (T - I)^{-1} + 2(T_1 - I)^{-1}(T - I)^{-1} = 0.$$

After multiplying the last equality by the operator $T - I$, we obtain

$$(T_1 - I)^{-1}(T - I) + I + 2(T_1 - I)^{-1} = 0$$

or

$$(T_1 - I)^{-1}(-T - I) = I.$$

This implies that if $g \in D_T$ and $(-T - I)g = f$, then

$$(T_1 - I)^{-1} f = g,$$

and therefore $g \in D_{T_1}$, where $T_1 g = -Tg$.

An operator $\mathscr{K}$ defind by the equality

$$\mathscr{K} f = \int_{-\infty}^{\infty} k(t)\, U_t f\, dt \qquad (f \in \mathfrak{B}) \tag{9.3}$$

can be put into correspondence with each function $k(t) \in \bar{L}_1$.

2. Let us denote by $\mathscr{C}$ the set of all operators of the form $\mu I + \mathscr{K}$, where μ is an arbitrary complex number and $k(t)$ is an arbitrary function of $\bar{L}_1$; and by $\bar{\mathscr{C}}$ the closure of the linear manifold $\mathscr{C}$ in the operator norm. We note in addition that if $k_1(t), k_2(t) \in \bar{L}_1$, then

$$|\mathscr{K}_1 - \mathscr{K}_2| \leq \int_{-\infty}^{\infty} |k_1(t) - k_2(t)|\, dt. \tag{9.4}$$

Let us form the space $\mathfrak{R}(V)$. From (9.4) it follows that the set of linear combinations of the operators $V^{(j)}$ $(j = 0, \pm 1, \cdots)$ forms a dense set in $\mathscr{C}$. This implies that $\bar{\mathscr{C}} = \mathfrak{R}(V)$.

The operators V and $V^{(-1)}$ satisfy conditions (Σ I) and (Σ II) of Theorem 3.1. Condition (Σ I) follows immediately from propositions 1° and 2°. Fulfillment of (Σ II) is established exactly as in Lemma 8.1.

Let $A(\zeta) \in \mathfrak{R}(\zeta)$ be the symbol of the operator $A \in \mathfrak{R}(V)$. If we make the substitution $\zeta = (\lambda - i)/(\lambda + i)$ $(-\infty < \lambda < \infty)$, we obtain the function $\mathscr{A}(\lambda)$ $(= A(\zeta))$, the symbol of A on the λ-axis.

It is natural to regard the operator A as a value of the function $\mathscr{A}(\lambda)$ of the operator $-iT$. Let us note that if A has the form

$$Af = f - \int_{-\infty}^{\infty} k(t) U_t \, dt \qquad (f \in \mathfrak{B}),$$

where $k(t) \in L_1$, then

$$\mathscr{A}(\lambda) = 1 - \int_{-\infty}^{\infty} k(t) e^{i\lambda t} \, dt.$$

Now let us introduce the following refinement of Theorem 3.1 for the case under consideration.

THEOREM 9.1. *The invertibility of an operator $A \in \mathscr{C}$ is determined by its symbol $\mathscr{A}(\lambda)$. If the symbol $\mathscr{A}(\lambda) \neq 0$, then for*

$$\kappa = \operatorname{ind} \mathscr{A}(\lambda) = \frac{1}{2\pi} [\arg \mathscr{A}(\lambda)]_{\lambda=-\infty}^{\infty} < 0 \tag{9.5}$$

the equality $\dim \operatorname{Ker} A = \mathfrak{m} |\kappa|$ *holds.*

If, in addition, the algebra $\mathfrak{R}(\zeta)$ is decomposing, then an $\mathfrak{m}$-dimensional linear manifold $\mathfrak{L}$ exists such that $\mathfrak{L} \subset D_{T^{|\kappa|-1}}$, and

$$\operatorname{Ker} A = \mathfrak{L} \dotplus T\mathfrak{L} \dotplus \cdots \dotplus T^{|\kappa|-1} \mathfrak{L},$$

where $T^{|\kappa|-1} \mathfrak{L} \cap D_T = \{0\}$.

PROOF. Only the last assertion of the theorem needs proof.

Let us denote by $\mathfrak{L}_k$ the subspace $\operatorname{Ker} V^{(-k)}$ $(k = 1,2,\cdots)$. Since $V^{(-1)} = I + 2(T_1 - I)^{-1}$, we have for every vector $\varphi \in \mathfrak{L}_k$

$$\varphi + c_1 (T_1 - I)^{-1} \varphi + \cdots + c_k (T_1 - I)^{-k} \varphi = 0,$$

where the c_j are specific numbers. It is easy to deduce from this that $\mathfrak{L}_k \subset D_{T^k}$. For any $f \in D_{T^k}$

$$V^{(-k)} f = ((I + T_1)(I - T_1)^{-1})^k f = (I - T_1)^{-k} (I + T_1)^k f,$$

and consequently $\mathfrak{L}_k \subset \operatorname{Ker} (I + T_1)^k$.

On the other hand, if $(I + T_1)^k f = 0$, then

$$(I - T_1)^{-k} (I + T_1)^k f = (I + T_1)^k (I - T_1)^{-k} f = 0.$$

Thus $\mathfrak{L}_k = \operatorname{Ker} (I + T_1)_k$.

Taking into account that $(T_1 + I)(T - I)^{-1} = (-T + I)(T - I)^{-1} = -I$, and repeating the argument in the proof of Theorem 6.1, we obtain

$$\mathfrak{L}_{|\kappa|} = \mathfrak{L}_1 \dotplus (T - I)^{-1} \mathfrak{L}_1 \dotplus \cdots \dotplus (T - I)^{\kappa+1} \mathfrak{L}_1.$$

Continuing to follow the proof of Theorem 6.1, we arrive at the equality

$$\operatorname{Ker} A = A_+^{-1} \mathfrak{L}_1 \dotplus A_+^{-1} (T - I)^{-1} \mathfrak{L}_1 \dotplus \cdots \dotplus A_+^{-1} (T - I)^{\kappa+1} \mathfrak{L}_1. \tag{9.6}$$

Without difficulty we establish that

$$A_+^{-1} (T - I)^{-1} = (T - I)^{-1} A_+^{-1};$$

consequently equality (9.6) takes the form

$$\operatorname{Ker} A = \mathfrak{L} \dotplus (T - I)\mathfrak{L} \dotplus \cdots \dotplus (T - I)^{|\kappa|-1} \mathfrak{L}, \tag{9.7}$$

where $\mathfrak{L} = (T - I)^{\kappa+1} A_+^{-1} \mathfrak{L}_1$.

The subspace $\tilde{\mathfrak{L}} = \mathfrak{L} \cap D_{T^{|\kappa|}} = \{0\}$. Indeed, in the opposite case we would have $(\{0\} \neq) \tilde{\mathfrak{L}} \subset D_{(T-I)^{|\kappa|}}$, which would give us

$$\tilde{\mathfrak{L}}_1 = A_+(T-I)^{-\kappa-1}\tilde{\mathfrak{L}} \subset D_{(T-I)}.$$

Since T_1 is an extension of $-T$, the last inclusion implies $(-T+I)\mathfrak{L} = 0$, which is impossible. The fact that

$$(T-I)^r \mathfrak{L} \cap D_{(T-I)^{|\kappa|-r}} = \{0\} \qquad (r = 1,2,\cdots, |\kappa|-1)$$

is proved analogously. From this we infer that the linear manifolds $T^r\mathfrak{L}$ $(r = 0,\cdots, |\kappa|-1)$ intersect only at zero.

Every vector φ $(\in \operatorname{Ker} A)$ representable in the form

$$\varphi = \varphi_1 + (T-I)\varphi_2 + \cdots + (T-I)^{|\kappa|-1}\varphi_{|\kappa|} \tag{9.8}$$

where $\varphi_j \in \mathfrak{L}$, can obviously be represnted in the form

$$\varphi = \psi_1 + T\psi_2 + \cdots + T^{|\kappa|-1}\psi_{|\kappa|}, \tag{9.9}$$

where the ψ_j also belong to $\mathfrak{L}$. It is easy to see that, conversely, every vector of the form (9.9) can be represented in the form (9.8) .Thus

$$\operatorname{Ker} A = \mathfrak{L} \dotplus T\mathfrak{L} \dotplus \cdots \dotplus T^{|\kappa|-1}\mathfrak{L}.$$

The theorem is proved.

Let us consider separately the case of Hilbert space $\mathfrak{B} = \mathfrak{H}$. Let H be an arbitrary maximal symmetric operator with deficiency index $(0, \mathfrak{m})$. As is well known, the operator $T = iH$ is the generating operator for some continuous semigroup U_t $(0 \leqq t < \infty)$ of isometric operators. The system $U_{-t} = U_t^*$ $(0 \leqq t < \infty)$ also forms a strongly continuous semigroup, where $U_{-t}U_t = I$ $(0 \leqq t < \infty)$. The operator $T_1 = -iH^*$ generates the semigroup U_t $(-\infty < t \leqq 0)$.

Thus, by virtue of the definitions introduced above, if

$$K(\lambda) = \int_{-\infty}^{\infty} k(t)e^{i\lambda t}\,dt,$$

then $K(H)$ is given by the equality

$$K(H)f = \int_{-\infty}^{\infty} k(t)U_t f\,dt \qquad (f \in \mathfrak{B}).$$

3. Let us now consider in greater detail the example cited at the beginning of the section. Let $E(0,\infty)$ be an arbitrary Banach space of functions measurable on the segment $[0,\infty)$ in which the semigroup U_t $(0 \leqq t < \infty)$ defined by the equality

$$(U_t f)(s) = \begin{cases} f(s-t), & s \geqq t, \\ 0, & s < t \end{cases} \qquad (f \in E),$$

is a strongly continuous semigroup of isometric operators, while the semigroup U_{-t} $(0 \leqq t < \infty)$ defined by the equality $(U_{-t}f)(s) = f(s+t)$ is strongly continuous and $|U_{-t}| = 1$.

Obviously any of the spaces $L_p(0,\infty)$ $(1 \leqq p < \infty)$ or $C_0(0,\infty)$ may be taken as the space $E(0,\infty)$.

If $k(t) \in \bar{L}_1$, the formula prescribing the operator

$$(\mathscr{K} f)(t) = \int_{-\infty}^{\infty} k(s)U_s f\,ds$$

can be transformed thus:

$$(\mathscr{K} f)(t) = \int_{-\infty}^{t} k(s)f(t-s)\,ds,$$

and therefore

$$(\mathscr{K} f)(t) = \int_0^\infty k(t - s) f(s)\, ds \qquad (0 \leqq t < \infty).$$

It is easy to see that the semigroup U_t is generated by the operator $T = -iD$, where $Df = df/dt$ for $f(0) = 0$.

Thus, by virtue of the preceding definitions, $\mathscr{K} = K(-iD)$, where $K(\lambda)$ is the Fourier transform of the function $k(t)$.

Let us note that Theorem 8.2 on the structure of the kernel of a Wiener-Hopf operator is a corollary of Theorem 9.1 proved above.

§ 10. Finite difference equations

Let an isometric operator V be defined in Hilbert space $L_2(0, \infty)$ by the equality

$$(Vf)(t) = \begin{cases} f(t-1), & t > 1, \\ 0, & t < 1. \end{cases}$$

The operator $V^{(-1)}$ defined by the equality $(V^{(-1)}f)(t) = f(t+1)$ is a left inverse of V, and $|V| = |V^{(-1)}| = 1$.

It is easy to see that $\mathfrak{R}(V)$ consists of all operators of the form

$$(Rf)(t) = \sum_{j=-n}^{n} a_j f(t - j),$$

where the a_j are arbitrary complex numbers. Here, as in the sequel, it is assumed that $f(t-j) = 0,\ 0 \leqq t \leqq j$.

Let $a(\zeta)$ be an arbitrary function continuous on the unit circle, and a_j $(j = 0, \pm 1, \cdots)$ its Fourier coefficients. By virtue of proposition 2° of § 2 there exists a unique operator $A \in \mathfrak{R}(V)$ whose symbol $A(\zeta)$ coincides with $a(\zeta)$: $A(\zeta) = a(\zeta)$. We shall write this operator in the form

$$(Af)(t) = \sum_{j=-\infty}^{\infty} a_j f(t - j). \tag{10.1}$$

The series on the right side of (10.1) converges to $(Af)(t)$ in the L_2 norm on any finite segment. The proof of this fact is cited in § 6, Chapter IV.

Let us note that by virtue of relation (2.5) the operators $P_n(V)$, where $P_n(\zeta)$ is the arithmetic average of the partial sums of the Fourier series for $a(\zeta)$, converge uniformly to A. The convergence of series (10.1) can be understood in this sense, also.

Applying Theorem 3.1 to the operator A, we obtain the following proposition.

THEOREM 10.1. *Let $a(\zeta)$ be an arbitrary function continuous on the unit circle, and a_j $(j = 0, \pm 1, \cdots)$ its Fourier coefficients. The invertibility of the operator A prescribed in L_2 by equality* (10.1) *is determined by its symbol $a(\zeta)$.*

Let us note that by virtue of equalities (3.5) and (3.6), for $\kappa > 0$

$$\dim \operatorname{Coker} A = \infty, \tag{10.2}$$

while for $\kappa < 0$

$$\dim \operatorname{Ker} A = \infty. \tag{10.2'}$$

The equations

$$\sum_{j=-\infty}^{\infty} a_j \varphi(t - j + \kappa) = f(t + \kappa) \qquad (0 < t < \infty) \tag{10.3}$$

for $\kappa > 0$ and

$$\sum_{j=-\infty}^{\infty} a_j \psi(t - j + \kappa) = f(t) \qquad (0 < t < \infty) \tag{10.4}$$

for $\kappa < 0$ will be cancellations for the equation

$$\sum_{j=-\infty}^{\infty} a_j \varphi(t - j) = f(t) \qquad (0 < t < \infty), \tag{10.5}$$

considered in L_2.

A canceled equation is always uniquely solvable.

In the first case ($\kappa > 0$) its solution is a solution of the original equation (10.5) if the additional condition

$$\sum_{j=-\infty}^{\kappa} a_j \varphi(t - j) = f(t) \qquad (0 < t < \kappa)$$

is fulfilled.

In the second case ($\kappa < 0$) the solution $\psi(t)$ of equation (10.4) is one of the solutions of equation (10.5); the general solution of the homogeneous equation

$$\sum_{j=-\infty}^{\infty} a_j \varphi(t - j) = 0 \qquad (0 < t < \infty)$$

is given by the equality

$$\varphi(t) = \chi(t + \kappa) + g(t),$$

where $g(t)$ ($\in L_2$) is an arbitrary function equal to zero for $t > -\kappa$, while $\chi(t)$ is the unique solution of the equation

$$\sum_{j=-\infty}^{\infty} a_j \chi(t - j + \kappa) = - \sum_{j=\kappa}^{\infty} a_j g(t - j).$$

These assertions have been obtained as a realization of propositions 6° and 8°, § 1.

Analogous results are valid in the spaces L_p ($p \geqq 1$), etc. Operators of a more general type will be considered in Chapter VII.

§ 11. General propositions concerning normally solvable operators and their indices

In the sequel we shall need certain results of the general theory of linear operators, a detailed presentation of which can be found in Gohberg and Kreĭn **[1]**. They will be cited here without proof.

As before, let $\mathfrak{A} = \mathfrak{A}(\mathfrak{B})$ be the algebra of all linear bounded operators in the Banach space $\mathfrak{B}$.

If $A \in \mathfrak{A}$, then by $\alpha(A)$ and $\beta(A)$ we denote the dimensions of the subspaces Ker A and Ker A^*, respectively, where A^* is the conjugate operator (operating in the conjugate space $\mathfrak{B}^*$).

Let us recall that if the set of values Im A of A is a subspace having a direct complement in $\mathfrak{B}$, then $\beta(A) = \dim \operatorname{Coker} A$.

If the numbers $\alpha(A)$ and $\beta(A)$ are finite, their difference $\kappa(A) = \alpha(A) - \beta(A)$ is called the *index* of A.

A is called *normally solvable* if the equation $Ax = y$ is solvable for every $y \in \mathfrak{B}$ if and only if $f(y) = 0$ for all $f \in$ Ker A^*. Normal solvability of A is equivalent to the closure of its set of values.

A normally solvable operator A is called a Φ_+ (Φ_-)*-operator* if the number $\alpha(A)$ $(\beta(A))$ is finite, and a Φ*-operator* if both of the numbers $\alpha(A)$ and $\beta(A)$ are finite.

The following propositions hold:

A) *For every Φ-operator A a positive number ρ exists such that for all operators B $(\in \mathfrak{A})$ for which $|B| < \rho$ the operator $A + B$ is a Φ-operator, where $\kappa(A + B) = \kappa(A)$ and $\alpha(A + B) \leqq \alpha(A)$.*

B) *Let A be some Φ-operator. Then, for any completely continuous operator T $(\in \mathfrak{A})$, the operator $A + T$ is a Φ-operator, and $\kappa(A + T) = \kappa(A)$.*

C) *The product AB of two Φ-operators is also a Φ-operator, and $\kappa(AB) = \kappa(A) + \kappa(B)$.*

Propositions A) and B) can be generalized to the case of Φ_+- and Φ_--operators.

We need two more auxiliary assertions.

D) *Let $A, B \in \mathfrak{A}$. If the product AB is a Φ_+ (Φ_-)-operator, the operator $B(A)$ is also a Φ_+ (Φ_-)-operator.* (Cf. B. Yood **[1]**.)

Let us note in addition that if the standard condition (3.2) in Theorem 3.1 (or the corresponding conditions in Theorems 7.1, 8.1, 9.1 or 10.1) is not fulfilled, then the operator A is neither a Φ_+-nor a Φ_--operator. By virtue of propositions A) and B) it is sufficient to show this for the case $A \in \mathfrak{R}(V)$.

Let us assume that $A(\zeta_0) = 0$ $(|\zeta_0| = 1)$ and A is, for example, a Φ_+-operator. As in Theorem 1.1, let us represent A in the form

$$A = A_1(V - \zeta_0 I) \qquad (A_1 \in \mathfrak{R}(V)).$$

By virtue of proposition D), the operator $V - \zeta_0 I$ is a Φ_+-operator. But this contradicts the theorem on the stability of the index of Φ_+-operators.

E) *For the operator A to be a Φ-operator it is necessary and sufficient that an operator $B \in \mathfrak{A}$ exist such that the operators $AB - I$ and $BA - I$ are completely continuous* (cf. F. V. Atkinson **[2]**).

In other words, *for an operator A to be a Φ-operator it is necessary and sufficient that the residue class containing A be invertible in the factor algebra $\mathfrak{A}/\mathfrak{T}$, where $\mathfrak{T}$ is the ideal of completely continuous operators* (see, for example, Gohberg, Markus and Fel'dman **[1]**).

CHAPTER II

GALERKIN'S METHOD AND PROJECTION METHODS OF SOLVING LINEAR EQUATIONS

In the present chapter general statements concerning the Galerkin method and projection methods are set forth. The relationship between these methods is explained. Criteria are established for their applicability. The theorems on the stability of projection methods for small in norm or completely continuous perturbations are important for the sequel.

It is proved (the theorem has the character of an existence theorem) that every linear equation with an invertible operator in Hilbert space can be solved by some projection method.

A class of invertible operators is described to which the Galerkin method is applicable relative to *any* orthogonal basis. The same problem is solved also with respect to operators to which the Galerkin method is applicable relative to any basis equivalent to an orthogonal one (i.e. a Riesz basis).

Only the results of the first three sections of this chapter are utilized in the subsequent chapters.

§ 1. Galerkin's method and its generalization

Let $\mathfrak{B}_1$, $\mathfrak{B}_2$ be two Banach spaces, and $\mathfrak{A}(\mathfrak{B}_1, \mathfrak{B}_2)$ the set of all bounded linear operators acting from $\mathfrak{B}_1$ into $\mathfrak{B}_2$. Let us suppose that $\{\varphi_j\}_1^\infty$ is a vector system complete in $\mathfrak{B}_1$ and $\{f_j\}_1^\infty$ is a total system of functionals of $\mathfrak{B}_2^*$. The following method of approximate solution of the equation

$$Ax = y \qquad (y \in \mathfrak{B}_2), \tag{1.1}$$

where $A \in \mathfrak{A}(\mathfrak{B}_1, \mathfrak{B}_2)$, is called the *Galerkin* (or *Galerkin-Petrov*) method.

An approximate solution x_n of equation (1.1) is sought in the form of a linear combination of the vectors $\{\varphi_j\}_1^n$:

$$x_n = \sum_1^n \xi_j \varphi_j. \tag{1.2}$$

The numbers ξ_j are determined from the system of equations $f_j(Ax_n - y) = 0$ $(j = 1,\cdots,n)$, which can be written in the following form:

$$\sum_{k=1}^{n} f_j(A\varphi_k)\xi_k = f_j(y) \qquad (j = 1,\cdots,n). \tag{1.3}$$

We shall say that the Galerkin method is applicable to the operator A relative to the system (φ_j, f_j) if for any $y \in \mathfrak{B}_2$ the system (1.3), beginning with some n, has a unique solution and the vectors x_n defined by (1.2) tend to a solution of equation (1.1) as $n \to \infty$.

The Galerkin method admits the following generalization. Let Ω be some unbounded set of positive numbers, and let $\{\mathfrak{L}_\tau\}$ and $\{\mathfrak{N}_\tau\}$ $(\tau \in \Omega)$ be families of subspaces of $\mathfrak{B}_1$ and $\mathfrak{B}_2$, respectively. The generalized method consists of finding an approximate solution x_τ $(\tau \in \Omega)$ of equation (1.1) in accordance with the following conditions:

$$x_\tau \in \mathfrak{L}_\tau, \qquad Ax_\tau - y \in \mathfrak{N}_\tau. \tag{1.4}$$

If for any $y \in \mathfrak{B}_2$ and any τ beginning with a certain one a unique vector x_τ exists satisfying conditions (1.4) and as $\tau \to \infty$ the vectors x_τ tend to a solution of equation (1.1), we shall say that the generalized Galerkin method relative to the system $(\mathfrak{L}_\tau, \mathfrak{N}_\tau)$ is applicable to the operator A. Let us denote by $\Gamma\{\mathfrak{L}_\tau, \mathfrak{N}_\tau) \subset \mathfrak{A}(\mathfrak{B}_1, \mathfrak{B}_2)$ the class of all operators to which the generalized Galerkin method $(\mathfrak{L}_\tau, \mathfrak{N}_\tau)$ is applicable.

Obviously the method (φ_j, f_j) coincides with the method $\{\mathfrak{L}_\tau, \mathfrak{N}_\tau\}$ if it is assumed that Ω is the set of natural numbers, $\mathfrak{L}_n$ is the linear span of the vectors $\{\varphi_j\}_1^n$, and $\mathfrak{N}_n$ is the subspace of all common zeros of the functionals $\{f_j\}_1^n$.

The following criterion holds for the applicability of the generalized Galerkin method.

THEOREM 1.1. *For the generalized Galerkin method* $(\mathfrak{L}_\tau, \mathfrak{N}_\tau)$ *to be applicable to an invertible operator* A *it is necessary and sufficient that the following conditions be fulfilled*:

1) $\lim_{\tau\to\infty} \rho(x, \mathfrak{L}_\tau) \to 0$ *for any* $x \in \mathfrak{B}_1$.

2) *Beginning with some* τ_0, *the space* $\mathfrak{B}_2$ *decomposes into the direct sum* $\mathfrak{B}_2 = A\mathfrak{L}_\tau \dotplus \mathfrak{N}_\tau$ $(\tau \geqq \tau_0)$.

3) $\sup_{\tau\geqq\tau_0} |P_\tau| < \infty$, *where* P_τ *is an operator projecting* $\mathfrak{B}_2$ *onto* $A\mathfrak{L}_\tau$ *parallel to* $\mathfrak{N}_\tau$.

Let us elucidate these conditions. Condition 1) signifies that $\bigcup_{\tau\in\Omega} \mathfrak{L}_\tau$ is dense in $\mathfrak{B}_1$. Condition 2) by virtue of the invertibility of A implies in particular that $\bigcap_{\tau\in\Omega} \mathfrak{N}_\tau = \{0\}$. Finally, condition 3) means that for large τ there is no unbounded "rapprochement" of the spaces $A\mathfrak{L}_\tau$ and $\mathfrak{N}_\tau$.

PROOF. Let us first prove the necessity of the conditions. Let x be an arbitrary vector of $\mathfrak{B}_1$ and $y = Ax$. Then the vectors x_τ $(\in \mathfrak{L}_\tau)$, which are found by the $(\mathfrak{L}_\tau, \mathfrak{N}_\tau)$ method, converge to x; consequently condition 1) is fulfilled. Fulfillment of condition 2) follows from the fact that for any $y \in \mathfrak{B}_2$ there exists a unique vector

$x_\tau \in \mathfrak{L}_\tau$ such that $z_\tau = Ax_\tau - y \in \mathfrak{N}_\tau$ $(\tau \geqq \tau_0)$. Finally, by virtue of the last condition, $P_\tau y = Ax_\tau$, and since $Ax_\tau \to y$, we have $P_\tau y \to y$, from which condition 3) follows.

Let us turn to the proof of sufficiency. Let y be an arbitrary vector of $\mathfrak{B}_2$. Condition 2) implies that for $\tau > \tau_0$ there exists a unique vector $x_\tau \in \mathfrak{L}_\tau$ such that $y - Ax_\tau = z_\tau \in \mathfrak{N}_\tau$. To complete the proof it is obviously sufficient to show that $z_\tau \to 0$ as $\tau \to \infty$. The latter follows from the equality

$$z_\tau = (I - P_\tau)_y = (I - P_\tau)(y - u_\tau),$$

where u_τ is an arbitrary vector of $A\mathfrak{L}_\tau$, and conditions 1) and 3). As a matter of fact, according to condition 1), the vectors u_τ can be so chosen that $y - u_\tau \to 0$, while according to condition 3) the norms of the operators $I - P_\tau$ $(\tau \geqq \tau_0)$ are bounded in the aggregate.

§ 2. **Projection methods**

Let $\{P_\tau\}$ and $\{Q_\tau\}$ $(P_\tau \in \mathfrak{A}(\mathfrak{B}_1),\ Q_\tau \in \mathfrak{A}(\mathfrak{B}_2),\ \tau \in \Omega)$ be sets of projections, and let A be an operator of $\mathfrak{A}(\mathfrak{B}_1, \mathfrak{B}_2)$. We call the approximate method of solving equation (1.1) which consists of finding a solution x_τ belonging to $P_\tau \mathfrak{B}_1$ of the equation

$$Q_\tau A P_\tau x = Q_\tau y \tag{2.1}$$

the *projection* method. We call equation (2.1) itself *truncated*.

If beginning with some τ_0 equation (2.1) has a unique solution x_τ for any $y \in \mathfrak{B}_2$ and as $\tau \to \infty$ the vectors x_τ tend to a solution of equation (1.1), we say that the projection method relative to the system (P_τ, Q_τ) is *applicable* to the operator A. Let us denote by $\Pi(P_\tau, Q_\tau) \subset \mathfrak{A}(\mathfrak{B}_1, \mathfrak{B}_2)$ the class of all operators to which the projection method (P_τ, Q_τ) is applicable.

Let us explain the relationship between the Galerkin method $(\mathfrak{L}_\tau, \mathfrak{N}_\tau)$ and the projection method (P_τ, Q_τ). It is easy to see that these two methods coincide if and only if

$$\mathfrak{L}_\tau = P_\tau \mathfrak{B}_1 \quad \text{and} \quad \mathfrak{N}_\tau = (I - Q_\tau)\mathfrak{B}_2. \tag{2.2}$$

This implies that the generalized Galerkin method $(\mathfrak{L}_\tau, \mathfrak{N}_\tau)$ coincides with some projection method if and only if all subspaces $\mathfrak{L}_\tau$ have direct complements in $\mathfrak{B}_1$ and the $\mathfrak{N}_\tau$ have direct complements in $\mathfrak{B}_2$. (The corresponding projections P_τ and Q_τ are so chosen that equalities (2.2) are fulfilled.)

Thus the class of generalized Galerkin methods is formally broader than the class of projection methods.

However, if the generalized Galerkin method $(\mathfrak{L}_\tau, \mathfrak{N}_\tau)$ is applicable to at least one *invertible* operator A, that method coincides with some projection method.

This assertion follows immediately from Theorem 1.1, according to which the subspace $A^{-1}\mathfrak{N}_\tau$ in the case under consideration is the direct complement of $\mathfrak{L}_\tau$ and $A\mathfrak{L}_\tau$ is the direct complement of $\mathfrak{N}_\tau$.

Let us note in addition that the projection method (P_τ, Q_τ) does not depend on the subspaces of the zeros of the projections P_τ or the subspaces of values of the projections Q_τ. Equalities (2.2) imply that *such a method is determined entirely by the subspaces of values of the projections P_τ and the subspaces of zeros of the projections Q_τ.*

We shall assume in the sequel that as $\tau \to \infty$ the projections P_τ and Q_τ converge strongly to the identity operators in $\mathfrak{B}_1$ and $\mathfrak{B}_2$, respectively. In that case the applicability of the projection method (P_τ, Q_τ) to an operator A is equivalent to the fact that operator A is invertible and, beginning with some τ, the operators $Q_\tau AP_\tau$, as operators operating from $P_\tau \mathfrak{B}_1$ into $Q_\tau \mathfrak{B}_2$, are invertible and the operators $(Q_\tau AP_\tau)^{-1} Q_\tau$ converge strongly to A^{-1} as $\tau \to \infty$.

The criterion for the applicability of the projection method to an operator A could be derived from Theorem 1.1 for such a system of projections (P_τ, Q_τ). However, it is easy to establish it immediately.

THEOREM 2.1. *For an invertible operator A to belong to $\Pi\{P_\tau, Q_\tau\}$ it is necessary and sufficient that, beginning with some τ, the relations*

$$|Q_\tau AP_\tau x| \geqq c|P_\tau x| \qquad (x \in \mathfrak{L}_1, c > 0), \tag{2.3}$$

$$Q_\tau AP_\tau \mathfrak{B}_1 = Q_\tau \mathfrak{B}_2 \tag{2.4}$$

be fulfilled.

PROOF. Let $A \in \Pi\{P_\tau, Q_\tau\}$. Then, as we noted above, the operators $(Q_\tau AP_\tau)^{-1} Q_\tau$ converge strongly to the operator A^{-1} as $\tau \to \infty$. (Let us emphasize once more that $(Q_\tau AP_\tau)^{-1}$ is inverse to $Q_\tau AP_\tau$, but only when $Q_\tau AP_\tau$ is considered as acting from $P_\tau \mathfrak{B}_1$ into $Q_\tau \mathfrak{B}_2$.) Hence there exists a $\tau_0 \in \Omega$ such that

$$\sup_{\tau > \tau_0} |(Q_\tau AP_\tau)^{-1} Q_\tau| = \gamma < \infty;$$

consequently

$$|P_\tau x| = |(Q_\tau AP_\tau)^{-1} Q_\tau(Q_\tau AP_\tau x)| \leqq \gamma |Q_\tau AP_\tau x|,$$

which implies (2.3). The fulfillment of condition (2.4) is obvious.

Conversely, let conditions (2.3) and (2.4) be fulfilled. Then, beginning with some τ, the truncated equation (2.1) has the unique solution x_τ for any $y \in \mathfrak{B}_2$. Let us denote by x a solution of the full equation (1.1). Then we obtain the equality

$$Q_\tau AP_\tau(x_\tau - P_\tau x) = Q_\tau A(I - P_\tau)x \tag{2.5}$$

from (1.1) and (2.1).

Since the right side of (2.5) tends to zero as $\tau \to \infty$, condition (2.3) implies that

$$\lim_{\tau\to\infty} x_\tau = \lim_{\tau\to\infty} P_\tau x = x.$$

REMARK. Relation (2.5) implies that the estimate

$$|x - x_\tau| = O(|x - P_\tau x|) \qquad (\tau \to \infty)$$

holds for the error $|x - x_\tau|$.

It is not difficult to derive from Theorem 2.1 the following sufficient condition for the applicability of the projection method.

1°. *Let an operator A have the form $A = H + iG$, where H is a positive definite, and G a selfadjoint, operator in Hilbert space $\mathfrak{H}$, and let the orthogonal projections P_τ $(\tau \in \Omega)$ converge strongly to the identity operator in $\mathfrak{H}$ as $\tau \to \infty$. Then $A \in \Pi\{P_\tau, P_\tau\}$.*

Indeed, the relations

$$\begin{aligned} |P_\tau x||P_\tau A P_\tau x| &\geqq |(P_\tau(H + iG)P_\tau x, P_\tau x)| \\ &\geqq (HP_\tau x, P_\tau x) \geqq \delta |P_\tau x|^2 \qquad (\delta > 0) \end{aligned}$$

imply that $|P_\tau A P_\tau x| \geqq \delta\, |P_\tau x|$ $(x \in \mathfrak{H})$. In addition, the operator $P_\tau A P_\tau$ is invertible in the subspace $P_\tau \mathfrak{H}$, since the values of the quadratic form $(P_\tau A P_\tau x, x)$ $(x \in P_\tau \mathfrak{H};\ |x| = 1)$ are inside the right half-plane.

The class $\Pi\{P_\tau, Q_\tau\}$ is mapped into itself under multiplication by operators of a certain type. The following theorem holds.

THEOREM 2.2. *Let $A \in \mathfrak{A}(\mathfrak{B}_2)$, $B \in \mathfrak{A}(\mathfrak{B}_1, \mathfrak{B}_2)$ and $C \in \mathfrak{A}(\mathfrak{B}_1)$ be invertible operators, and let the following conditions hold*:

1) *$Q_\tau A = Q_\tau A Q_\tau$; the equality $Q_\tau A Q_\tau x = 0$ implies $Q_\tau x = 0$.*
2) *$B \in \Pi\{P_\tau, Q_\tau\}$.*
3) *$CP_\tau = P_\tau C P_\tau$ and $P_\tau C P_\tau \mathfrak{B}_1 = P_\tau \mathfrak{B}_1$.*

Then $ABC \in \Pi\{P_\tau, Q_\tau\}$.

Let us explain condition 1). The first equality in it signifies the invariance of the subspace $\operatorname{Im}(I - Q_\tau)$ relative to A. The second requirement signifies that A maps this subspace precisely onto itself.

PROOF. Let us show first that $A \in \Pi\{Q_\tau, Q_\tau\}$ and $C \in \Pi\{P_\tau, P_\tau\}$. Indeed, by virtue of conditions 1), the equation $Q_\tau A Q_\tau y = Q_\tau f$ has the unique solution $y_\tau = Q_\tau A^{-1} f$ in the subspace $Q_\tau \mathfrak{B}_2$, whatever the vector $f \in \mathfrak{B}_2$ may be. Since $\lim y_\tau = A^{-1} f$, $A \in \Pi\{Q_\tau, Q_\tau\}$. By virtue of conditions 3), the equation $P_\tau C P_\tau x = P_\tau \varphi$ has the unique solution $x_\tau = C^{-1} P_\tau \varphi$ in the subspace $P_\tau \mathfrak{B}_1$, whatever the vector $\varphi \in \mathfrak{B}_1$ may be. Thus $C \in \Pi\{P_\tau, P_\tau\}$.

Furthermore, the equality

$$Q_\tau ABCP_\tau = (Q_\tau A Q_\tau)(Q_\tau B P_\tau)(P_\tau C P_\tau)$$

and Theorem 2.1 imply the relations

$$\begin{gathered} |Q_\tau ABCP_\tau x| \geqq \delta |P_\tau x| \qquad (x \in \mathfrak{B}_1, \delta > 0), \\ Q_\tau ABCP_\tau \mathfrak{B}_1 = Q_\tau \mathfrak{B}_2. \end{gathered}$$

Hence, by virtue of the same Theorem 2.1, $ABC \in \Pi\{P_\tau, Q_\tau\}$.

In the sequel we shall need in addition the following proposition concerning the block applicability of the projection method.

2°. *Let Q_1 and Q_2 be two complementary projections of $\mathfrak{A}(\mathfrak{B})$ ($Q_2 = I - Q_1$), and let them commute with the projections P_τ ($\tau \in \Omega$). Furthermore, let the subspaces* Im Q_1 *and* Im Q_2 *be invariant relative to an invertible operator $A \in \mathfrak{A}(\mathfrak{B})$. Denote by A_j and $P_{j\tau}$ the restrictions of the operators A and P_τ to the subspace* Im Q_j, $j = 1,2$.

For $A \in \Pi\{P_\tau, P_\tau\}$ it is necessary and sufficient that $A_j \in \Pi\{P_{j\tau}, P_{j\tau}\}$, $j = 1,2$.

§ 3. Stability of projection methods

Two propositions are established in this section concerning perturbation of operators of the class $\Pi\{P_\tau, Q_\tau\}$ by small in norm, or completely continuous, operators. The second of these propositions, Theorem 3.1, plays a vital role in the sequel.

1°. *Let $A \in \Pi\{P_\tau, Q_\tau\}$. Then there exists a constant $\gamma > 0$ such that if $B \in \mathfrak{A}(\mathfrak{B}_1, \mathfrak{B}_2)$ and $|B| < \gamma$, then $A + B \in \Pi\{P_\tau, Q_\tau\}$. In other words, the class $\Pi\{P_\tau, Q_\tau\}$ is an open set.*

PROOF. Let us set $\gamma = (c - \varepsilon)/m$, where c is the constant of condition (2.3), $0 < \varepsilon < c/2$ and $m = \sup_\tau |P_\tau| \cdot |Q_\tau|$. Then the estimate

$$|Q_\tau(A + B)P_\tau x| \geqq c\,|P_\tau x| - |Q_\tau BP_\tau x| \geqq \varepsilon |P_\tau x|$$

is valid for $|B| < \gamma$. In addition, the operator $Q_\tau(A + B)P_\tau$ mapping from $P_\tau\mathfrak{B}_1$ into $Q_\tau\mathfrak{B}_2$ is invertible, since

$$Q_\tau(A + B)P_\tau = Q_\tau AP_\tau\,[I_\tau + (Q_\tau AP_\tau)^{-1}\,Q_\tau BP_\tau],$$

where I_τ is the identity operator is $P_\tau\mathfrak{B}_1$, and

$$|(Q_\tau AP_\tau)^{-1}\,Q_\tau BP_\tau| < \frac{c - \varepsilon}{c} < 1.$$

Thus the conditions of Theorem 2.1 are fulfilled for the operator $A + B$; therefore $A + B \in \Pi\{P_\tau, Q_\tau\}$.

THEOREM 3.1. *Let $A \in \Pi\{P_\tau, Q_\tau\}$, and let the operator $T \in \mathfrak{A}(\mathfrak{B}_1, \mathfrak{B}_2)$ be completely continuous. If the operator $A + T$ is invertible, it also belongs to the class $\Pi\{P_\tau, Q_\tau\}$.*

PROOF. By virtue of estimate (2.3), the following relations are valid:

$$\begin{aligned} |Q_\tau(A + T)P_\tau x| &= |Q_\tau AP_\tau\,[P_\tau x + (Q_\tau AP_\tau)^{-1}\,Q_\tau TP_\tau x]| \\ &\geqq c|P_\tau x + (Q_\tau AP_\tau)^{-1}\,Q_\tau TP_\tau x| \\ &\geqq c|P_\tau x + A^{-1}TP_\tau x| - c|\,[(Q_\tau AP_\tau)^{-1}\,Q_\tau T - A^{-1}T]\,P_\tau x|. \end{aligned}$$

Since the operator $I + A^{-1}T$ ($\in \mathfrak{A}(\mathfrak{B}_1)$) is invertible, there exists a constant

$\gamma > 0$ such that

$$|P_\tau x + A^{-1}TP_\tau x| \geqq \gamma|P_\tau x| \qquad (x \in \mathfrak{B}_1, \tau \in \Omega).$$

By utilizing the complete continuity of the operator T, it is easy to obtain the result that the operators $(Q_\tau AP_\tau)^{-1} Q_\tau T$ converge uniformly to $A^{-1}T$. This follows from the compactness of the set AS, where S is the unit sphere of the space $\mathfrak{B}_1$, and from the strong convergence of the operators $(Q_\tau AP_\tau)^{-1} Q_\tau$.

Thus τ_0 can be so chosen that

$$|(Q_\tau AP_\tau)^{-1} Q_\tau T - A^{-1}T| < \tfrac{1}{2}\gamma \qquad (\tau \geqq \tau_0).$$

The inequality

$$|Q_\tau (A+T)P_\tau x| \geqq \tfrac{1}{2} c\gamma|P_\tau x| \qquad (\tau \geqq \tau_0) \tag{3.1}$$

follows from the estimates just obtained.

To complete the proof of the theorem let us note that the operator $Q_\tau(A + T)P_\tau$ acting from $P_\tau\mathfrak{B}_1$ into $Q_\tau\mathfrak{B}_2$ is invertible. Indeed, by virtue of (3.1), $Q_\tau(A + T)P_\tau$ operates in a one-to-one manner, and since it equals the sum of the invertible operator $Q_\tau AP_\tau$ and the completely continuous operator $Q_\tau TP_\tau$, it is invertible.

REMARK. The above results imply the stability of the generalized Galerkin method under small and completely continuous perturbations (preserving the invertibility of the operator).

§ 4. Existence theorem

Let $\{\varphi_j\}_1^\infty$ be a basis of the separable Hilbert space $\mathfrak{H}$. We shall say that the operator A admits *reduction* relative to the basis $\{\varphi_j\}_1^\infty$ if $A \in \Pi\{P_n, P_n\}$, where P_n is a projection onto the linear span of the vectors $\varphi_1,\cdots,\varphi_n$ which annihilates the vectors $\varphi_{n+1}, \varphi_{n+2},\cdots$. The operator A admits reduction relative to the basis $\{\varphi_j\}_1^\infty$ if and only if the Galerkin method (φ_j, ψ_j) is applicable to it, where $\{\psi_j\}_1^\infty$ is a basis biorthogonal to the basis $\{\varphi_j\}_1^\infty$.

Let $A \in \mathfrak{A}(\mathfrak{H})$ be an invertible operator, and $\|a_{jk}\|_1^\infty$ the matrix of A in the basis $\{\varphi_j\}_1^\infty$. The operator A admits reduction relative to the basis $\{\varphi_j\}_1^\infty$ if, beginning with some n, the matrices $A^{(n)} = \|a_{jk}\|_1^n$ are invertible and the vectors

$$x^{(n)} = \sum_1^n \xi_j^{(n)} \varphi_j, \qquad \text{where} \quad \{\xi_j^{(n)}\}_1^n = (A^{(n)})^{-1}\{\eta_j\}_1^n, \qquad y = \sum_1^\infty \eta_j\varphi_j,$$

tend in norm to a solution of the equation $Ax = y$ as $n \to \infty$.

THEOREM 4.1. *For every invertible operator $A \in \mathfrak{A}(\mathfrak{H})$ there exists an orthonormal basis $\{\varphi_j\}_1^\infty$ of the space $\mathfrak{H}$ relative to which A admits reduction.*

PROOF. Let us use the polar representation of the operator A: $A = HU$, where U is a unitary, and H is a positive, operator.

Let us decompose the spectrum of the operator U into a sum of closed sets σ_k, $k = 1,\cdots,q$, such that the intersection $\sigma_j \cap \sigma_k$ $(j \neq k)$ consists of not more than one point and the diameter of each set σ_j does not exceed ε (the number $\varepsilon > 0$ will be chosen later). Then the space $\mathfrak{H}$ can be represented in the form of an orthogonal sum of the subspaces $\mathfrak{H}_k$, $k = 1,\cdots,q$, invariant relative to U, where the spectrum of the operator $U_k = U|\mathfrak{H}_k$ (U_k is the restriction of U to $\mathfrak{H}_k$) is contained in σ_k.

Let I_k be the identity operator in $\mathfrak{H}_k$ and $t_k \in \sigma_k$. Then obviously

$$|U_k - t_k I_k| < \varepsilon \qquad (k = 1,\cdots,q),$$

and so for any orthogonal projector R in $\mathfrak{H}_k$ and any vector $x \in \mathfrak{H}_k$

$$|RU_kRx| \geqq |Rx| - |R(U_k - t_kI_k)Rx| \geqq (1 - \varepsilon)|Rx|. \tag{4.1}$$

Let $\{\varphi_j\}_1^\infty$ be an orthonormal basis of $\mathfrak{H}$ which is obtained by joining the bases of the subspaces $\mathfrak{H}_k$, $k = 1,\cdots,q$, and let $P_nx = \sum_1^n (x, \varphi_j)\varphi_j$.

By means of inequality (4.1) it is easy to obtain the fact that for any vector $x \in \mathfrak{H}$

$$|P_nUP_nx| \geqq (1 - \varepsilon)|P_nx| \qquad (n = 1, 2, \cdots). \tag{4.2}$$

Setting $Q_n = I - P_n$, we obtain

$$\begin{aligned} |Q_nUP_nx|^2 &= |UP_nx|^2 - |P_nUP_nx|^2 \\ &\leqq |P_nx|^2 - (1 - \varepsilon)^2 |P_nx|^2 = (2\varepsilon - \varepsilon^2)|P_nx|^2. \end{aligned} \tag{4.3}$$

Since for every orthogonal projection P $(\in \mathfrak{A}(\mathfrak{H}))$ and every vector $y \in \mathfrak{H}$

$$|PHPy| \geqq m|Py| \qquad (m = |H^{-1}|^{-1} = |A^{-1}|^{-1}),$$

by setting $M = |H|$ we obtain, by virtue of (4.2) and (4.3),

$$\begin{aligned} |P_nHUP_nx| &\geqq |P_nHP_nUP_nx| - |P_nHQ_nUP_nx| \\ &\geqq m|P_nUP_nx| - M|Q_nUP_nx| \\ &\geqq [m(1 - \varepsilon) - M\sqrt{2\varepsilon - \varepsilon^2}]|P_nx|. \end{aligned}$$

Assuming that the number ε $(0 < \varepsilon < 1)$ is so chosen that the expression in square brackets is positive, we obtain the fact that A admits reduction relative to the orthonormal basis $\{\varphi_j\}_1^\infty$

The theorem is proved.

§ 5. Operators admitting reduction relative to any orthonormal basis

Proposition 1° of § 2 and Theorem 3.1 imply that every invertible operator $A \in \mathfrak{A}(\mathfrak{H})$ of the form

$$A = c(H + iG) + T, \tag{5.1}$$

where c is a complex number, H a positive definite, G a selfadjoint, and T a completely continuous operator, admits reduction relative to *any* orthonormal basis. It turns out that the converse assertion holds.

Before proceeding to its proof, let us note that representation (5.1) is equivalent to the following:

$$A = \alpha (I + S) + T, \tag{5.2}$$

where α is a complex number, $S \in \mathfrak{A}(\mathfrak{H})$, $|S| < 1$, and T is a completely continuous operator. This follows from the following proposition.

LEMMA 5.1. *An operator $A \in \mathfrak{A}$ has the form*

$$A = \alpha (I + S), \tag{5.3}$$

where α ($\neq 0$) is a complex number and S is an operator with $|S| < 1$, if and only if the closure of the numerical domain

$$W(A) = \{(Ax, x) : |x| = 1\}$$

of A does not contain the point $\lambda = 0$.

PROOF. If A has the form (5.3), then $W(A)$ is obviously contained in the disk $\Gamma = \{\lambda : |\lambda - \alpha| \leqq |\alpha|\,|S|\}$.

To prove sufficiency, let us recall that the set $W(A)$ is convex, and therefore if $0 \in \overline{W(A)}$ there exists a complex number z such that $\operatorname{Re}(zA) \geqq \delta I$ ($\delta > 0$).

Let $B = zA$ and $\varepsilon > 0$. Then

$$(I - \varepsilon B)(I - \varepsilon B)^* = I - \varepsilon (2 \operatorname{Re} B - \varepsilon BB^*).$$

Since $BB^* > 0$ and $\operatorname{Re} B > \delta I$, for sufficiently small $\varepsilon > 0$ the operator

$$C = 2 \operatorname{Re} B - \varepsilon BB^*$$

will be positive, where $|C| < 2|\operatorname{Re} B|$. Consequently for sufficiently small $\varepsilon > 0$ we have $|I - \varepsilon C| < 1$, and therefore $|I - \varepsilon B| < 1$. Denoting the operator $\varepsilon B - I$ by S, we obtain

$$A = (1/2\varepsilon)(I + S).$$

The lemma is proved.

We need one more auxiliary assertion.

LEMMA 5.2. *If the operator A cannot be represented in the form* (5.2), *there exists an orthonormal sequence $\{g_j\}_1^\infty$ such that* $\lim_{j\to\infty} (Ag_j, g_j) = 0$.

PROOF. Let $\mathfrak{N}$ be some subspace of finite codimenseion, P an orthogonal projection onto $\mathfrak{N}$, and $\hat{A} = PA\,|\mathfrak{N}$. Since the operator $A - \hat{A}P$ is finite-dimensional, the operator $\hat{A}$ also cannot be represented in the form $\alpha(\hat{I} + \hat{S})$, where $\hat{I}$ is the identity operator in $\mathfrak{N}$, $S \in \mathfrak{A}(\mathfrak{N})$ and $|\hat{S}| < 1$. By virtue of Lemma 5.1, $0 \in \overline{W(\hat{A})}$;

therefore for any $\varepsilon > 0$ there is a vector $x \in \mathfrak{N}$ such that $|x| = 1$ and $|(\hat{A}x, x)| < \varepsilon$, i. e. $|(Ax, x)| < \varepsilon$.

Let us choose a vector $g_1 \in \mathfrak{H}$ so that $|(Ag_1, g_1)| < 1$. Let us develop the construction of the remaining vectors of the sequence g_j by induction. Let the vectors $g_1,\cdots,g_n$ be already chosen so that the relations

$$(g_j, g_k) = \delta_{jk} \qquad (j, k = 1,\cdots,n),$$
$$|(Ag_j, g_j)| < 1/j \qquad (j = 1,\cdots,n)$$

hold. Let us denote by $\mathfrak{N}_n$ the orthogonal complement of the subspace spanned by the vectors g_j $(j = 1,\cdots,n)$. By virtue of what was established above, there is a vector $g_{n+1} \in \mathfrak{N}_n$ such that $|g_{n+1}| = 1$ and $|(Ag_{n+1}, g_{n+1})| < 1/(n+1)$.

The lemma is proved.

THEOREM 5.1. *An invertible operator $A \in \mathfrak{A}(\mathfrak{H})$ admits reduction relative to any orthonormal basis $\{\varphi_j\}_1^\infty$ of the space $\mathfrak{H}$ if and only if that operator admits a representation of the form* (5.2).

PROOF. The sufficiency of condition (5.2) has already been established. Let us pass to the proof of its necessity.

Let us suppose that A cannot be represented in the form (5.2), and let $\{g_j\}_1^\infty$ be the orthonormal system whose existence was established in Lemma 5.2. A subsequence $\{\tilde{g}_j\}_1^\infty$ can be chosen such that

$$\lim_{n\to\infty} |R_n A\tilde{g}_n| = 0, \tag{5.4}$$

where $R_n x = \sum_1^n (x, \tilde{g}_j)\tilde{g}_j$.

Indeed, let us set $\tilde{g}_1 = g_1$. Since $(Ag_j, \tilde{g}_1) = (g_j, A^*\tilde{g}_1) \to 0$, a vector $\tilde{g}_2$ can be so chosen from the sequence $\{g_j\}_2^\infty$ such that $|(A\tilde{g}_2, \tilde{g}_1)| < 1/2$. By repeating this construction we obtain a subsequence $\{\tilde{g}_j\}_1^\infty$ of the sequence $\{g_j\}_1^\infty$ such that

$$|(A\tilde{g}_j, \tilde{g}_k)| < 1/j \qquad (k < j, j = 2, 3,\cdots).$$

Then

$$|R_n A\tilde{g}_n|^2 \leqq \frac{n-1}{n^2} + |(A\tilde{g}_n, \tilde{g}_n)|^2.$$

Since $(A\tilde{g}_n, \tilde{g}_n) \to 0$, (5.4) is established.

Let $\{\psi_j\}_1^\infty$ be some sequence extending $\{\tilde{g}_j\}_1^\infty$ to an orthonormal basis of the space $\mathfrak{H}$. Since

$$\lim_{j\to\infty} (A\tilde{g}_j, \psi_k) = \lim_{j\to\infty} (\tilde{g}_j, A^*\psi_k) = 0 \qquad (k = 1, 2,\cdots),$$

there is a subsequence $\{j_n\}$ of the natural numbers such that

$$|(A\tilde{g}_{j_n}, \psi_k)| < 1/n \qquad (k \leqq n, n = 1, 2,\cdots). \tag{5.5}$$

Now let us arrange the system $\{\tilde{g}_j\} \cup \{\psi_k\}$ as follows:

$$\tilde{g}_1, \cdots, \tilde{g}_{j_1}, \psi_1, \tilde{g}_{j_1+1}, \cdots, \tilde{g}_{j_2}, \psi_2, \tilde{g}_{j_2+1}, \cdots,$$

and dentoe by φ_k the vector in the kth place of this sequence.

Let us set $P_n x = \sum_1^n (x, \varphi_j) \varphi_j$. As is easy to see,

$$P_{j_n+n-1} A\varphi_{j_n+n-1} = R_{j_n} A\tilde{g}_{j_n} + \sum_{k=1}^{n-1} (A\tilde{g}_{jn}, \psi_k) \psi_k.$$

By virtue of inequalities (5.4) and (5.5) this implies

$$\lim_{n\to\infty} |P_{j_n+n-1} A\varphi_{j_n+n-1}| = 0.$$

Thus condition (2.3) is not fulfilled for the operator A and the system of projections P_n, $n = 1, 2, \cdots$; consequently A does not admit reduction relative to the basis $\{\varphi_j\}_1^\infty$.

§ 6. Operators admitting reduction relative to any basis equivalent to an orthonormal one

Two bases $\{\varphi_j\}_1^\infty$ and $\{\psi_j\}_1^\infty$ of the space $\mathfrak{H}$ are called *equivalent* if there exists an invertible operators $B \in \mathfrak{A}$ such that $B\varphi_j = \psi_j$ $(j = 1, 2, \cdots)$. According to the Lorch-Gel'fand theorem (cf. Gohberg and Kreĭn **[4]**, Chapter VI, Theorem 2.2), *a basis* $\{\psi_j\}_1^\infty$ *is equivalent to an orthonormal one if and only if it is absolute and*

$$\sup_j |\psi_j| < \infty, \qquad \inf_j |\psi_j| > 0.$$

In this section a description is given of a class of operators admitting reduction relative to any basis equivalent to an orthonormal one. Auxiliary Theorem 6.1 plays a vital role here. The following notation is introduced for its formulation.

Let $A \in \mathfrak{A}$, $x \in \mathfrak{H}$, $|x| = 1$, and let $\mathfrak{M}$ be some subspace of $\mathfrak{H}$. We set

$$\eta_A(x) = |Ax - (Ax, x)x|,$$
$$\eta_A(\mathfrak{M}) = \sup\{\eta_A(x) : x \in \mathfrak{M}, |x| = 1\}.$$

THEOREM 6.1. *For an operator $A \in \mathfrak{A}$ to have the form $A = \alpha I + T$, where α is a number and T is a completely continuous operator, it is necessary and sufficient that* $\inf \eta_A(\mathfrak{M}) = 0$, *where the greatest lower bound is taken with respect to all subspaces of finite codimension.*

The proof of this theorem is based on several lemmas.

Let $x, y \in \mathfrak{H}$ and $|x| = 1$. By $\theta_x(y)$ we denote the number

$$\theta_x(y) = |y - (y, x)x|.$$

The equality $\theta_x(y) = 0$ obviously indicates the linear dependence of the vectors x and y; if, however, $\theta_x(y) > 0$, then the vectors x and $e = (y - (y, x)x)/\theta_x(y)$

form an orthonormal basis of the linear span of the vectors x and y.

The following properties of the function $\theta_x(y)$ are also obvious:

$$\theta_x^2(y) = |y|^2 - |(y, x)|^2, \qquad \theta_x(y + \alpha x) = \theta_x(y).$$

LEMMA 6.1. *Let $\mathfrak{P}_1$ and $\mathfrak{P}_2$ be two-dimensional spaces, and T a linear operator from $\mathfrak{P}_1$ into $\mathfrak{P}_2$ which maps some orthonormal basis of $\mathfrak{P}_1$ into the linearly independent vectors $x, y \in \mathfrak{P}_2$; $|x| = 1$. Then*

$$|T| \leqq (1 + |y|^2)^{1/2} \qquad \textit{and} \qquad |T^{-1}| \leqq (1 + |y|^2)^{1/2}/\theta_x(y).$$

PROOF. Obviously it is sufficient to establish the lemma for the case when $\mathfrak{P}_1 = \mathfrak{P}_2$ and the orthonormal basis x, e is transformed by the operator T into the vectors x and y, respectively. Since

$$y = (y, x)x + \theta_x(y)e,$$

the matrices of the operators T and T^{-1} have the following form in the basis x, e:

$$T = \begin{Vmatrix} 1 & (y, x) \\ 0 & \theta_x(y) \end{Vmatrix}, \qquad T^{-1} = \begin{Vmatrix} 1 & -(y, x)/\theta_x(y) \\ 0 & 1/\theta_x(y) \end{Vmatrix}.$$

Hence the lemma's assertion follows.

The quantities η_A and θ_x, obviously, are connected by the following relation:

$$\eta_A(x) = \theta_x(A_x).$$

This implies that $\eta_A(x) = 0$ if and only if x is an eigenvector of A. If $\mathfrak{M}$ is a subspace of $\mathfrak{H}$, then, as is easy to see,

$$\eta_A(\mathfrak{M}) \leqq |A|\mathfrak{M}|, \qquad \eta_{A+\lambda I}(\mathfrak{M}) = \eta_A(\mathfrak{M}), \tag{6.1}$$

where $|A|\mathfrak{M}| = \sup\{|Ax| : x \in \mathfrak{M}, |x| = 1\}$.

Let us denote by $W_A(\mathfrak{M})$ the numerical domain of the operator A on the subspace $\mathfrak{M}$, i. e.

$$W_A(\mathfrak{M}) = \{(Ax, x) : x \in \mathfrak{M}, |x| = 1\}.$$

As is well known, $W_A(\mathfrak{M})$ is a convex set.

LEMMA 6.2. *Let $A \in \mathfrak{A}$, and let $\mathfrak{M}$ be a subspace of $\mathfrak{H}$. Then*

$$\operatorname{diam} W_A(\mathfrak{M}) \leqq 8\,|A|^{1/2}\eta_A^{1/2}(\mathfrak{M}).$$

PROOF. Let x and y be arbitrary unit vectors of $\mathfrak{M}$, and set $\lambda = (Ax, x)$ and $\mu = (Ay, y)$. Let us denote the quantity $\eta_A(\mathfrak{M})$ by η_0. It is sufficient to show the validity of the inequality

$$|\lambda - \mu|^2 \leqq 64|A|\eta_0. \tag{6.2}$$

If the vectors x and y are linearly dependent, $\lambda = \mu$. We shall assume their linear independence. Let us introduce the notation $\alpha = (y, x)$ and $\beta = \theta_x(y)$. Then $\beta > 0$; it can be assumed that $\alpha \geqq 0$ since this can be attained by multiplying x by a number whose modulus equals unity. The vectors x and $e = (y - \alpha x)/\beta$ form an orthonormal basis of the subspace $\mathfrak{P}$ spanned by x and y. Since $y = \alpha x + \beta e$, it follows that $\alpha^2 + \beta^2 = 1$, and the relations

$$|w|^2 = (1 - \alpha)^2 + \beta^2 = 2(1 - \alpha) \leqq 2\beta^2$$

are fulfilled for the vector $w = x - y$. The equalities

$$\begin{aligned}\lambda = (Ax, x) &= (Ay + Aw, y + w)\\ &= \mu + (Aw, y) + (Ay, w) + (Aw, w)\end{aligned}$$

imply

$$|\lambda - \mu| \leqq 2|A|\,|w| + |A|\,|w|^2 \leqq 4|A|\,|w| \leqq 4\sqrt{2}\,|A|\,\beta. \tag{6.3}$$

Now let us set $z = (x + y)/|x + y|$ and $\nu = (Az, z)$. Then

$$\eta_A(z) = |Az - \nu z| = \frac{|A(x + y) - \nu(x + y)|}{|x + y|} \leqq \eta_0$$

and

$$|A(x + y) - \nu(x + y)| < 2\eta_0.$$

Since $|A_x - \lambda_x| < \eta_0$ and $|Ay - \mu y| < \eta_0$, it follows that

$$|(\lambda - \nu)x + (\mu - \nu)y| = |\lambda x + \mu y - \nu(x + y)| \leqq 4\eta_0.$$

Let us consider an operator T which converts a pair of complex numbers (γ, δ) into a vector $\gamma x + \delta y \in \mathfrak{P}$. By virtue of Lemma 6.1, $|T^{-1}| \leqq \sqrt{2}/\beta$; therefore

$$(|\lambda - \nu|^2 + |\mu - \nu|^2)^{1/2} \leqq (\sqrt{2}/\beta)|(\lambda - \nu)x + (\mu - \nu)y| \leqq 4\sqrt{2}\,\eta_0/\beta,$$

and consequently

$$|\lambda - \mu| \leqq 8\sqrt{2}\,\eta_0/\beta. \tag{6.4}$$

(6.3) and (6.4) imply (6.2). The lemma is proved.

We still need the following proposition.

LEMMA 6.3. *Let $A \in \mathfrak{A}$, let $\mathfrak{M}$ be a subspace of $\mathfrak{H}$ and let the point λ_0 belong to the closure of the set $W_A(\mathfrak{M})$. Then*

$$|(A - \lambda_0 I)|\mathfrak{M}| \leqq 65\,|A|\eta_A(\mathfrak{M}).$$

PROOF. The closure of the set $W_{A-\lambda_0 I}(\mathfrak{M}) = W_A(\mathfrak{M}) - \lambda_0$ contains the point $\lambda = 0$; therefore, by virtue of Lemma 6.2,

$$\begin{aligned}|(Ax - \lambda_0 x, x)|^2 &\leqq [\operatorname{diam} W_{A-\lambda_0 I}(\mathfrak{M})]^2\\ &= [\operatorname{diam} W_A(\mathfrak{M})]^2 \leqq 64|A|\eta_A(\mathfrak{M})\end{aligned}$$

for any unit vector $x \in \mathfrak{M}$. Consequently

$$|(A - \lambda_0 I)x|^2 \leqq |(Ax - \lambda_0 x, x)|^2 + \eta_A^2(x) \leqq 65|A|\eta_A(\mathfrak{M}).$$

PROOF OF THEOREM 6.1. Let $A = \alpha I + T$, where T is a completely continuous operator. Then, by virtue of (6.1), $\eta_A(\mathfrak{M}) = \eta_T(\mathfrak{M})$ for any subspace $\mathfrak{M}$. Let P_n, $n = 1, 2, \cdots$, be a sequence of finite-dimensional orthoprojections converging strongly to I. Then

$$\lim_{n\to\infty} |T(I - P_n)| = 0.$$

Let us set $\mathfrak{M}_n = (I - P_n)\mathfrak{H}$. By virtue of (6.1),

$$\eta_A(\mathfrak{M}_n) = \eta_T(\mathfrak{M}_n) \leqq |T|\mathfrak{M}_n| = |T(I - P_n)| \to 0.$$

Conversely, let $\inf \eta_A(\mathfrak{M}) = 0$, where the greatest lower bound is taken with respect to all subspaces of finite codimension. Then there is a sequence of subspaces $\mathfrak{M}_n$, each of which is of finite codimension, such that

$$\lim_{n\to\infty} \eta_A(\mathfrak{M}_n) = 0.$$

Obviously it can be assumed that $\mathfrak{M}_{j+1} \subset \mathfrak{M}_j$ $(j = 1, 2, \cdots)$.

Then we obtain

$$\overline{W_A(\mathfrak{M}_{j+1})} \subset \overline{W_A(\mathfrak{M}_j)} \qquad (j = 1, 2, \cdots)$$

for the sequence of closed convex sets $\overline{W_A(\mathfrak{M}_n)}$

By virtue of Lemma 6.2,

$$\lim_{n\to\infty} \operatorname{diam} \overline{W_A(\mathfrak{M}_n)} = 0.$$

Therefore there exists a unique point α which belongs to all $\overline{W_A(\mathfrak{M}_n)}$. Let $T = A - \alpha I$, let Q_n be an orthoprojection onto $\mathfrak{M}_n$, and $P_n = I - Q_n$. Then, by virtue of Lemma 6.3,

$$|TQ_n|^2 = |T|\mathfrak{M}_n|^2 \leqq 65\, |A|\, \eta_A(\mathfrak{M}_n) \to 0,$$

and the sequence of finite-dimensional operators TP_n consequently converges uniformly to T. Therefore the operator T is completely continuous.

THEOREM 6.2. *An invertible operator $A \in \mathfrak{A}$ admits reduction relative to any basis equivalent to an orthonormal one if and only if*

$$A = \alpha I + T, \tag{6.5}$$

where α is a number and T is a completely continuous operator.

PROOF. The sufficiency of condition (6.5) follows from Theorem 3.1. Let us prove its necessity.

Let us suppose that the operator A cannot be represented in the form (6.5). Then by virtue of Theorem 6.1 there exists a number $\delta > 0$ such that in each subspace $\mathfrak{N}$ of finite codimension there is a *normalized* vector for which

$$|Ax - (Ax, x)x| > \delta.$$

Let $x_1 \in \mathfrak{H}$, $|x_1| = 1$, be some vector for which

$$|Ax_1 - (Ax_1, x_1)x_1| > \delta,$$

let $\mathfrak{M}_1$ be the linear span of the vectors x_1 and Ax_1, and let $\mathfrak{H}_1 = \mathfrak{H} \ominus \mathfrak{M}$, and $\mathfrak{N}_1 = A^{-1}\mathfrak{H}_1 \cap \mathfrak{H}_1$. Since $\mathfrak{N}_1$ is a subspace of finite codimension, there is a vector $x_2 \in \mathfrak{N}_1$, $|x_2| = 1$, such that

$$|Ax_2 - (Ax_2, x_2)x_2| > \delta.$$

Let us denote by $\mathfrak{M}_2$ the linear span of the vectors x_2 and Ax_2. Obviously $\mathfrak{M}_2 \supset \mathfrak{H}_1$; therefore $\mathfrak{M}_2$ and $\mathfrak{M}_1$ are orthogonal.

Continuing this construction we obtain a sequence of pairwise orthogonal two-dimensional subspaces $\mathfrak{M}_j, j = 1, 2, \cdots$, spanned by the vectors x_j and Ax_j, where

$$|x_j| = 1, \qquad |Ax_j - (Ax_j, x_j)\, x_j| > \delta \qquad (j = 1, 2, \cdots). \tag{6.6}$$

Let us denote by $\mathfrak{M}$ the orthogonal sum of the subspaces $\mathfrak{M}_j, j = 1, 2, \cdots$.

Let us set $\varphi_{2j-1} = x_j$ and $\varphi_{2j} = Ax_j$, $j = 1, 2, \cdots$, and show that $\{\varphi_j\}_1^\infty$ is a basis of the subspace $\mathfrak{M}$ which is equivalent to an orthonormal one.

Obviously for any natural n and any complex numbers $a_j, j = 1, \cdots, 2n$, we have

$$\Big| \sum_{j=1}^{2n} a_j \varphi_j \Big|^2 = \sum_{j=1}^{n} |a_{2j-1}x_j + a_{2j}\, Ax_j\, |^2. \tag{6.7}$$

The parallelogram equality

$$\begin{aligned} |a_{2j-1}x_j + a_{2j}Ax_j|^2 + |a_{2j-1}x_j - a_{2j}Ax_j|^2 \\ = 2\,(|a_{2j-1}x_j|^2 + |a_{2j}Ax_j|^2) \end{aligned} \tag{6.8}$$

implies that

$$\Big| \sum_{j=1}^{2n} a_j\varphi_j \Big|^2 \leqq 2 \sum_{j=1}^{n} (|a_{2j-1}|^2 + |a_{2j}|^2\, |A|^2) \leqq c \sum_{j=1}^{2n} |a_j|^2. \tag{6.9}$$

On the other hand,

$$|a_{2j-1}x_j - a_{2j}Ax_j| \leqq |a_{2j-1}x_j + a_{2j}Ax_j| + 2|a_{2j}Ax_j|. \tag{6.10}$$

Since the number $\rho_j = |Ax_j - (Ax_j, x_j)\, x_j|$ equals the distance from the vector Ax_j to the one-dimensional subspace spanned by x_j, we have

$$|a_{2j}Ax_j + a_{2j-1}x| \geqq |a_{2j}|\rho_j > |a_{2j}|\,\delta \geqq \frac{\delta}{|A|}\,|a_{2j}Ax_j| \tag{6.11}$$

by virtue of (6.6).

Relations (6.7), (6.8), (6.10) and (6.11) imply that

$$\begin{aligned}\sum_{j=1}^{2n} |a_j|^2 &\leqq \sum_{j=1}^{n} (|a_{2j-1}x_j|^2 + |A^{-1}|^2 |a_{2j}Ax_j|^2) \\ &\leqq c_1 \sum_{j=1}^{n} (|a_{2j-1}x_j|^2 + |a_{2j}Ax_j|^2) \\ &\leqq c_2 \sum_{j=1}^{n} |a_{2j-1}x_j + a_{2j}Ax_j|^2 = c_2 \Big| \sum_{j=1}^{2n} a_j\varphi_j \Big|^2. \end{aligned} \tag{6.12}$$

By virtue of N. K. Bari's well-known result (cf. Gohberg and Kreĭn [4], Theorem VI.2.1), inequalities (6.9) and (6.12) imply that the sequence $\{\varphi_j\}_1^\infty$ is a basis of the subspace $\mathfrak{M}$ which is equivalent to an orthonormal one.

Let us denote by $\mathfrak{N}$ the orthogonal complement of $\mathfrak{M}$ in $\mathfrak{H}$. Let us suppose first that the subspace $\mathfrak{N}$ is infinite-dimensional, and let $\{g_j\}_1^\infty$ be some orthonormal basis of $\mathfrak{N}$. Let us set

$$\psi_{3j-2} = x_j, \qquad \psi_{3j-1} = Ax_j, \quad \psi_{3j} = g_j \qquad (j = 1, 2, \cdots).$$

Obviously $\{\psi_j\}_1^\infty$ is a basis of $\mathfrak{H}$ which is equivalent to an orthonormal one. Let $\{f_j\}_1^\infty$ be a sequence biorthogonal to it, and let $P_nx = \sum_1^n(x, f_j)\psi_j$. Then

$$P_{3j-2}AP_{3j-2}\psi_{3j-2} = 0 \qquad (j = 1, 2, \cdots);$$

consequently A does not admit reduction relative to the basis $\{\psi_j\}_1^\infty$.

If, however, $\dim \mathfrak{N} = r < \infty$, let us set

$$\psi_j = g_j \quad (j \leqq r), \qquad \psi_j = \varphi_{j-r} \quad (j > r).$$

Then

$$P_{r+2k-1}AP_{r+2k-1}\psi_{r+2k-1} = 0 \qquad (k = 1, 2, \cdots),$$

and A again does not admit reduction relative to the basis $\{\psi_j\}_1^\infty$.

The theorem is proved.

CHAPTER III

PROJECTION METHODS OF SOLVING THE WIENER-HOPF EQUATION AND ITS DISCRETE ANALOGUE

The justification of several projection methods (in particular the reduction method) is presented in this chapter through application to the Wiener-Hopf integral operator and its discrete analogue. The justification of these methods is first developed for a function of abstract unilaterally invertible operators, and only later is their application to various specific types of Wiener-Hopf equations derived. We particularly call the reader's attention to the method of inverting a finite Toeplitz matrix and its continuous analogue which is set forth in this chapter, as well as to an iterative method of calculating the index of a polynomial.

In connection with two unsolved problems we are also concerned (by way of survey) with the case of the multivariate discrete Wiener-Hopf equation.

§ 1. Projection method for functions of unilaterally invertible operators

The abstract foundation of this chapter is presented in the present section.

The applicability of several projection methods for solving a Wiener-Hopf integral equation and its discrete analogue is derived from the results of this section.

§§ 2 and 3 are devoted to these applications.

THEOREM 1.1. *Let* V $(\in \mathfrak{A}(\mathfrak{B}))$ *be an operator invertible only on the left, and let* $V^{(-1)}$ *be an operator inverse to it on the left. Suppose that the projections* P_τ $(\tau \in \Omega)$, *which converge strongly to the identity operator as* $\tau \to \infty$, *are related to the operators* V *and* $V^{(-1)}$ *by the following conditions*:

1) *The operators* V *and* $V^{(-1)}$ *possess properties* (Σ I) *and* (Σ II) *of* § 3, *Chapter* I.
2) dim Coker $V < \infty$.
3) $P_\tau V P_\tau = P_\tau V$ *and* $P_\tau V^{(-1)} P_\tau = V^{(-1)} P_\tau$ *for* $\tau \in \Omega$.

If the symbol of an operator $A \in \mathfrak{R}(V)$ *does not vanish and has index zero,*

$$A(\zeta) \neq 0 \quad (|\zeta| = 1) \qquad \text{and} \qquad \kappa = \operatorname{ind} A(\zeta) = 0,$$

then $A \in \Pi\{P_\tau, P_\tau\}$.

PROOF. The operator A satisfies the basic invertibility theorem (Theorem 3.1, Chapter I). Therefore (cf. equality (3.5), Chapter I) A can be represented in the form $A = R_-(I + C)R_+$, where $R_\pm$ are invertible operators of $\mathfrak{R}^\pm(V)$, respectively, with $R_\pm^{-1} \in \mathfrak{R}^\pm(V)$, and the operator $C \in \mathfrak{R}(V)$, it being possible to choose the latter so small in norm that $I + C \in \Pi\{P_\tau, P_\tau\}$.

Let us take an auxiliary operator $B = R_+(I + C)R_-$ and show that it is contained in the class $\Pi\{P_\tau, P_\tau\}$. The equalities $P_\tau R_+ P_\tau = P_\tau R_+$ and $P_\tau R_- P_\tau = R_- P_\tau$ $(\tau \in \Omega)$ and analogous equalities for the operators $R_\pm^{-1}$ follow immediately from condition 3). Therefore $(P_\tau R_\pm P_\tau)^{-1} = P_\tau R_\pm^{-1} P_\tau$ and, by virtue of the multiplication theorem (Theorem 2.2, Chapter II), the operator $B \in \Pi\{P_\tau, P_\tau\}$.

Since A is invertible, it follows from the theorem concerning perturbation by a completely continuous operator (Theorem 3.1, Chapter II) that in order to complete the proof it is sufficient to ascertain that the operator $A - B$ is completely continuous. We shall now establish a somewhat more general fact, namely, that *for any operators* $A_1, A_2 \in \mathfrak{R}(V)$ *the operator* $A_1A_2 - A_2A_1$ *is completely continuous.* Let us show first that the operator $N = V^{(j)}V^{(k)} - V^{(k)}V^{(j)}$ is finite-dimensional (j and k are *any* integers). If the numbers j and k are of the same sign, then obviously $N = 0$. Now for the sake of definiteness let $j < 0$ and $k > 0$. Then $V^{(j)}V^k = V^{(j+k)}V^k = V^{(j+k)}V^{-j}$ and

$$N = V^{(j+k)} - V^{(j+k)}V^{-j}V^{(j)} = V^{(j+k)}(I - V^{-j}V^{(j)}).$$

By virtue of condition 2), the operator $I - V^{-j}V^{(j)}$ is a finite-dimensional projection; therefore the operator N is also finite-dimensional. This implies that for any $A_1, A_2 \in \bar{\mathfrak{R}}(V)$ the operator $A_1A_2 - A_2A_1$ is finite-dimensional. The proof in the general case $A_1, A_2 \in \mathfrak{R}(V)$ is obtained by passage to the limit.

The theorem is proved. It permits projection methods to be applied to the inversion of operators A of $\mathfrak{R}(V)$ in the "basic" case, i.e. when the index $\kappa = 0$.

If, however, $\kappa \neq 0$, it is necessary to cancel the operator's index, i.e. to introduce the operator

$$B = \begin{cases} V^{(-\kappa)}A & \text{for } \kappa > 0, \\ AV^{-\kappa} & \text{for } \kappa < 0, \end{cases}$$

with index equal to 0.

By virtue of the theorem just proved, the operators $P_\tau BP_\tau$, beginning with some τ, are invertible on the subspace $P_\tau\mathfrak{B}$, and their inverses converge strongly to the operator B^{-1}.

Therefore the operators[1] $(P_\tau BP_\tau)^{-1}P_\tau V^{(-\kappa)}$ for $\kappa > 0$ or the operators $V^{-\kappa}(P_\tau BP_\tau)^{-1}P_\tau$ for $\kappa < 0$ converge strongly to an appropriate (left or right) inverse of A.

REMARK. If in condition 1) of Theorem 1.1 the requirement (Σ II) is discarded by replacing it with the condition $A \in \hat{\mathfrak{R}}(U)$ (cf. the second scheme, § 3.2, Chapter I), Theorem 1.1 remains valid.

[1] If $A \in \mathfrak{A}(\mathfrak{B})$ and the operator $P_\tau AP_\tau$ is invertible in the subspace $P_\tau\mathfrak{B}$, then $(P_\tau AP_\tau)^{-1}$, as before, denotes the operator inverse to the operator $P_\tau AP_\tau$ on the subspace $P_\tau\mathfrak{B}$.

An operator A ($\in \mathfrak{A}(\mathfrak{B})$) is called an operator of *regular* type if the set of its values $A\mathfrak{B}$ is a closed subspace and Ker $A = \{0\}$. It follows immediately from Banach's inverse operator theorem that A is an operator of regular type if and only if there exists a constant $\gamma > 0$ such that for all x in $\mathfrak{B}$ the relation $|Ax| \geqq \gamma\, |x|$ is valid.

LEMMA 1.1. *Let a sequence of invertible operators* A_n ($\in \mathfrak{A}(\mathfrak{B})$) *converge strongly to the operator* A, *and let* $\sup_n |A_n^{-1}| < \infty$. *Then* A *is an operator of regular type.*

If in addition the sequence A_n^* *converges strongly to the operator* A^*, *then* A *is invertible.*

PROOF. The inequality $|x| \leqq |A_n^{-1}|\,|A_nx|$ ($x \in \mathfrak{B}$) implies the relation

$$|A_nx| \geqq |x|\,/\,M, \tag{1.1}$$

where $M = \sup_n |A_n^{1-}|$. By passing to the limit in (1.1) we obtain the inequality $|Ax| \geqq |x|/M$, which proves the first assertion of the lemma.

If, moreover, the additional requirement is fulfilled, it can be shown by an analogous argument that the operator A^* is an operator of regular type. The invertibility of A follows immediately from this.

Let us introduce the following definition. If a Banach space $\mathfrak{B}$ is dual to some Banach space $\mathfrak{B}_1$, we say that $\mathfrak{B}_1$ is *predual* to $\mathfrak{B}$ and denote it by the symbol $^*\mathfrak{B}$. In exactly the same way, if an operator $A \in \mathfrak{A}(\mathfrak{B})$ is adjoint to some operator A_1 of $\mathfrak{A}(^*\mathfrak{B})$, we say that A_1 is preadjoint to A and write $A_1 = {}^*A$.

LEMMA 1.2. *Lemma* 1.1 *remains valid if the additional condition is changed as follows.*

Preadjoints $^*A_n$ *and* $^*A \in \mathfrak{A}(^*\mathfrak{B})$ *exist for the operators* A_n *and* A, *and the sequence* $^*A_n$ *converges strongly to* $^*A$.

The proof is carried out analogously. It is necessary to establish only that $^*A$ is an operator of regular type. As in Lemma 1.1, the invertibility of A will follow immediately from this.

THEOREM 1.2. *Let a system of projections* P_τ ($\tau \in \Omega$) *converge strongly to the identity operator as* $\tau \to \infty$, *and let*

$$\lim_{\tau\to\infty} |(P_\tau A P_\tau)^{-1}| < \infty^{2)}$$

for an operator $A \in \mathfrak{R}(V)$.

Let at least one of the following two conditions be fulfilled:

1) *The projections* P_τ^* *converge strongly to the identity operator in* $\mathfrak{B}^*$.

---

2)If the operator $P_\tau A P_\tau$ is not invertible in the subspace $P_\tau\mathfrak{B}$, we assume $|P_\tau A P_\tau|^{-1} = \infty$.

2) *The predual space* $^*\mathfrak{B}$ *exists, and the operators* $^*A, ^*P_\tau \in \mathfrak{A}(\mathfrak{B})$ *are such that the* $^*P_\tau$ *converge strongly to the identity operator in* $^*\mathfrak{B}$ *as* $\tau \to \infty$.

Then

$$A(\zeta) \neq 0 \quad (|\zeta| = 1) \qquad \textit{and} \qquad \operatorname{ind} A(\zeta) = 0. \tag{1.2}$$

PROOF. Let $\tau_n \to \infty$ ($\tau_n \in \Omega$) be a sequence such that the operators $A_n = P_{\tau_n} A P_{\tau_n} + Q_{\tau_n}$ are invertible ($Q_{\tau_n} = I - P_{\tau_n}$) and $\sup_n |A_n^{-1}| < \infty$. By applying Lemma 1.1 or Lemma 1.2 to the sequence A_n and the operator A we obtain the fact that A is invertible, which, by virtue of the basic invertibility theorem (Theorem 3.1, Chapter I), is equivalent to fulfillment of condition (1.2).

REMARK. If the space $\mathfrak{B}$ is reflexive, $P_\tau \to I$, and the equality $P_\tau x = 0$ implies $P_\nu x = 0$ for all $\nu < \tau$, then $P_\tau^* \to I^*$.

As a matter of fact, let

$$\mathfrak{L} = \bigcup_{\tau \in \Omega} P_\tau^* \mathfrak{B}^*.$$

Since $P_\nu^* \mathfrak{B}^* \subset P_\tau^* \mathfrak{B}^*$ for $\nu < \tau$, it follows that $P_\tau^* \varphi \to \varphi$ for any φ in $\mathfrak{L}$. (The equality $P_\tau^* \varphi = \varphi$ will be fulfilled beginning with some τ.) From this it is easy to obtain the fact that $P_\tau^* \varphi \to \varphi$ for all φ in $\tilde{\mathfrak{L}}$. Now let us show that $\tilde{\mathfrak{L}} = \mathfrak{B}^*$. Let us assume that this is not so, and let the functional $\varphi (\in \mathfrak{B}^*)$ not belong to $\tilde{\mathfrak{L}}$. Then there is an element $f \in \mathfrak{B}^{**}$ such that $f(\psi) = 0$ for all $\psi \in \tilde{\mathfrak{L}}$ and $f(\varphi) = 1$. This implies that $P_\tau^{**} f = 0$ ($\tau \in \Omega$), and, since $P_\tau^{**} f \to f$, we have $f = 0$, which contradicts the equality $f(\varphi) = 1$.

§ 2. Solution of discrete equations by the reduction method

1. Now let us pass to applications of the general theorems. The following theorem is a corollary of Theorems 1.1 and 1.2.

THEOREM 2.1. *Let* $a(\zeta)$ *be an arbitrary function continuous on the unit circle, and* $\{a_j\}_{-\infty}^{\infty}$ *the sequence of its Fourier coefficients. If the relation*

$$\liminf_{n \to \infty} |A_n^{-1}| < \infty^{3)}$$

is valid for the matrices $A_n = \|a_{j-k}\|_{j,k=1}^n$, *then*

$$a(\zeta) \neq 0 \quad (|\zeta| = 1) \qquad \textit{and} \qquad \kappa = \operatorname{ind} a(\zeta) = 0. \tag{2.1}$$

Conversely, under fulfillment of conditions (2.1) *the system of equations*

$$\sum_{k=1}^{n} a_{j-k} \xi_j = \eta_j \qquad (j = 1, \cdots, n),$$

[3] If the matrix A_n is singular, we assume $|A_n^{-1}| = \infty$; the norm $|\ |$ is understood to be in l_2.

beginning with some n, will have the solution $\{\xi_j^{(n)}\}_{j=1}^n$, *and as* $n \to \infty$ *the vectors* $\xi^{(n)} = \{\xi_1^{(n)},\cdots,\xi_n^{(n)},0,0,\cdots\}$ *will converge in the norm of* l_2 *to a solution of the system*

$$\sum_{k=1}^{\infty} a_{j-k}\xi_k = \eta_j \qquad (j = 1,2,\cdots),$$

whatever the vector $\{\eta_j\}_1^\infty \in l_2$ *may be.*

PROOF. Let us realize the space $\mathfrak{B}$ and the operators V and $V^{(-1)}$ just as in § 7, Chapter I; i.e. let us set $\mathfrak{B} = l_2$ and

$$V\{\xi_j\}_1^\infty = \{0, \xi_1, \xi_2,\cdots\}, \qquad V^{(-1)}\{\xi_j\}_1^\infty = \{\xi_2, \xi_3,\cdots\}.$$

Let us consider the projections P_n, $n = 1,2,\cdots$, defined by the equalities

$$P_n\{\xi_j\}_1^\infty = \{\xi_1,\cdots,\xi_n,0,0,\cdots\}.$$

It is easy to see that the operators V and $V^{(-1)}$ and the projections P_n satisfy all of the conditions of Theorems 1.1 and 1.2 ,while the operator A defined in l_2 by the matrix $\|a_{j-k}\|_{j,k=1}^\infty$ belongs, as was shown in § 7, Chapter I, to the set $\mathfrak{R}(V)$, and $A(\zeta) = a(\zeta)$.

The theorem is proved.

Let us note that Theorems 1.1 and 1.2 imply the validity of Theorem 2.1 for the case of the spaces l_p $(p \geqq 1)$ and c_0 only if the continuous function $a(\zeta)$ is such that $a(V) \in \mathfrak{R}(V)$ (i.e. $a(\zeta) \in \mathfrak{R}(\zeta)$). The latter in particular holds for all l_p $(p \geqq 1)$ and c_0 if $a(\zeta)$ expands into an absolutely convergent Fourier series. The second part of Theorem 2.1 has been established by G. Baxter for the space l_1 under this condition. Let us give his proof.

BAXTER'S PROOF. Let $\{\xi_0,\cdots,\xi_n\}$ be an arbitrary vector, and set

$$\eta_j = \sum_{k=0}^{n} a_{j-k}\xi_k \qquad (j = 0,\cdots,n). \tag{2.2}$$

Assuming ξ_j and η_j equal to zero for $j < 0$ and $j > n$, let us write equalities (2.2) in the form

$$\sum_{k=-\infty}^{\infty} a_{j-k}\xi_k = \eta_j + g_{1j} + g_{2j} \qquad (j = 0, \pm 1,\cdots), \tag{2.3}$$

where

$$g_{1j} = \begin{cases} \sum_{k=0}^{n} a_{j-k}\xi_k, & j > n, \\ 0, & j \leqq n, \end{cases} \qquad g_{2j} = \begin{cases} \sum_{k=0}^{n} a_{j-k}\xi_k, & j < 0, \\ 0, & j \geqq 0. \end{cases}$$

The vectors $\{g_{1j}\}_{-\infty}^\infty$ and $\{g_{2j}\}_{-\infty}^\infty$ obviously belong to the space $\tilde{l}_1$. Multiplying the equation with index j in (2.3) by ζ^j and summing over j from $-\infty$ to ∞, we obtain

$$a(\zeta)\xi(\zeta) = \eta(\zeta) + g_1(\zeta) + g_2(\zeta).$$

All of the functions entering into the latter equality belong to the Wiener algebra W of functions expanding into absolutely convergent Fourier series; a norm is introduced into it as follows: if $c(\zeta) = \sum_{-\infty}^{\infty} c_j\zeta^j$, then $|c| = \sum_{-\infty}^{\infty}|c_j|$.

By virtue of the theorem concerning factorization of functions of a decomposing algebra (Theorem 5.1, Chapter I), the function $a(\zeta)$ admits the factorization $a(\zeta) = a_-(\zeta)a_+(\zeta)$, where $a_+^{\pm 1}(\zeta) \in W^+$ and $a_-^{\pm 1}(\zeta) \in W^-$.

Let us introduce in the algebra W the projections P_m $(m = 0, \pm 1, \cdots)$ defined by the equality

$$P_m\Big(\sum_{-\infty}^{\infty} c_j\zeta^j\Big) = \sum_{-\infty}^{m} c_j\zeta^j.$$

Since $\xi(\zeta)a_+(\zeta) \in W^+$ and $g_2(\zeta)a_-^{-1}(\zeta) \in W^-$, the equality

$$\xi a_+ = \eta a_-^{-1} + g_1 a_-^{-1} + g_2 a_-^{-1} \tag{2.4}$$

implies $g_2 a_-^{-1} = -P_{-1}(\eta a_-^{-1} + g_1 a_-^{-1})$. It is easy to ascertain that $P_{-1}(g_1 a_-^{-1}) = P_{-1}(g_1 P_{-n} a_-^{-1})$. Therefore

$$g_2 a_-^{-1} = -P_{-1}(\eta a_-^{-1}) - P_{-1}(g_1 P_{-n} a_-^{-1}),$$
$$|g_2 a_-^{-1}| \leqq |\eta|\,|a_-^{-1}| + |g_1 a_+^{-1}|\,|a_+ P_{-n} a_-^{-1}|.$$

Whatever $\varepsilon > 0$ may be, the relation $|a_+ P_{-n} a_-^{-1}| < \varepsilon$ is fulfilled for sufficiently large n; consequently

$$|g_2 a_-^{-1}| \leqq c|\eta| + \varepsilon|g_1 a_+^{-1}|, \tag{2.5}$$

where $c = \max(|a_-^{-1}|, |a_+^{-1}|)$. The fact that for sufficiently large n

$$|g_1 a_+^{-1}| \leqq c|\eta| + \varepsilon|g_2 a_-^{-1}| \tag{2.6}$$

is established analogously.

If n is chosen so large that ε is less than 1 in relations (2.5) and (2.6), we obtain

$$|g_2 a_-^{-1}| \leqq \frac{c}{1-\varepsilon}|\eta|, \qquad |g_1 a_+^{-1}| \leqq \frac{c}{1-\varepsilon}|\eta|.$$

The latter relations and equality (2.4) imply

$$|\xi| \leqq \frac{c^2(3-\varepsilon)}{1-\varepsilon}|\eta|,$$

which is equivalent to the inequality

$$\sup_n |A_n^{-1}| < \infty,$$

which implies the second assertion of Theorem 2.1.

Let us cite one more proof of the second part of Theorem 2.1.

THIRD PROOF. Let us denote by A the operator defined in l_2 by the matrix $\|a_{j-k}\|_{j,k=1}^{\infty}$; by B the operator defined in the space $\tilde{l}_2$ of sequences $\{\xi_j\}_{-\infty}^{\infty}$ by the matrix $\|a_{j-k}\|_{j,k=-\infty}^{\infty}$; and by Q_n the projection operating in $\tilde{l}_2$ according to the rule

$$Q_n\{\xi_j\}_{-\infty}^{\infty} = \{\cdots, \xi_{-1}, \xi_0, \xi_1, \cdots, \xi_n, 0, \cdots\} \qquad (n = 0,1,\cdots).$$

The operators Q_nBQ_n $(n = 0,1,\cdots)$ are obviously invertible in the subspace $Q_n\tilde{l}_2$, because in bases natural for them one and the same matrix $A' = \|a_{k-j}\|_{j,k=1}^{\infty}$ always corresponds to them; therefore $|(Q_nBQ_n)^{-1}| = |(A')^{-1}|$. The operator $B \in \Pi\{Q_n,Q_n\}$ by virtue of the theorem on the applicability of the projection method (Theorem 2.1, Chapter II).

The operator $T = PBQ + QBP$ is completely continuous by virtue of Lemma 1.1, Chapter VI. (Here $Q = Q_0$ and $P = I - Q_0$.) The operator $C = QBQ + PBP$ is invertible, and since $C = B - T$, this operator, by virtue of Theorem 3.1, Chapter II, belongs to the class $\Pi\{Q_n, Q_n\}$.

The equality $Q_nCQ_n = QBQ + Q_nPBPQ_n$ implies that, beginning with some $n > 0$, the operators defined by the matrices $\|a_{j-k}\|_{j,k=1}^{n}$ are invertible in the subspace P_nl_2, and the norms of their inverses are bounded in the aggregate, from which $A \in \Pi\{P_n,P_n\}$ follows.

REMARK. In the case when ind $a(\zeta) \neq 0$, an approximate method of solving a Wiener-Hopf system is obtained easily by means of the method of § 7, Chapter I.

THEOREM 2.2. *Let $a(\zeta)$ $(|\zeta| = 1)$ be an arbitrary continuous function satisfying the conditions $a(\zeta) \neq 0$ $(|\zeta| = 1)$ and* ind $a(\zeta) = 0$, *and let a_j $(j = 0, \pm 1,\cdots)$ be its Fourier coefficients. Then, beginning with some n, the determinants $D_n(a) =$* det $\|a_{j-k}\|_{j,k=1}^{n}$ *do not vanish, and*

$$\lim_{n\to\infty} \frac{D_n(a)}{D_{n-1}(a)} = \exp\left\{\frac{1}{2\pi}\int_{-\pi}^{\pi} \log a(e^{i\varphi})\, d\varphi\right\}.$$

PROOF. The nonvanishing of the determinants $D_n(a)$ beginning with some n follows from Theorem 2.1.

Let us consider the system

$$\sum_{k=1}^{n} a_{j-k}x_k = \delta_{j1} \qquad (j = 1, 2,\cdots, n), \tag{2.7}$$

where n is so great that D_n and all subsequent determinants do not vanish. Let $\{x_j^{(n)}\}_{j=1}^{n}$ be a solution of this system. Obviously $x_1^{(n)} = D_{n-1} / D_n$. Theorem 2.1 implies that as $n \to \infty$, $x_1^{(n)}$ tends to the first coordinate of a solution $\{x_j\}_1^{\infty}$ of the system

$$\sum_{k=1}^{\infty} a_{j-k}x_k = \delta_{j1} \qquad (j = 1, 2, \cdots). \tag{2.8}$$

Thus $\lim(D_{n-1}/D_n)$ exists and equals x_1. Now let us establish the equality

$$x_1 = \exp\left\{-\frac{1}{2\pi}\int_{-\pi}^{\pi} \log a(e^{i\varphi})\, d\varphi\right\}. \tag{2.9}$$

Let us suppose at the outset that the function $a(\zeta)$ expands into an absolutely convergent Fourier series. The system (2.8) is equivalent to the following equation:

$$a(\zeta) \sum_{1}^{\infty} x_j \zeta^{j-1} = 1 + \varphi(\zeta), \tag{2.10}$$

where the function $\varphi(\zeta)$ expands into an absolutely convergent Fourier series in which all coefficients for nonnegative powers of ζ equal zero. By a theorem of Wiener, the function $b(\zeta) = \log a(\zeta)$ expands into the absolutely convergent Fourier series $\sum_{-\infty}^{\infty} b_j \zeta^j$. It is easy to obtain from (2.10) that

$$(\exp b_+(\zeta)) \sum_{1}^{\infty} x_j \zeta^{j-1} = 1,$$

where $b_+(\zeta) = \sum_0^{\infty} b_j \zeta^j$. This implies that $(\exp b_0)x_1 = 1$, but

$$b_0 = (1/2\pi) \int_{-\pi}^{\pi} \log a(e^{i\varphi})\, d\varphi,$$

and equality (2.9) is established.

Now let $a(\zeta)$ be an arbitrary continuous function for which the theorem's conditions are fulfilled, $p_m(\zeta)$ a sequence of polynomials converging uniformly on the unit circle to $a(\zeta)$, and $\{p_j^{(m)}\}_{j=-\infty}^{\infty}$ the Fourier coefficients of $p_m(\zeta)$. Then, beginning with some m, the system

$$\sum_{k=1}^{\infty} p_{j-k}^{(m)} y_k = \delta_{j1} \qquad (j = 1, 2, \cdots)$$

has the unique solution $\{y_k^{(m)}\}_{k=1}^{\infty}$, where $y_1^{(m)} \to x_1$ as $m \to \infty$.

By virtue of the part of the theorem which has been proved,

$$y_1^{(m)} = \exp\left\{-\frac{1}{2\pi}\int_{-\pi}^{\pi} \log p_m(e^{i\varphi})\, d\varphi\right\}.$$

By passing to the limit in this equality as $m \to \infty$ we obtain the theorem's assertion.

§ 3. Projection methods for solving integral equations

The continuous analogue of the basic theorem of § 2 is derived in the present section. In addition, justification of the Galerkin method (relative to a special system of functions) is presented for the Wiener-Hopf integral equation.

1. Let us denote by P_τ $(0 < \tau < \infty)$ the projection defined[4)] in the space $L_p(0,\infty)$ $(1 \leq p < \infty)$ by the equality

$$(P_\tau \varphi)(t) = \begin{cases} \varphi(t), & 0 < t < \tau, \\ 0, & \tau < t < \infty. \end{cases}$$

[4)]The case $p = \infty$ is excluded since the projections P_τ in that case do not converge (strongly) to the identity operator.

The family of projections P_τ converges strongly to the identity operator as $\tau \to \infty$ and, together with the operators V and $V^{(-1)}$ introduced in § 8, Chapter I, satisfies conditions 1)—3) of Theorem 1.1.

Theorems 1.1 and 1.2 are applicable to the operator $A \in \mathfrak{R}(V)$ defined by the left side of the equation

$$\varphi(t) - \int_0^\infty k(t-s)\varphi(s)\,ds = f(t) \qquad (0 < t < \infty;\, k(t) \in \tilde{L}_1), \tag{3.1}$$

and to the familty of projections P_τ.

If the operator

$$P_\tau A P_\tau \varphi = \varphi(t) - \int_0^\tau k(t-s)\varphi(s)\,ds \qquad (0 < t < \tau)$$

is invertible in the subspace $P_\tau L_p$, then, as before, we denote by $(P_\tau A P_\tau)^{-1}$ its inverse on the subspace $P_\tau L_p$. If the operator $P_\tau A P_\tau$ is not invertible, let us set $|(P_\tau A P_\tau)^{-1}| = \infty$.

THEOREM 3.1. *Let $k(t) \in \tilde{L}_1$. If $\liminf_{\tau\to\infty} |(P_\tau A P_\tau)^{-1}| < \infty$, then*

$$\mathscr{A}(\lambda) = 1 - \int_{-\infty}^{\infty} k(t)e^{i\lambda t}\,dt \neq 0 \qquad (-\infty < \lambda < \infty), \tag{3.2}$$

$$x = \operatorname{ind} \mathscr{A}(\lambda) = \frac{1}{2\pi}[\arg \mathscr{A}(\lambda)]_{-\infty}^{\infty} = 0.$$

Under fulfillment of these conditions the equation

$$\varphi(t) - \int_0^\tau k(t-s)\varphi(s)\,ds = f(t) \qquad (0 < t < \tau),$$

beginning with some τ, has the unique solution $\varphi_\tau(t)$, and as $\tau \to \infty$ the functions

$$\tilde{\varphi}_\tau(t) = \begin{cases} \varphi_\tau(t), & 0 < t < \tau, \\ 0, & \tau < t < \infty \end{cases}$$

converge in the norm of the space L_p to a solution of equation (3.1).

The remarks made after Theorem 1.1 permit an approximate method of solving equation (3.1) to be justified also in the case when $\kappa \neq 0$. Let us note, moreover, that the second assertion of Theorem 3.1 can be proved in two additional ways analogous to those cited in § 2 for the discrete case.

2. In the case of *Hilbert* space L_2 justification of the Galerkin method for integral equation (3.1) relative to some system of functions can also be derived from the general theorem concerning the projection method (Theorem 1.1).

In the case of L_2 the operator V is strictly isometric, and the operator $V^{(-1)}$ is its adjoint. Let us define a system of functions $\psi_j(t)$ by setting

$$\psi_0(t) = \sqrt{2}\, e^{-t}, \qquad \psi_{j+1}(t) = (V\psi_j)(t) \quad (j = 0, 1, \cdots; 0 < t < \infty). \quad (3.3)$$

System (3.3) is an orthonormal basis in L_2. Indeed, the equality $|\psi_j| = 1$ follows from the fact that V is isometric, while the orthogonality of ψ_j and ψ_k $(k > j)$ follows from the equalities

$$\psi_m = V^m\psi_0, \quad V^{(-1)}\psi_0 = 0,$$
$$(V^j\psi_0, \ V^k\psi_0) = (V^{j-k-1}\psi_0, \ V^{(-1)}\psi_0) = 0.$$

It is not difficult to establish the following relationship between the functions $\psi_j(t)$ and the normalized Laguerre polynomials $\Lambda_j(t)$:

$$\psi_n(t) = \sqrt{2}\, e^{-t}\Lambda_n(2t) \qquad (n = 0, 1, \cdots). \quad (3.4)$$

Indeed, since the function $\psi_n(t)$ has the form $p_n(t)e^{-t}$, where $p_n(t)$ is a polynomial whose degree equals precisely n and the system (3.3) is orthonormal, it follows that

$$\int_0^\infty e^{-2t}\, p_j(t)p_k(t)\, dt = \delta_{jk} \qquad (j, k = 1, 0, \cdots).$$

On the other hand, the polynomials $\Lambda_j(t)$, as is well known, are orthogonal, with weight e^{-t}:

$$\int_0^\infty e^{-t}\Lambda_j(t)\Lambda_k(t)\, dt = \delta_{jk} \qquad (j, k = 0, 1, \cdots).$$

Comparison of the last two equalities leads to (3.4).

As noted in § 8, Chapter I, the operator

$$P_n = I - V^nV^{(-n)} \qquad (n = 1, 2, \cdots) \quad (3.5)$$

is an orthogonal projection projecting the space L_2 onto the linear span of the functions $\psi_j(t)$, $j = 0, \cdots, n-1$.

Let us note in addition that the formula

$$\Psi_n(\lambda) = \int_0^\infty e^{i\lambda t}\psi_n(t)\, dt = \left(\frac{\lambda - i}{\lambda + i}\right)^n \frac{i\sqrt{2}}{\lambda + i} \qquad (n = 0, 1, \cdots)$$

is valid for the Fourier transform $\Psi_n(\lambda)$ of the function $\psi_n(t)$ continued as zero on the negative semi-axis.

This follows from equality (3.3) when the Fourier transform is applied to it.

The operators V, $V^{(-1)}$ and the system of projections P_n defined by equalities (3.5) obviously satisfy the conditions of Theorems 1.1 and 1.2. Before concretizing these theorems apropos the present case, let us note that the equation $P_nAP_n\varphi = P_nf$ is equivalent to the algebraic system

$$\sum_{k=0}^{n-1} (A\psi_k, \psi_j)c_k = (f, \psi_j) \qquad (j = 0, \cdots, n-1). \quad (3.6)$$

THEOREM 3.2. *Let $k(t) \in \tilde{L}_1$, and let the operator A be defined by equation* (3.1).

If $\liminf_{n\to\infty}|A_n^{-1}| < \infty$, *where* $A_n = \|(A\psi_k, \psi_j)\|_{j,k=0}^{n-1}$, *then conditions* (3.2) *are fulfilled.*

Under fulfillment of these conditions system (3.6), *beginning with some n, has the unique solution* $\{c_j^n\}_{j=0}^{n-1}$, *and as* $n \to \infty$ *the functions*

$$\varphi_n(t) = \sum_{j=0}^{n-1} c_j^{(n)} \psi_j(t)$$

converge in the norm of L_2 *to a solution of equation* (3.1) *whatever the function* $f(t) \in L_2$ *may be.*

REMARK. Under fulfillment of conditions (3.2) the Galerkin method is also applicable to equation (3.1) relative to the system of functions $\chi_j(t) = t^j e^{-t}$ $(j = 0, 1, \cdots)$, since the function

$$\tilde{\varphi}_n(t) = \sum_{j=0}^{n-1} \tilde{c}_j^{(n)} \chi_j(t),$$

where the vector $\{\tilde{c}_j^{(n)}\}_{j=0}^{n-1}$ is a solution of the system

$$\sum_{k=0}^{n-1} (A\chi_k, \chi_j) c_k = (f, \chi_j) \qquad (j = 0, \cdots, n-1), \tag{3.7}$$

coincides with the function $\varphi_n(t)$.

System (3.6) has an advantage over system (3.7) in that the elements of its matrix depend only on indicial difference. Indeed,

$$\begin{aligned} a_{j+m,\,k+m} &= (A\psi_{k+m}, \psi_{j+m}) = (AV^m\psi_k, V^m\psi_j) \\ &= (V^{(-m)} AV^m\psi_k, \psi_j) = a_{jk}, \end{aligned}$$

because $V^{(-m)} AV^{(m)} = A$.

On the other hand, the functions $\chi_j(t)$ are simpler than the functions $\psi_j(t)$.

The following formulas may turn out useful when finding the coefficients of systems (3.6) and (3.7):

$$\begin{aligned} (A\psi_k, \psi_j) &= \frac{1}{\pi} \int_{-\infty}^{\infty} \frac{1 - K(\lambda)}{\lambda^2 + 1} \left(\frac{\lambda - i}{\lambda + i}\right)^{k-j} d\lambda, \\ (f, \psi_j) &= -\frac{i}{\pi\sqrt{2}} \left(F(\lambda), \left(\frac{\lambda - i}{\lambda + i}\right)^j \frac{1}{\lambda + i}\right), \\ (A\chi_k, \chi_j) &= \frac{i!\,k!\,i^{k-j}}{2\pi} \int_{-\infty}^{\infty} \frac{1 - K(\lambda)}{(\lambda + i)^{k+1}(\lambda - i)^{j+1}}\, d\lambda \\ (f, \chi_j) &= \frac{i!(-i)^{j+1}}{2\pi} \left(F(\lambda), \frac{1}{(\lambda + i)^{j+1}}\right), \end{aligned}$$

where $K(\lambda)$ and $F(\lambda)$ are the Fourier transforms of the functions $k(t)$ and $f(t)$.

By means of these formulas it is easy to obtain the fact that equation (3.1) is equivalent to the following infinite system:

$$\sum_{k=0}^{\infty} a_{j-k}\xi_k = b_j \qquad (j = 0, 1, \cdots), \tag{3.8}$$

where $\{a_j\}_{-\infty}^{\infty}$ and $\{b_j\}_0^{\infty}$ are the Fourier coefficients of the functions

$$1 - K\left(i\frac{1+\zeta}{1-\zeta}\right) \quad \text{and} \quad F\left(i\frac{1+\zeta}{1-\zeta}\right) \qquad (|\zeta| = 1),$$

where if $\{\xi_j\}_0^{\infty}$ is a solution of system (3.8), the function

$$\varphi(t) = \sum_0^{\infty} \xi_j \psi_j(t)$$

is a solution of equation (3.1), and conversely.

System (3.6) of the Galerkin method coincides with the "truncated" system (3.8)

$$\sum_{k=0}^{n-1} a_{j-k}\xi_k = b_j \qquad (j = 0, \cdots, n-1),$$

and we thus obtain one more proof of the convergence of the Galerkin method.

It is not difficult now to understand how the Galerkin method is applied to the solution of equation (3.1) in the case when only the first of conditions (3.2) is fulfilled, i.e. when $\kappa = \operatorname{ind} \mathscr{A}(\lambda) \neq 0$.

For example, *if* $\kappa > 0$, *then, beginning with some n, the system*

$$\sum_{k=0}^{n-1} (A\psi_k, \psi_j)c_k = (f, \psi_j) \qquad (j, = \kappa, \cdots, \kappa + n - 1)$$

has the unique solution $\{c_j^{(n)}\}_{j=0}^{n-1}$, *and as* $n \to \infty$ *the functions*

$$\varphi_n(t) = \sum_{j=0}^{n-1} c_j^{(n)} \psi_j(t)$$

converge to some function $\varphi(t)$. *If equation* (3.1) *is solvable, then* $\varphi(t)$ *is a solution of it.*

§ 4. Multivariate discrete equations

In the present section a survey of results is presented, and two problems are formulated.

1. Let us denote by R the set of all vectors $j = (j_0, \cdots, j_n)$ with integral coordinates, and by $R+$ the subset of R which is the "half-space" defined by the inequality $j_0 \geqq 0$.

Let us denote by $l_p(R^+)$ $(1 \leqq p < \infty)$ the space of sequences $\xi = \{\xi_j\}_{j \in R^+}$ of complex numbers ξ_j with norm

$$|\xi|_p = (\sum_{j \in R^+} |\xi_j|^p)^{1/p}.$$

The space $l_p(R)$ is introduced analogously. Finally, let us denote by T_{n^+} the set of all points $\zeta = \{\zeta_0, \cdots, \zeta_n\}$ of the $(n+1)$-dimensional complex euclidean space such that $|\zeta_j| = 1$ $(j = 0, \cdots, n)$.

Let us put into correspondence with every sequence $a = \{a_j\}_{j \in R} \in l_1(R)$ a function

$$a(\zeta) = \sum_{j \in R} a_j \zeta^j \qquad (\zeta \in T_{n+1}; i = (i_0, \cdots, i_n); \zeta^j = \zeta_0^{j_0}, \cdots, \zeta_n^{j_n})$$

and a bounded linear operator A defined in the space $l_p(R^+)$ by a multidimensional matrix $\| a_{j-k} \|_{j,k\in R^+}$.

The equality

$$\eta = A\xi \ (\xi = \{\xi_j\}_{j\in R^+}, \ \eta = \{\eta_j\}_{j\in R^+} \in l_p(R^+)),$$

signifies that $\eta_j = \sum_{k\in R^+} a_{j-k}\xi_k \ (j \in R^+)$.

THEOREM 4.1. *For an operator A to be invertible on at least one side, it is necessary and sufficient that the conditions $a(\zeta) \neq 0 \ (\zeta \in T_{n+1})$ be fulfilled.*

If this condition is fulfilled, then A is invertible, left invertible, or right invertible if the number

$$\kappa_0 = \frac{1}{2\pi}[\arg a(e^{i\varphi}, \zeta_1, \cdots, \zeta_n)]_{\varphi=0}^{2\pi} \qquad (|\zeta_1| = \cdots = |\zeta_n| = 1)$$

equals zero, is positive, or is negative, respectively.

Sufficiency has been proved by L.S. Gol'denšteĭn and I. C. Gohberg **[1]**, and necessity by Gol'denšteĭn **[1]**.

Various generalizations of this theorem have been obtained by L. S. Gol'denšteĭn **[1, 2]**. He obtained in particular the multivariate generalization of the first abstract scheme presented in Chapter I.

Let us denote by $P_m \ (m = 1, 2, \cdots)$ the projector defined by the equality $P_m\{\xi_j\} = \{\eta_j\}, j \in R^+$, where $\eta_j = \xi_j$ for $j_0 = 0, \cdots, m$, and $\eta_j = 0$ for $j_0 = m+1, m+2, \cdots$.

THEOREM 4.2. *Let $\{a_j\} \in l_1(R)$, and let $a(\zeta)$ and A be the function and operator corresponding to this sequence. For the operators P_mAP_m, beginning with some m, to be invertible on the subspace $P_ml_p(R^+)$ and for the inverse operators $(P_mAP_m)^{-1}$ to fulfill the condition*

$$\liminf_{m\to\infty} |(P_mAP_m)^{-1}| < \infty,$$

it is necessary and sufficient that $a(\zeta) \neq 0 \ (\zeta \in T_{n+1})$ and $\kappa_0 = 0$.

If this condition is fulfilled, $A \in \Pi\{P_m, P_m\}$.

The sufficiency has been proved by Gol'denšteĭn **[2]**. This proof is analogous to Baxter's proof of the sufficiency of the conditions of Theorem 2.1.

The necessity is established just as in the case $n = 0$ (cf. § 2).

Theorem 4.2 remains valid if in its formulation A is replaced by the operator $\tilde{A}$ defined in the space $l_p(R)$ by the matrix $\| a_{j-k} \|_{j,k\in R}$, and the projections P_m by the projections $\tilde{P}_m$ defined by the equalities $\tilde{P}_m\{\xi_j\} = \{\eta_j\}, j \in R$, where $\eta_j = \xi_j$ for $|j_0| \leq m$ and $\eta_j = 0$ for $|j_0| > m$.

Some of the other propositions of Chapters I and III can be generalized to the multivariate case. Continuous analogues are valid for those propositions in particular.

2. Remaining *open* is the question: What are necessary and sufficient conditions for an operator A to belong to the class $\Pi\{\tilde{P}_m, \tilde{P}_m\}$, where the projection $\tilde{P}_m$ is defined by the equality $\tilde{P}_m\{\xi_j\} = \{\eta_j\}, j \in R^+$, in which $\eta_j = \xi_i$ for $j_k \leq m \ (k = 0, \cdots, n)$ and $\eta_j = 0$ for $j_k > m$?

It is possible that

$$a(\zeta) \neq 0 \quad (\zeta \in T_{n+1}), \qquad \kappa_j = 0 \quad (i = 0, 1, \cdots, n)$$

are such conditions, where

$$\kappa_j = \frac{1}{2\pi}[\arg a(\zeta_0, \cdots, \zeta_{j-1}, e^{i\varphi}, \zeta_{j+1}, \cdots, \zeta_n)]_{\varphi=0}^{2\pi}.$$

Exactly the same question arises for the operator $\tilde{A}$ as well.

§ 5. Iterative method for calculating the index of a function

1. As is evident from the preceding sections, the applicability of one or another projection method to the Wiener-Hopf integral equation is determined by the index of the function $1 - K(\lambda)$:

$$\kappa = \operatorname{ind}(1 - K(\lambda)) = (1/2\pi)\,[\arg(1 - K(\lambda))]_{\lambda=-\infty}^{\infty},$$

where $K(\lambda)$ is the Fourier transform of the kernel $k(t)$.

Since the function's index does not vary under small perturbations and the function $1-K(\lambda)$ can be approximated quite well by a rational function, the problem of calculating the index of an "arbitrary" function can be reduced to the problem of evaluating the index of a rational function

$$R(\lambda) = p_1(\lambda)/p_2(\lambda),$$

where the $p_j(\lambda)$ are polynomials.

As is well known, ind $R(\lambda) = r_1 - r_2$, where r_j is the number of zeros (taking account of their multiplicities) of the polynomial $p_j(\lambda)$ in the upper half-plane.

Thus the problem reduces to calculating the number of roots of a polynomial in the upper half-plane.

2. Let the polynomial $p(\lambda) = a_0\lambda^n + a_1\lambda^{n-1} + \cdots + a_n$, $a_0 \neq 0$, satisfy the condition

$$p(\lambda) \neq 0, \qquad -\infty < \lambda < \infty. \tag{5.1}$$

It is not difficult to ascertain that the function $p(\lambda)p(-1/\lambda)$ is an nth degree polynomial in the variable $\lambda^{(1)} = \frac{1}{2}(\lambda - 1/\lambda)$. Let us denote this polynomial by $p_1(\lambda)$ (having redesignated the argument $\lambda^{(1)}$ as λ):

$$p_1(\lambda) = a_0^{(1)}\lambda^n + a_1^{(1)}\lambda^{n-1} + \cdots + a_n^{(1)}.$$

The coefficients $a_j^{(1)}$ can be expressed in terms of the coefficients a_j by the following formula:

$$a_j^{(1)} = 2^{n-j}\sum_{k=0}^{j} a_k a_{n-j+k}(-1)^{j-k} - \sum_{k=1}^{[j/2]} C_{n-j+2k}^{k}\, 2^{-2k}\, a_{j-2k}^{(1)}(-1)^k \qquad (j = 0, 1, \cdots, n).$$

Passing from the polynomial $p(\lambda)$ to the polynomial $p_1(\lambda)$ constitutes one step of the iterative process. Continuing this process, we obtain the sequence of polynomials

$$p_k(\lambda) = a_0^{(k)}\lambda^n + a_1^{(k)}\lambda^{n-1} + \cdots + a_n^{(k)} \qquad (k = 1, 2, \cdots).$$

THEOREM 5.1. *Let the polynomial $p(\lambda)$ satisfy condition* (5.1). *Then the sequence of numbers $c_k = a_1^{(k)}/a_0^{(k)}$ $(k = 1, 2, \cdots)$ has a limit, and the number r of roots of the polynomial $p(\lambda)$ in the upper half-plane is calculated by the formula*

$$r = \frac{1}{2}\left(n + i\lim_{k\to\infty} c_k\right).$$

PROOF. Let us denote by $\lambda_j^{(k)}, j = 1,\cdots, n$, the roots of the polynomial $p_k(\lambda)$. It is easy to see that the numbers

$$\lambda_j^{(k+1)} = \frac{1}{2}\left(\lambda_j^{(k)} - \frac{1}{\lambda_j^{(k)}}\right) \qquad (j = 1,\cdots, n), \tag{5.2}$$

and only they, are the roots of the polynomial $p_{k+1}(\lambda)$. Therefore the number of roots of $p_k(\lambda)$ in the upper (lower) half-plane does not depend on k and consequently equals r (equals $n - r$).

Let us show that

$$\lim_{k\to\infty} \lambda_j^{(k)} = \alpha_j \qquad (j = 1,\cdots, n), \tag{5.3}$$

where

$$\alpha_j = \begin{cases} i, & \operatorname{Im}\lambda_j > 0, \\ -i, & \operatorname{Im}\lambda_j < 0. \end{cases}$$

Let λ_j be a root of $p(\lambda)$ such that at some step k_j the root $\lambda_j^{(k_j)} = -i$. Then equality (5.2) implies that $\lambda_j^{(k_j+l)} = -i$ $(l = 1, 2,\cdots)$ at all subsequent steps; consequently (5.3) holds.

Now let λ_j be a root such that $\lambda_j^{(k,)} \neq -i$, $k = 1, 2,\cdots$. Then equality (5.2) implies the relations

$$\lambda_j^{(k+1)} - i = \frac{1}{2\lambda_j^{(k)}}(\lambda_j^{(k)} - i)^2$$

and

$$\lambda_j^{(k+1)} + i = \frac{1}{2\lambda_j^{(k)}}(\lambda_j^{(k)} + i)^2,$$

and consequently

$$\frac{\lambda_j^{(k+1)} - i}{\lambda_j^{(k+1)} + i} = \left(\frac{\lambda_j^{(k)} - i}{\lambda_j^{(k)} + i}\right)^2.$$

Introducing the notation

$$q_j = \left|\frac{\lambda_j - i}{\lambda_j + i}\right|,$$

we obtain

$$\left| \frac{\lambda_j^{(k)} - i}{\lambda_j^{(k)} + i} \right| = q_j^{2^k}. \tag{5.4}$$

If $\operatorname{Im} \lambda_j > 0$, then $q_j < 1$, and equality (5.4) implies that $\lim_{k\to\infty} \lambda_j^{(k)} = i$. If, however, $\operatorname{Im} \lambda_j < 0$, then $q_j > 1$, and the same equality (5.4) implies that $\lim_{k\to\infty} \lambda_j^{(k)} = -i$. Thus relation (5.3) has been established for all roots. Since

$$c_k = - \sum_{\operatorname{Im} \lambda_j > 0} \lambda_j^{(k)} - \sum_{\operatorname{Im} \lambda_j < 0} \lambda_j^{(k)}$$

and

$$\lim_{k\to\infty} \sum_{\operatorname{Im} \lambda_j > 0} \lambda_j^{(k)} = ri, \qquad \lim_{k\to\infty} \sum_{\operatorname{Im} \lambda_j < 0} \lambda_j^{(k)} = -(n-r)i,$$

the theorem is proved.

Calculating the index of a rational function relative to the unit circle reduces to the case under consideration in an obvious way.

§ 6. Inversion of finite Toeplitz matrices*

This section is devoted to the problem of inverting Toeplitz matrices, i.e. matrices of the form $\|a_{j-k}\|_{j,k=0}^n$.

1. The following theorem is basic in this subsection.

THEOREM 6.1. *If the Toeplitz matrix* $\mathscr{A}_n = \|a_{j-k}\|_{j,k=0}^n$ *is such that each of the systems of equations*

$$\sum_{k=0}^n a_{j-k} x_k = \delta_{j0} \qquad (j = 0, \cdots, n), \tag{6.1}$$

$$\sum_{k=0}^n a_{j-k} y_{k-n} = \delta_{jn} \qquad (j = 0, \cdots, n) \tag{6.2}$$

is solvable and the condition $x_0 \neq 0$ *is fulfilled, then the matrix* $\mathscr{A}_n$ *is invertible, and its inverse is formed according to the formula*

$$\mathscr{A}_n^{-1} = x_0^{-1} \left\{ \begin{Vmatrix} x_0 & 0 & \cdots & 0 \\ x_1 & x_0 & \cdots & 0 \\ \cdot & \cdot & \cdot & \cdot \\ x_n & x_{n-1} & \cdots & x_0 \end{Vmatrix} \begin{Vmatrix} y_0 & y_{-1} & \cdots & y_{-n} \\ 0 & y_0 & \cdots & y_{-n+1} \\ \cdot & \cdot & \cdot & \cdot \\ 0 & 0 & \cdots & y_0 \end{Vmatrix} \right.$$
$$\left. - \begin{Vmatrix} 0 & 0 & 0 & \cdots & 0 & 0 \\ y_{-n} & 0 & 0 & \cdots & 0 & 0 \\ y_{-n+1} & y_{-n} & 0 & \cdots & 0 & 0 \\ \cdot & \cdot & \cdot & \cdot & \cdot & \cdot \\ y_{-1} & y_{-2} & y_{-3} & \cdots & y_{-n} & 0 \end{Vmatrix} \begin{Vmatrix} 0 & x_n & x_{n-1} & \cdots & x_1 \\ 0 & 0 & x_n & \cdots & x_2 \\ \cdot & \cdot & \cdot & \cdot & \cdot \\ 0 & 0 & 0 & \cdots & x_n \\ 0 & 0 & 0 & \cdots & 0 \end{Vmatrix} \right\}. \tag{6.3}$$

**Editor's note.* This and the following two sections have been revised in translation.

PROOF. Let us prove the invertibility of the matrix $\mathscr{A}_n$ by contradiction.

Let $\det \mathscr{A}_n = 0$. Then the rows $\Gamma_k = (a_k, a_{k-1}, \cdots, a_{k-n})$ of $\mathscr{A}_n$ are linearly dependent. The solvability of system (6.1) implies that $\Gamma_0 \neq 0$; hence one of the rows of $\mathscr{A}_n$ is expressable linearly in terms of those preceding it. Let us denote by $p\ (\geqq 1)$ the largest number for which

$$\Gamma_p = \sum_0^{p-1} \alpha_j \Gamma_j,$$

where $\alpha_0, \cdots, \alpha_{p-1}$ are complex numbers. Let us show that $p = n$. Let $p < n$. Since

$$a_{p-k} = \sum_{j=0}^{p-1} \alpha_j a_{j-k} \qquad (k = 0, \cdots, n),$$

it follows that

$$a_{p+1-k} = \sum_{j=1}^{p} \alpha_{j-1} a_{j-k} \qquad (k = 1, \cdots, n). \tag{6.4}$$

By virtue of the solvability of system (6.1), we have

$$\begin{aligned} 0 &= \sum_{k=0}^{n} a_{p+1-k} x_k - \sum_{j=1}^{p} \alpha_{j-1} \sum_{k=0}^{n} a_{j-k} x_k \\ &= \left(a_{p+1} - \sum_{j=1}^{p} \alpha_{j-1} a_j\right) x_0 + \sum_{k=1}^{n} \left(a_{p+1-k} - \sum_{j=1}^{p} \alpha_{j-1} a_{j-k}\right) x_k. \end{aligned}$$

Taking equalities (6.4) and the condition $x_0 \neq 0$ into consideration, we obtain

$$a_{p+1} = \sum_{j=1}^{p} \alpha_{j-1} a_j. \tag{6.5}$$

Then, combining (6.4) and (6.5), we have

$$\Gamma_{p+1} = \sum_{j=1}^{p} \alpha_{j-1} \Gamma_j,$$

which contradicts the choice of p.

Thus row Γ_n is expressable linearly in terms of those preceding it. The latter contradicts the solvability of system (6.2).

The invertibility of $\mathscr{A}_n$ implies that

$$x_0 = y_0 = \det \mathscr{A}_{n-1} / \det \mathscr{A}_n. \tag{6.6}$$

Let us denote by $\mathfrak{B} = \|b_{jk}\|_0^n$ the matrix in the right side of (6.3). It is easy to see that the elements of this matrix can be calculated from the formula

$$b_{jk} = x_0^{-1} \sum_{s=0}^{\min(j,k)} (x_{j-s} y_{s-k} - y_{-n-1+j-s} x_{n+1+s-k})$$

for all $j, k = 0, \cdots, n$. It is assumed here that $x_{n+1} = y_{-n-1} = 0$.

The equalities

$$b_{jk} = b_{j-1,\,k-1} + x_0^{-1}\{x_j y_{-k} - y_{-n-1+j}x_{n+1-k}\} \qquad (j, k = 1, \cdots, n), \tag{6.7}$$
$$b_{0k} = y_{-k}, \qquad b_{k0} = x_k \qquad (k = 0, 1, \cdots, n) \tag{6.8}$$

follow immediately from this formula and (6.6).

The equalities

$$b_{nk} = x_{n-k} \quad \text{and} \quad b_{kn} = y_{k-n} \qquad (k = 0, \cdots, n) \tag{6.9}$$

are also easily derived.

Let us show that the matrix $\mathscr{C} = \mathscr{A}_n\mathscr{B} = \|c_{jk}\|_0^n$ is Toeplitz. Indeed, we have

$$c_{jk} = a_j b_{0k} + \sum_{s=1}^{n} a_{j-s}\,[b_{s-1,k-1} - x_0^{-1}(x_s y_{-k} - y_{n-1+s}x_{n+1-k})]$$

by virtue of the equalities (6.7) for $j, k = 1,\cdots,n$. By virtue of (6.6), (6.8) and (6.9), this implies

$$c_{jk} = \sum_{s=0}^{n} a_{j-1-s}b_{s,k-1} - x_0^{-1}b_{n,k-1}a_{j-1-n}y_0 + x_0^{-1}b_{0k}\sum_{s=1}^{n} a_{j-s}x_s$$
$$+ x_0^{-1}b_{0k}a_j x_0 - x_0^{-1}b_{n,k-1}\sum_{s=0}^{n-1} a_{j-1-s}y_{s-n}.$$

Thus

$$c_{jk} = c_{j-1,k-1} + x_0^{-1}b_{0k}\sum_{s=0}^{n} a_{j-s}x_s - x_0^{-1}b_{n,k-1}\sum_{s=0}^{n} a_{j-1-s}y_{s-n}.$$

Taking (6.1) and (6.2) into account, we arrive at the equalities $c_{j-1,k-1} = c_{jk}$ $(j, k = 1,\cdots,n)$. Therefore $\mathscr{C}$ is Toeplitz. By virtue of (6.8) and (6.9), the first and last columns of $\mathscr{C}$ coincide with the corresponding columns of the identity matrix. Consequently the Toeplitz matrix $\mathscr{C}$ coincides with the identity matrix.

The theorem is proved.

That the condition $x_0 \neq 0$ in Theorem 6.1 is essential can be determined without difficulty. If systems (6.1) and (6.2) are solvable and $x_0 = 0$, then $\mathscr{A}_n$ can turn out to be singular, while if it is nonsingular, its inverse may not be determined by the solutions of systems (6.1) and (6.2)

Let us note that a different formula for inverting the Toeplitz matrix $\mathscr{A}_n = \|a_{j-k}\|_0^n$ is obtained in the following subsection.

2. It is easy to see that the matrix $\mathscr{A}_{n-1} = \|a_{j-k}\|_{j,k=0}^{n-1}$ is also invertible under fulfillment of the conditions of Theorem 6.1. What is more, it turns out that in this case its inverse matrix $\mathscr{A}_{n-1}^{-1}$ can be formed from solutions of systems (6.1) and (6.2).

THEOREM 6.2. *If solutions $x_0,\cdots,x_n$ and $y_{-n},\cdots,y_0$ of systems* (6.1) *and* (6.2) *exist and the condition $x_0 \neq 0$ is fulfilled, then the matrix $\mathscr{A}_{n-1} = \|a_{j-k}\|_{j,k=0}^{n-1}$ is invertible, and its inverse is formed according to the formula*

$$\mathscr{A}_{n-1}^{-1} = x_0^{-1}\left\{\begin{Vmatrix} x_0 & 0 & \cdots & 0 \\ x_1 & x_0 & \cdots & 0 \\ \cdot & \cdot & \cdot & \cdot \\ x_{n-1} & x_{n-2} & \cdots & x_0 \end{Vmatrix}\begin{Vmatrix} y_0 & y_{-1} & \cdots & y_{-n+1} \\ 0 & y_0 & \cdots & y_{-n+2} \\ \cdot & \cdot & \cdot & \cdot \\ 0 & 0 & \cdots & y_0 \end{Vmatrix} - \begin{Vmatrix} y_{-n} & 0 & \cdots & 0 \\ y_{-n+1} & y_{-n} & \cdots & 0 \\ \cdot & \cdot & \cdot & \cdot \\ y_{-1} & y_{-2} & \cdots & y_{-n} \end{Vmatrix}\begin{Vmatrix} x_n & x_{n-1} & \cdots & x_1 \\ 0 & x_n & \cdots & x_2 \\ \cdot & \cdot & \cdot & \cdot \\ 0 & 0 & \cdots & x_n \end{Vmatrix}\right\}. \tag{6.10}$$

PROOF. Let us denote by $\mathscr{D} = \|d_{jk}\|_0^{n-1}$ the matrix in the right side of (6.10). It is easy to see that

$$d_{jk} = x_0^{-1} \sum_{s=0}^{\min(j,k)} (x_{j-s}y_{s-k} - y_{n+j-s}x_{n+s-k})$$

for all j, $k=0,\cdots,n-1$. Hence, taking into account the equality $x_0=y_0$, it follows immediately that

$$d_{k0} = x_k - x_0^{-1}x_n y_{k-n}; \qquad d_{0k} = y_{-k} - x_0^{-1}y_{-n}x_{n-k} \qquad (k = 0,\cdots,n-1), \tag{6.11}$$

$$d_{jk} = d_{j-1,k-1} + x_0^{-1}(x_j y_{-k} - y_{-n+j}x_{n-k}) \qquad (j,k = 1,\cdots,n-1). \tag{6.12}$$

In addition, the equalities

$$d_{k,n-1} = y_{-n+k+1} - x_0^{-1}y_{-n}x_{k+1}; \qquad d_{n-1,k} = x_{n-k-1} - x_0^{-1}x_n y_{-k-1} \tag{6.13}$$

are easily obtained for $k = 0,\cdots,n-1$.

Let us ascertain that the first and last columns of the matrix $\mathscr{E} = \mathscr{A}_{n-1}\mathscr{D} = \|e_{jk}\|_0^{n-1}$ coincide with the corresponding columns of the identity matrix. Indeed, taking (6.11), (6.13) and (6.6) into consideration, we obtain the equalities

$$\begin{aligned} e_{j0} &= \sum_{k=0}^{n-1} a_{j-k}d_{k0} = \sum_{k=0}^{n-1} a_{j-k}x_k - x_0^{-1}x_n \sum_{k=0}^{n-1} a_{j-k}y_{k-n} \\ &= \sum_{k=0}^{n} a_{j-k}x_k - x_0^{-1}x_n \sum_{k=0}^{n} a_{j-k}y_{k-n} = \delta_{j0} \qquad (j = 0,\cdots, n-1), \end{aligned}$$

as well as the equalities

$$\begin{aligned} e_{j,n-1} &= \sum_{k=0}^{n-1} a_{j-k}d_{k,n-1} = \sum_{k=0}^{n-1} a_{j-k}(y_{-n+k+1} - x_0^{-1}y_{-n}x_{k+1}) \\ &= \sum_{k=1}^{n} a_{j+1-k}y_{k-n} - x_0^{-1}y_{-n}\sum_{k=1}^{n} a_{j+1-k}x_k \\ &= \sum_{k=0}^{n} a_{j+1-k}y_{k-n} - x_0^{-1}y_{-n}\sum_{k=0}^{n} a_{j+1-k}x_k = \delta_{j,n-1} \qquad (j=0,\cdots,n-1). \end{aligned}$$

To complete the proof, it remains to show that the matrix $\mathscr{E}$ is Toeplitz. Using (6.11), (6.12), (6.13) and (6.6) we have for j, $k = 1,\cdots, n-1$

$$\begin{aligned}
e_{jk} &= \sum_{s=0}^{n-1} a_{j-s}d_{sk} = a_j d_{0k} + \sum_{s=1}^{n-1} a_{j-s}[d_{s-1,k-1} + x_0^{-1}(x_s y_{-k} - y_{-n+s}x_{n-k})] \\
&= \sum_{s=0}^{n-2} a_{j-1-s}d_{s,k-1} + a_j d_{0k} + x_0^{-1}y_{-k}\sum_{s=1}^{n-1} a_{j-s}x_s - x_0^{-1}x_{n-k}\sum_{s=1}^{n-1} a_{j-s}y_{s-n} \\
&= \sum_{s=0}^{n-1} a_{j-1-s}d_{s,k-1} - a_{j-n}(x_{n-k} - x_0^{-1}x_n y_{-k}) + a_j(y_{-k} - x_0^{-1}y_{-n}x_{n-k}) \\
&\quad + x_0^{-1}y_{-k}\sum_{s=1}^{n-1} a_{j-s}x_s - x_0^{-1}x_{n-k}\sum_{s=1}^{n-1} a_{j-s}y_{s-n}.
\end{aligned}$$

Consequently, by virtue of (6.1) and (6.2), $e_{jk} = e_{j-1,k-1}$ for $j, k = 1,\cdots,n-1$. The theorem is proved.

3. Let us cite one more method of inverting a nonsingular Toeplitz matrix $\mathscr{A}_n = \|a_{j-k}\|_{j,k=0}^n$. This method is also suitable for the case when $\det \mathscr{A}_{n-1} = 0$. Let us consider the matrix

$$\mathscr{A}_{n+1}(\zeta) = \left\| \begin{matrix} a_0 & a_{-1} & \cdots & a_{-n} & \zeta \\ a_1 & a_0 & \cdots & a_{-n+1} & a_{-n} \\ \cdots & \cdots & \cdots & \cdots & \cdots \\ a_n & a_{n-1} & \cdots & a_0 & a_{-1} \\ a_{n+1} & a_n & \cdots & a_1 & a_0 \end{matrix} \right\|,$$

where a_{n+1} is some fixed number and ζ is a complex parameter. Let us show that if $\mathscr{A}_n$ is nonsingular, the matrix $\mathscr{A}_{n+1}(\zeta)$ will also be nonsingular for all values of ζ, with the exception, perhaps, of one.

Indeed, let $\det \mathscr{A}_{n+1}(\zeta_0) = 0$ be fulfilled for some $\zeta = \zeta_0$. It can be deduced from the condition $\det \mathscr{A}_n \neq 0$ that the first row of $\mathscr{A}_{n+1}(\zeta_0)$ is expressable linearly in terms of the remaining rows. Consequently the first row of the matrix $\mathscr{M} = \|a_{j-k}\|$, where $j = 0, \cdots, n+1$ and $k = 0, \cdots, n$, possesses the same property. The rank of $\mathscr{M}$ equals $n+1$, by virtue of the condition $\det \mathscr{A}_n \neq 0$. Hence $\det \|a_{1+j-k}\|_{j,k=0}^n \neq 0$.

The determinant of the matrix $\mathscr{A}_{n+1}(\zeta)$ is a linear function of ζ:

$$\det \mathscr{A}_{n+1}(\zeta) = a\zeta + b,$$

where $a = \det \|a_{1+j-k}\|_{j,k=0}^n \neq 0$. Consequently this determinant does not vanish for $\zeta \neq \zeta_0$.

Thus for all values of ζ, with the exception, perhaps, of one, $\mathscr{A}_{n+1}(\zeta)$ satisfies all of the conditions of Theorem 6.2, and hence $\mathscr{A}_n$ can be inverted by formula (6.10).

4. In Theorem 6.1 a method is indicated for forming the inverse matrix of a given Toeplitz matrix from the first and last columns of the inverse matrix. Here we consider the problem of reconstructing the original Toeplitz matrix from the first and last columns of its inverse matrix.

THEOREM 6.3. *Let $x_0,\cdots,x_n$ and $y_{-n},\cdots,y_0$ be given systems of complex numbers, and $x_0 \neq 0$.*

For a Toeplitz matrix $\mathscr{A}_n = \|a_{j-k}\|_{j,k=0}^n$ *such that*

$$\sum_{k=0}^{n} a_{j-k}x_k = \delta_{j0} \quad \textit{and} \quad \sum_{k=0}^{n} a_{j-k}y_{k-n} = \delta_{jn} \quad (j = 0,\cdots, n) \tag{6.14}$$

to exist, it is necessary and sufficient that the condition

$$x_0 = y_0 \tag{6.15}$$

be fulfilled, and that the matrix

$$\mathscr{P} = \left\| \begin{matrix} y_0 & y_{-1} & \cdots & y_{-n+1} & y_{-n} & 0 & \cdots & 0 \\ 0 & y_0 & \cdots & y_{-n+2} & y_{-n+1} & y_{-n} & \cdots & 0 \\ \cdots & \cdots & \cdots & \cdots & \cdots & \cdots & \cdots & \cdots \\ 0 & 0 & \cdots & y_0 & y_{-1} & y_{-2} & \cdots & y_{-n} \\ x_n & x_{n-1} & \cdots & x_1 & x_0 & 0 & \cdots & 0 \\ 0 & x_n & \cdots & x_2 & x_1 & x_0 & \cdots & 0 \\ \cdots & \cdots & \cdots & \cdots & \cdots & \cdots & \cdots & \cdots \\ 0 & 0 & \cdots & x_n & x_{n-1} & x_{n-2} & \cdots & x_0 \end{matrix} \right\|$$

be nonsingular.

If these conditions are fulfilled, then $\mathscr{A}_n$ *is nonsingular and can be uniquely recovered by means of formula* (6.3).

PROOF. The necessity of condition (6.15) is a simple corollary of Theorem 6.1. Let us show that $\mathscr{P}$ is nonsingular. Let us subdivide this matrix into the four square blocks

$$P = \left\| \begin{matrix} \mathscr{Y} & \mathscr{U} \\ \mathscr{V} & \mathscr{X} \end{matrix} \right\|,$$

where

$$\mathscr{X} = \left\| \begin{matrix} x_0 & 0 & \cdots & 0 \\ x_1 & x_0 & \cdots & 0 \\ \cdots & \cdots & \cdots & \cdots \\ x_{n-1} & x_{n-2} & \cdots & x_0 \end{matrix} \right\|, \quad \mathscr{Y} = \left\| \begin{matrix} y_0 & y_{-1} & \cdots & y_{-n+1} \\ 0 & y_0 & \cdots & y_{-n+2} \\ \cdots & \cdots & \cdots & \cdots \\ 0 & 0 & \cdots & y_0 \end{matrix} \right\|,$$

$$\mathscr{U} = \left\| \begin{matrix} y_{-n} & 0 & \cdots & 0 \\ y_{-n+1} & y_{-n} & \cdots & 0 \\ \cdots & \cdots & \cdots & \cdots \\ y_{-1} & y_{-2} & \cdots & y_{-n} \end{matrix} \right\|, \quad \mathscr{V} = \left\| \begin{matrix} x_n & x_{n-1} & \cdots & x_1 \\ 0 & x_n & \cdots & x_2 \\ \cdots & \cdots & \cdots & \cdots \\ 0 & 0 & \cdots & x_n \end{matrix} \right\|.$$

By virtue of a proposition in the theory of determinants (cf. F. R. Gantmaher [**1**], Chapter II, § 5.3),

$$\det \mathscr{P} = \det\,(\mathscr{Y} - \mathscr{U}\mathscr{X}^{-1}\mathscr{V}) \det \mathscr{X}.$$

The matrices $\mathscr{X}$ and $\mathscr{U}$ are commutative; therefore

$$\det \mathscr{P} = \det(\mathscr{X}\mathscr{Y} - \mathscr{U}\mathscr{V}).$$

This implies that $\mathscr{P}$ is nonsingular if and only if $\mathscr{X}\mathscr{Y} - \mathscr{U}\mathscr{V}$ is nonsingular. The nonsingularity of the latter follows from Theorem 6.2 (cf. (6.10)).

Let us pass to the proof of the sufficiency of the theorem's conditions. Let us rewrite equalities (6.14) as a system of equations with the unknowns $a_{-n}, \cdots, a_n$:

$$\sum_{k=j}^{j+n} y_{j-k}a_k = \delta_{j0} \qquad (j = -n, \cdots, 0),$$

$$\sum_{k=j-n}^{j} x_{j-k}a_k = \delta_{j0} \qquad (j = 0, \cdots, n). \tag{6.16}$$

The matrix of this system has the form

$$\mathscr{Q} = \left\| \begin{matrix} y_0 & y_{-1} & \cdots & y_{-n+1} & y_{-n} & 0 & \cdots & 0 & 0 \\ 0 & y_0 & \cdots & y_{-n+2} & y_{-n+1} & y_{-n} & \cdots & 0 & 0 \\ \cdots & \cdots & \cdots & \cdots & \cdots & \cdots & \cdots & \cdots & \cdots \\ 0 & 0 & \cdots & y_0 & y_{-1} & y_{-2} & \cdots & y_{-n} & 0 \\ 0 & 0 & \cdots & 0 & y_0 & y_{-1} & \cdots & y_{-n+1} & y_{-n} \\ x_n & x_{n-1} & \cdots & x_1 & x_0 & 0 & \cdots & 0 & 0 \\ 0 & x_n & \cdots & x_2 & x_1 & x_0 & \cdots & 0 & 0 \\ \cdots & \cdots & \cdots & \cdots & \cdots & \cdots & \cdots & \cdots & \cdots \\ 0 & 0 & \cdots & x_n & x_{n-1} & x_{n-2} & \cdots & x_0 & 0 \\ 0 & 0 & \cdots & 0 & x_n & x_{n-1} & \cdots & x_1 & x_0 \end{matrix} \right\|.$$

If the $(n + 1)$th row of $\mathscr{Q}$ is deleted, a matrix with determinant equal to $x_0 \det \mathscr{P}$ is obtained. Consequently $\mathscr{Q}$ has maximal rank $2n + 1$.

Let us denote by $\tilde{\mathscr{Q}}$ the augmented matrix of system (6.16). Expanding the determinant of $\tilde{\mathscr{Q}}$ with respect to the elements of the last column, we obtain $\det \tilde{\mathscr{Q}} = (-1)^n (y_0 - x_0) \det \mathscr{P}$. Therefore, by virtue of (6.15), we have $\det \tilde{\mathscr{Q}} = 0$.

Thus system (6.16) is uniquely solvable. The last assertion of the theorem follows from Theorem 6.1.

The theorem is proved.

§ 7. **Another formula for inverting Toeplitz matrices**

1. Let a_j $(j = 0, \cdots, \pm n)$ be complex numbers. For the square Toeplitz matrix $\mathscr{A}_n = \|a_{j-k}\|_{j,k=0}^{n}$, let us consider the two systems of linear equations

$$\sum_{k=0}^{n} a_{j-k}x_k = \delta_{j0} \qquad (j = 0, \cdots, n) \tag{7.1}$$

and

$$\sum_{k=0}^{n} a_{j-k}z_k = \delta_{j1} \qquad (j = 0, \cdots, n). \tag{7.2}$$

In this section we derive a formula which expresses the matrix $\mathscr{A}_n^{-1}$ in terms of the solutions $x_0,\cdots,x_n$ and $z_0,\cdots,z_n$ of equations (7.1) and (7.2). In addition, a number of results connected with this formula are cited.

2. The following theorem is basic.

THEOREM 7.1. *If solutions* $x_0,\cdots, x_n$ *and* $z_0,\cdots, z_n$ *of systems* (7.1) *and* (7.2) *exist and the condition* $x_n \neq 0$ *is fulfilled, then the Toeplitz matrix* $\mathscr{A}_n = \|a_{j-k}\|_{j,k=0}^n$ *is invertible, and its inverse is formed according to the formula*

$$\mathscr{A}_n^{-1} = x_n^{-1}\left\{\begin{Vmatrix} x_0x_n & x_0x_{n-1} & \cdots & x_0^2 \\ x_1x_n & x_1x_{n-1} & \cdots & x_1x_0 \\ \cdots & \cdots & \cdots & \cdots \\ x_n^2 & x_nx_{n-1} & \cdots & x_nx_0 \end{Vmatrix} + \begin{Vmatrix} z_0 & 0 & \cdots & 0 \\ z_1 & z_0 & \cdots & 0 \\ \cdots & \cdots & \cdots & \cdots \\ z_n & z_{n-1} & \cdots & z_0 \end{Vmatrix}\begin{Vmatrix} 0 & x_n & \cdots & x_1 \\ 0 & 0 & \cdots & x_2 \\ \cdots & \cdots & \cdots & \cdots \\ 0 & 0 & \cdots & x_n \\ 0 & 0 & \cdots & 0 \end{Vmatrix} \right.$$

$$\left. - \begin{Vmatrix} x_0 & 0 & \cdots & 0 \\ x_1 & x_0 & \cdots & 0 \\ \cdots & \cdots & \cdots & \cdots \\ x_n & x_{n-1} & \cdots & 0 \end{Vmatrix}\begin{Vmatrix} 0 & z_n & \cdots & z_1 \\ 0 & 0 & \cdots & z_2 \\ \cdots & \cdots & \cdots & \cdots \\ 0 & 0 & \cdots & z_n \\ 0 & 0 & \cdots & 0 \end{Vmatrix}\right\} \tag{7.3}$$

PROOF. Let us prove the invertibility of $\mathscr{A}_n$ by contradiction. Let us suppose that the theorem's conditions are fulfilled and $\mathscr{A}_n$ is not invertible. Then the rows $\Gamma_k = (a_k,\cdots, a_{k-n})$ $(k = 0,\cdots, n)$ of $\mathscr{A}_n$ are linearly dependent. Let p be the least number for which the vector Γ_p is a linear combination of the vectors $\Gamma_{p+1},\cdots, \Gamma_n$. It follows immediately from the existence of solutions of equations (7.1) and (7.2) that each of the first two rows Γ_0, Γ_1 is not a linear combination of the remaining rows. Consequently $p \geqq 2$. For some $p \geqq 2$ and some complex numbers $c_{p+1},\cdots, c_n$, let the equality $\Gamma_p = \sum_{p+1}^n c_k\Gamma_k$, or, what is the same thing,

$$a_{p-j} = \sum_{k=p+1}^{n} c_k a_{k-j} \qquad (j = 0,\cdots, n) \tag{7.4}$$

hold. Combining the equalities

$$c_{k+1} \sum_{j=0}^{n} a_{k-j}x_j = 0 \qquad (k = p,\cdots, n-1),$$

we obtain (by virtue of (7.4) and the condition $x_n \neq 0$)

$$a_{p-1-n} = \sum_{k=p+1}^{n} c_k a_{k-n-1}. \tag{7.5}$$

Equalities (7.4) and (7.5) imply that

$$a_{p-1-j} = \sum_{k=p}^{n-1} c_{k+1} a_{k-j} \qquad (j = 0,\cdots, n).$$

Hence

$$\Gamma_{p-1} = \sum_{p}^{n-1} c_{k+1}\Gamma_k.$$

The latter contradicts the hypothesis. Consequently $\mathscr{A}_n$ is invertible.

The derivation of formula (7.3) is based on Theorem 6.1.

Let us denote by $b_{j-k}(\varepsilon)$ the elements of the Toeplitz matrix $\mathscr{B}(\varepsilon) = \mathscr{A}_n - \varepsilon\mathscr{Y}$, where ε is a real parameter. Let us choose a number $\delta > 0$ so that the conditions

$$\det \mathscr{B}(\varepsilon) \neq 0 \text{ and } \det \left\| b_{j-k}(\varepsilon) \right\|_{j,k=0}^{n-1} \neq 0$$

are fulfilled for $0 < \varepsilon < \delta$.

By virtue of formula (6.3), the elements $d_{jk}(\varepsilon)$ $(j,\ k = 0, \cdots, n)$ of the matrix $\mathscr{B}^{-1}(\varepsilon)$ are determined for $0 < \varepsilon < \delta$ by the equalities

$$d_{jk}(\varepsilon) = x_0^{-1}(\varepsilon)\Big[\sum_{r=0}^{\min(j,k)} x_{j-r}(\varepsilon)y_{r-k}(\varepsilon) - \sum_{r=1}^{\min(j,k)} y_{j-n-r}(\varepsilon)x_{n-k+r}(\varepsilon),\Big] \tag{7.6}$$

where $x_j(\varepsilon)$ and $y_{-j}(\varepsilon)$ are solutions of the following systems of equations:

$$\sum_{k=0}^{n} b_{j-k}(\varepsilon)x_k(\varepsilon) = \delta_{j0} \qquad (j = 0,\cdots, n),$$

$$\sum_{k=0}^{n} b_{j-k}(\varepsilon)y_{k-n}(\varepsilon) = \delta_{jn} \qquad (j = 0,\cdots, n)$$

In particular, equalities (7.6) imply that

$$d_{01}(\varepsilon) = y_{-1}(\varepsilon) \qquad \text{and} \qquad x_0(\varepsilon) = y_0(\varepsilon), \tag{7.7}$$

and

$$d_{j1}(\varepsilon) = x_0^{-1}(\varepsilon)\,[x_j(\varepsilon)y_{-1}(\varepsilon) + x_{j-1}(\varepsilon)y_0(\varepsilon) + y_{j-n-1}(\varepsilon)x_n(\varepsilon)]. \tag{7.8}$$

Thus

$$d_{n+1-j,1}(\varepsilon) = x_0^{-1}(\varepsilon)\,[x_{n+1-j}(\varepsilon)d_{01}(\varepsilon) + x_{n-j}(\varepsilon)x_0(\varepsilon) - y_{-j}(\varepsilon)x_n(\varepsilon)].$$

Consequently

$$y_{-j}(\varepsilon) = x_n^{-1}(\varepsilon)[x_{n+1-j}(\varepsilon)\, d_{01}(\varepsilon) + x_{n-j}(\varepsilon)x_0(\varepsilon) - d_{n+1-j,1}(\varepsilon)x_0(\varepsilon)].$$

Let us note that the function $x_n(\varepsilon)$ is continuous; consequently it does not vanish in some neighborhood of zero.

Substituting the values $y_{-j}(\varepsilon)$ just obtained into equalities (7.6), we arrive at the equalities

$$d_{jk}(\varepsilon) = x_n^{-1}(\varepsilon)\Big[x_j(\varepsilon)x_{n-k}(\varepsilon) + \sum_{r=0}^{\min(j,k)} (x_{n-k+r+1}(\varepsilon)d_{j-r,1}(\varepsilon) - x_{j-r}(\varepsilon)d_{n-k+r+1,1}(\varepsilon))\Big] \tag{7.9}$$

$$(j,k = 0,\cdots, n).$$

It is easy to see that the functions $x_j(\varepsilon)$ and $d_{jk}(\varepsilon)$ $(j,k = 0,\cdots, n)$ are continuous functions in a neighborhood of the point $\varepsilon = 0$, the equalities $x_j(0) = x_j$ and $d_{j1}(0) = z_j$ holding for $j = 0, \cdots, n$.

Moreover, the numbers $d_{jk} = d_{jk}(0)$ are obviously elements of $\mathscr{A}_n^{-1}$. Setting $\varepsilon = 0$ in (7.9), we obtain

$$d_{jk} = x_n^{-1}\Big[x_j x_{n-k} + \sum_{r=0}^{\min(j,k-1)} (x_{n-k+r-1} z_{j-r} - x_{j-r} z_{n-k+r-1})\Big] \qquad (j, k = 0, \cdots, n).$$

These equalities are equivalent to (7.3).

The theorem is proved.

3. Let $\mathscr{A}_n = \|a_{j-k}\|_{j,k=0}^{n}$ be a Toeplitz matrix. Let us consider the matrix $\tilde{\mathscr{A}}_{n-1}$ defined by the equality

$$\tilde{\mathscr{A}}_{n-1} = \begin{Vmatrix} a_1 & a_0 & a_{-1} & \cdots & a_{-n+2} \\ a_2 & a_1 & a_0 & \cdots & a_{-n+3} \\ \cdots & \cdots & \cdots & \cdots & \cdots \\ a_n & a_{n-1} & a_{n-2} & \cdots & a_1 \end{Vmatrix}.$$

Theorem 7.1 implies that the solvability of equations (7.1) and (7.2), together with the condition $x_n \neq 0$, is equivalent to the following:

$$\det \mathscr{A}_n \neq 0 \qquad \text{and} \qquad \det \tilde{\mathscr{A}}_{n-1} \neq 0.$$

The following, more complete theorem concerning the matrix $\tilde{\mathscr{A}}_{n-1}$ holds.

THEOREM 7.2. *If solutions $x_0, \cdots, x_n$ and $z_0, \cdots, z_n$ of systems* (7.1) *and* (7.3) *exist and the condition $x_n \neq 0$ is fulfilled, then the Toeplitz matrix $\tilde{\mathscr{A}}_{n-1}$ is invertible, and its inverse is formed according to the formula*

$$\tilde{\mathscr{A}}_{n-1}^{-1} = x_n^{-1}\left\{ \begin{Vmatrix} z_0 & 0 & \cdots & 0 \\ z_1 & z_0 & \cdots & 0 \\ \cdots & \cdots & \cdots & \cdots \\ z_{n-1} & z_{n-2} & \cdots & z_0 \end{Vmatrix} \cdot \begin{Vmatrix} x_n & x_{n-1} & \cdots & x_1 \\ 0 & x_n & \cdots & x_2 \\ \cdots & \cdots & \cdots & \cdots \\ 0 & 0 & \cdots & x_n \end{Vmatrix} \right.$$
$$\left. - \begin{Vmatrix} x_0 & 0 & \cdots & 0 \\ x_1 & x_0 & \cdots & 0 \\ \cdots & \cdots & \cdots & \cdots \\ x_{n-1} & x_{n-2} & \cdots & x_0 \end{Vmatrix} \cdot \begin{Vmatrix} z_n & z_{n-1} & \cdots & z_1 \\ 0 & z_n & \cdots & z_2 \\ \cdots & \cdots & \cdots & \cdots \\ 0 & 0 & \cdots & z_n \end{Vmatrix} \right\}. \tag{7.10}$$

PROOF. Let us form the $(n+1)$th order square matrix

$$\mathscr{M} = \begin{Vmatrix} x_0 & 1 & 0 & \cdots & 0 \\ x_1 & 0 & 1 & \cdots & 0 \\ \cdot & \cdot & \cdot & \cdot & \cdot \\ x_{n-1} & 0 & 0 & \cdots & 1 \\ x_n & 0 & 0 & \cdots & 0 \end{Vmatrix}.$$

It is easy to see that

$$\mathscr{A}_n\mathscr{M} = \left\|\begin{matrix} 1 & a_0 \; a_{-1} \; \cdots \; a_{-n+1} \\ 0 & \\ \vdots & \tilde{\mathscr{A}}_{n-1} \\ 0 & \end{matrix}\right\|$$

Consequently

$$\mathscr{M}\mathscr{A}_n^{-1} = \left\|\begin{matrix} 1 & * \; \cdots \; * \\ 0 & \\ \vdots & \tilde{\mathscr{A}}_{n-1}^{-1} \\ 0 & \end{matrix}\right\| \tag{7.11}$$

where the asterisks denote elements not playing an essential role in the sequel.

The last equality will be utilized for the derivation of (7.10).

By Theorem 7.1, $\mathscr{A}_n^{-1}$ can be represented in the form

$$\mathscr{A}_n^{-1} = x_n^{-1}(\mathscr{K} + \mathscr{R} - \mathscr{S}),$$

where $\mathscr{K} = \|x_j x_{n-k}\|_{j,k=0}^n$ and

$$\mathscr{R} = \left\|\begin{matrix} z_0 & 0 & \cdots & 0 \\ z_1 & z_0 & \cdots & 0 \\ \cdot & \cdot & \cdot & \cdot \\ z_n & z_{n-1} & \cdots & z_0 \end{matrix}\right\| \cdot \left\|\begin{matrix} 0 & x_n & x_{n-1} & \cdots & x_1 \\ 0 & 0 & x_n & \cdots & x_2 \\ \cdot & \cdot & \cdot & \cdot & \cdot \\ 0 & 0 & 0 & \cdots & 0 \end{matrix}\right\|,$$

$$\mathscr{S} = \left\|\begin{matrix} x_0 & 0 & \cdots & 0 \\ x_1 & x_0 & \cdots & 0 \\ \cdot & \cdot & \cdot & \cdot \\ x_n & x_{n-1} & \cdots & x_0 \end{matrix}\right\| \cdot \left\|\begin{matrix} 0 & z_n & z_{n-1} & \cdots & z_1 \\ 0 & 0 & z_n & \cdots & z_2 \\ \cdot & \cdot & \cdot & \cdot & \cdot \\ 0 & 0 & 0 & \cdots & 0 \end{matrix}\right\|.$$

The matrix $\mathscr{M}^{-1}$ has the form

$$\mathscr{M}^{-1} = x_n^{-1} \left\|\begin{matrix} 0 & 0 & \cdots & 0 & 1 \\ 1 & 0 & \cdots & 0 & -x_0 \\ 0 & 1 & \cdots & 0 & -x_1 \\ \cdot & \cdot & \cdot & \cdot & \cdot \\ 0 & 0 & \cdots & 1 & -x_{n-1} \end{matrix}\right\|. \tag{7.12}$$

We verify directly that in the matrix $\mathscr{M}^{-1}\mathscr{K}$ the elements of all rows, except the first, equal zero.

It is also easily ascertained that the difference $\mathscr{R} - \mathscr{S}$ has the form

$$\mathscr{R} - \mathscr{S} = x_n \left\|\begin{matrix} 0 & \\ 0 & \mathscr{B} \\ \vdots & \\ 0 & 0 \; \cdots \; 0 \end{matrix}\right\|,$$

where $\mathscr{B}$ is an nth order square matrix coinciding with the right side of (7.10).

Taking into account equality (7.12) and the above established properties of the elements of $\mathscr{M}^{-1}\mathscr{K}$ and $\mathscr{R} - \mathscr{S}$, we obtain

$$\mathscr{M}^{-1}\mathscr{A}_n = \mathscr{M}^{-1}(\mathscr{R} - \mathscr{S}) + \mathscr{M}^{-1}\mathscr{K} = \begin{Vmatrix} * & * & \cdots & * \\ 0 & & & \\ \vdots & & \mathscr{B} & \\ 0 & & & \end{Vmatrix}.$$

This and (7.11) imply (7.10).

The theorem is proved.

4. Theorem 7.2 permits formula (7.10) to be used for the inversion of any nonsingular matrix $\mathscr{A}_n = \|a_{j-k}\|_{j,k=0}^{n}$ (and in particular when det $\tilde{\mathscr{A}}_{n-1} = 0$). Let us form the Toeplitz matrix

$$A_{n+1}(\zeta) = \begin{Vmatrix} a_{-1} & a_{-2} & \cdots & a_{-n-1} & \zeta \\ a_0 & a_{-1} & \cdots & a_{-n} & a_{-n-1} \\ \cdot & \cdot & \cdot & \cdot & \cdot \\ a_n & a_{n-1} & \cdots & a_0 & a_{-1} \end{Vmatrix},$$

where a_{-n-1} is some fixed complex number and ζ is a complex parameter. The function det $\mathscr{A}_{n+1}(\zeta)$ is a linear function of ζ and vanishes only at one value of ζ. Theorem 7.2 is applicable to the matrix $\mathscr{A}_{n+1}(\zeta)$ for all values of ζ besides the exceptional one; consequently $\mathscr{A}_n$ can be inverted by means of formula (7.10).

5. In the theorem given below, conditions are established under which, for prescribed systems of complex numbers $x_0, \cdots, x_n$ ($x_0 \neq 0$) and $z_0, \cdots, z_n$, a Toeplitz matrix $\mathscr{A}_n = \|a_{j-k}\|_{j,k=0}^{n}$ exists such that relations (7.1) and (7.2) hold.

THEOREM 7.3. *Let $x_0, \cdots, x_n$ and $z_0, \cdots, z_n$ be prescribed complex numbers, and $x_n \neq 0$.*

For a Toeplitz matrix $\mathscr{A}_n = \|a_{j-k}\|_{j,k=0}^{n}$ to exist such that

$$\sum_{k=0}^{n} a_{j-k}x_k = \delta_{j0} \quad \text{and} \quad \sum_{k=0}^{n} a_{j-k}z_k = \delta_{j1} \quad (j = 0, \cdots, n), \tag{7.13}$$

it is necessary and sufficient that the condition

$$z_n = x_{n-1} \tag{7.14}$$

be fulfilled and that the matrix

$$\mathscr{N} = \begin{Vmatrix} x_n & x_{n-1} & \cdots & x_1 & x_0 & 0 & \cdots & 0 \\ 0 & x_n & \cdots & x_2 & x_1 & x_0 & \cdots & 0 \\ \cdot & \cdot & \cdot & \cdot & \cdot & \cdot & \cdot & \cdot \\ 0 & 0 & & x_n & x_{n-1} & x_{n-2} & \cdots & x_0 \\ z_n & z_{n-1} & \cdots & z_1 & z_0 & 0 & \cdots & 0 \\ 0 & z_n & \cdots & z_2 & z_1 & z_0 & \cdots & 0 \\ \cdot & \cdot & \cdot & \cdot & \cdot & \cdot & \cdot & \cdot \\ 0 & 0 & \cdots & z_n & z_{n-1} & z_{n-2} & \cdots & z_0 \end{Vmatrix}$$

be nonsingular.

If these conditions are fulfilled, then $\mathscr{A}_n$ is invertible and can be uniquely reconstructed relative to the systems $x_0, \cdots, x_n$ and $z_0, \cdots, z_n$ by means of formula (7.3).

PROOF. The necessity of condition (7.14) is obvious. Let us show that the condition $\det \mathscr{N} \neq 0$ is also necessary. By Theorem 7.1, equalities (7.13) imply that the matrix is invertible and its inverse can be represented as

$$\mathscr{A}_n^{-1} = x_n^{-1}(\mathscr{B}\mathscr{C} - \mathscr{D}\mathscr{F}), \tag{7.15}$$

where

$$\mathscr{B} = \begin{Vmatrix} z_0 & 0 & \cdots & 0 \\ z_1 & z_0 & \cdots & 0 \\ \cdots & \cdots & \cdots & \cdots \\ z_n & z_{n-1} & \cdots & z_0 \end{Vmatrix}, \quad \mathscr{C} = \begin{Vmatrix} 0 & x_n & \cdots & x_1 \\ \cdots & \cdots & \cdots & \cdots \\ 0 & 0 & \cdots & x_n \\ 0 & 0 & \cdots & 0 \end{Vmatrix}, \quad \mathscr{D} = \begin{Vmatrix} x_0 & 0 & \cdots & 0 \\ x_1 & x_0 & \cdots & 0 \\ \cdots & \cdots & \cdots & \cdots \\ x_n & x_{n-1} & \cdots & x_0 \end{Vmatrix}$$

and

$$\mathscr{F} = \begin{Vmatrix} 0 & z_n & \cdots & z_1 \\ \cdots & \cdots & \cdots & \cdots \\ 0 & 0 & \cdots & z_n \\ 0 & 0 & \cdots & 0 \end{Vmatrix} - \begin{Vmatrix} x_n & x_{n-1} & \cdots & x_0 \\ 0 & 0 & \cdots & 0 \\ \cdots & \cdots & \cdots & \cdots \\ 0 & 0 & \cdots & 0 \end{Vmatrix}.$$

Since $\mathscr{B}\mathscr{D} = \mathscr{D}\mathscr{B}$, we have

$$\det(\mathscr{B}\mathscr{C} - \mathscr{D}\mathscr{F}) = \det \begin{Vmatrix} \mathscr{B} & \mathscr{F} \\ \mathscr{D} & \mathscr{C} \end{Vmatrix} \tag{7.16}$$

by a well-known proposition (cf. F. R. Gantmaher [1], Chapter II, § 5.3). It is easy to see that

$$\left| \det \begin{Vmatrix} \mathscr{B} & \mathscr{F} \\ \mathscr{D} & \mathscr{C} \end{Vmatrix} \right| = |x_n|^2 |\det \mathscr{N}|. \tag{7.17}$$

It follows immediately from (7.15), (7.16) and (7.17) that

$$|\det \mathscr{N}| = |x_n|^{-1} |\det \mathscr{A}_n^{-1}|.$$

Therefore $\det \mathscr{N} \neq 0$.

Let us pass to the proof of the sufficiency of the theorem's conditions. Let the conditions $\det \mathscr{N} \neq 0$ and $z_n = x_{n-1}$ be fulfilled. Let us consider equations (7.13) as a system of $2n + 2$ equations with $2n + 1$ unknowns $a_{-n}, \cdots, a_n$. The matrix of this system is of the form

$$\mathscr{G} = \left\| \begin{matrix} x_n & x_{n-1} & \cdots & x_1 & x_0 & 0 & \cdots & 0 \\ 0 & x_n & \cdots & x_2 & x_1 & x_0 & \cdots & 0 \\ \cdots & \cdots & \cdots & \cdots & \cdots & \cdots & \cdots & \cdots \\ 0 & 0 & \cdots & 0 & x_n & x_{n-1} & \cdots & x_0 \\ z_n & z_{n-1} & \cdots & z_1 & z_0 & 0 & \cdots & 0 \\ 0 & z_n & \cdots & z_2 & z_1 & z_0 & \cdots & 0 \\ \cdots & \cdots & \cdots & \cdots & \cdots & \cdots & \cdots & \cdots \\ 0 & 0 & \cdots & 0 & z_n & z_{n-1} & \cdots & z_0 \end{matrix} \right\|.$$

If the $(n+2)$th row is deleted in $\mathscr{G}$, the determinant of the remaining matrix will differ from the determinant of $\mathscr{N}$ by a factor equal to $\pm x_n$. Consequently the rank of $\mathscr{G}$ is maximal. Let us denote by $\tilde{\mathscr{G}}$ the matrix obtained from $\mathscr{G}$ by adding to it the column of the right sides of system (7.13). Since

$$\left\| \begin{matrix} x_n & x_{n-1} & \cdots & x_0 \\ 0 & x_n & \cdots & x_1 \\ \cdots & \cdots & \cdots & \cdots \\ 0 & 0 & \cdots & x_n \end{matrix} \right\| \cdot \left\| \begin{matrix} z_n & z_{n-1} & \cdots & z_0 \\ 0 & z_n & \cdots & z_1 \\ \cdots & \cdots & \cdots & \cdots \\ 0 & 0 & \cdots & z_n \end{matrix} \right\| = \left\| \begin{matrix} z_n & z_{n-1} & \cdots & z_0 \\ 0 & z_n & \cdots & z_1 \\ \cdots & \cdots & \cdots & \cdots \\ 0 & 0 & \cdots & z_n \end{matrix} \right\| \cdot \left\| \begin{matrix} x_n & x_{n-1} & \cdots & x_0 \\ 0 & x_n & \cdots & x_1 \\ \cdots & \cdots & \cdots & \cdots \\ 0 & 0 & \cdots & x_n \end{matrix} \right\|,$$

it follows that

$$\det \tilde{\mathscr{G}} = \det \left\{ \left\| \begin{matrix} x_n & x_{n-1} & \cdots & x_0 \\ 0 & x_n & \cdots & x_1 \\ \cdots & \cdots & \cdots & \cdots \\ 0 & 0 & \cdots & x_n \end{matrix} \right\| \cdot \left\| \begin{matrix} 0 & 0 & \cdots & 0 \\ z_0 & 0 & \cdots & 1 \\ z_1 & z_0 & \cdots & 0 \\ \cdots & \cdots & \cdots & \cdots \\ z_{n-1} & z_{n-2} & \cdots & 0 \end{matrix} \right\| - \left\| \begin{matrix} z_n & z_{n-1} & \cdots & z_0 \\ 0 & z_n & \cdots & z_1 \\ \cdots & \cdots & \cdots & \cdots \\ 0 & 0 & \cdots & z_n \end{matrix} \right\| \cdot \left\| \begin{matrix} 0 & 0 & \cdots & 1 \\ x_0 & 0 & \cdots & 0 \\ x_1 & x_0 & \cdots & 0 \\ \cdots & \cdots & \cdots & \cdots \\ x_{n-1} & x_{n-2} & \cdots & 0 \end{matrix} \right\| \right\},$$

by virtue of the proposition already cited (Gantmaher [1], loc. cit.). We verify directly that in the matrix in the right side of the latter equality all elements of the first row equal zero; consequently $\det \tilde{\mathscr{G}} = 0$.

Hence we have shown that system (7.13) is solvable relative to the unknowns $a_j \, (j = 0, \cdots, \pm n)$. We compose the Toeplitz matrix $\mathscr{A}_n = \|a_{j-k}\|_0^n$ from the numbers a_j. Theorem 7.1 implies that $\mathscr{A}_n$ is invertible and uniquely determined by formulas (7.3).

The theorem is proved.

§ 8. Inversion of truncated Wiener-Hopf integral operators

1. The continuous analogues of Theorems 6.1 and 6.3 are given in this section.

THEOREM 8.1. *Let the function $k(t) \in L_1(-\tau, \tau)$ be such that the equations*

$$\gamma_+(t) - \int_0^\tau k(t-s)\gamma_+(s)\,ds = k(t) \qquad (0 \leqq t \leqq \tau) \tag{8.1}$$

and

$$\gamma_-(-t) - \int_0^\tau k(s-t)\gamma_-(-s)\,ds = k(-t) \qquad (0 \leqq t \leqq \tau) \tag{8.2}$$

have solutions $\gamma_+(t) \in L_1(0, \tau)$ and $\gamma_-(t) \in L_1(-\tau, 0)$.
Then the equation

$$\varphi(t) - \int_0^\tau k(t-s)\,\varphi(s)\,ds = f(t) \qquad (0 \leqq t \leqq \pi) \tag{8.3}$$

has the unique solution $\varphi(t) \in L_p(0, \tau)$ $(1 \leqq p \leqq \infty)$ for any right side $f(t) \in L_p(0, \tau)$, and this solution can be found by the formula

$$\varphi(t) = f(t) + \int_0^\tau \gamma(t, s)\, f(s)\,ds,$$

where

$$\gamma(t,s) = \gamma_+(t-s) + \gamma_-(t-s) + \int_0^{\min(t,s)} \gamma_+(t-r)\,\gamma(r-s)\,dr$$
$$- \int_\tau^{\tau+\min(t,s)} \gamma_-(t-r)\gamma_+(r-s)\,dr \qquad (0 \leqq t, s \leqq \tau) \tag{8.4}$$

and $\gamma_+(-t) = \gamma_-(t) = 0, 0 \leqq t \leqq \tau$.

Let us note that the following equations with delta-functions will be the formal continuous analogues of equations (6.1) and (6.2):

$$x(t) = \int_0^\tau k(t-s)x(s)\,ds = \delta(t) \qquad (0 \leqq t \leqq \tau)$$

and

$$y(t-\tau) = \int_0^\tau k(t-s)y(s-t)\,ds = \delta(t-\tau) \qquad (0 \leqq t \leqq \tau).$$

Equations (8.1) and (8.2) are obtained from these equations if the functions $x(t)$ and $y(t)$ are sought in the following form:

$$x(t) = \delta(t) + \gamma_+(t), \qquad y(-t) = \delta(-t) + \gamma_-(-t) \qquad (0 \leqq t \leqq \tau).$$

PROOF OF THEOREM 8.1. The integral operator in equation (8.3) is completely continuous in the space $L_p(0, \tau)$ $(1 \leqq p < \infty)$. Therefore it is sufficient for the proof to ascertain that the function $\gamma(t, s)$ $(0 \leqq t, s \leqq \tau)$ defined by equality (8.4) satisfies the equation

$$\gamma(t, s) - \int_0^\tau k(t-u)\gamma(u, s)\, du = k(t-s) \qquad (0 \leqq t, s \leqq \tau). \tag{8.5}$$

Let us first consider the case $t \geqq s$. In this case

$$\gamma(t, s) = \gamma_+(t-s) + \chi_1(t, s) - \chi_2(t, s), \tag{8.6}$$

where

$$\chi_1(t, s) = \int_0^s \gamma_+(t-r)\, \gamma_-(r-s)\, dr$$

and

$$\chi_2(t, s) = \int_\tau^{\tau+s} \gamma_-(t-r)\, \gamma_+(r-s)\, dr.$$

Taking into account that the function $\gamma_+(t)$ is the solution of equation (8.1), we obtain

$$\chi_1(t, s) = \int_0^s \left[k(t-r) + \int_0^\tau k(t-r-u)\, \gamma_+(u)\, du \right] \gamma_-(r-s)\, dr$$
$$= \int_0^s k(t-r)\, \gamma_-(r-s)\, dr + \int_0^s \gamma_-(r-s)\, dr \int_r^{\tau+r} k(t-u)\, \gamma_+(u-r)\, du.$$

Switching the order of integration in the last integral, we obtain

$$\int_0^s \gamma_-(r-s)\, dr \int_r^{\tau+r} k(t-u)\, \gamma_+(u-r)\, du = \int_0^s k(t-u) \int_0^u \gamma_+(u-r)\, \gamma_-(r-s)\, dr$$
$$+ \int_s^\tau k(t-u)\, du \int_0^s \gamma_+(u-r)\, \gamma_-(r-s)\, dr$$
$$+ \int_\tau^{\tau+s} k(t-u)\, du \int_{-\tau+u}^{s} \gamma_+(ut-r)\, \gamma_-(r-s)\, dr,$$

or, what is the same thing,

$$\int_0^s \gamma_-(r-s)\, dr \int_r^{\tau+r} k(t-u)\, \gamma_+(u-r)\, du$$
$$= \int_0^\tau k(t-u)\, du \int_0^{\min(u,s)} \gamma_+(u-r)\, \gamma_-(r-s)\, dr$$
$$+ \int_\tau^{\tau+s} k(t-u)\, du \int_u^{\tau+s} \gamma_-(u-r)\, \gamma_+(r-s)\, dr.$$

Consequently

$$\chi_1(t, s) = \int_0^s k(t-u)\gamma_-(u-s)\,du + \int_0^\tau k(t-u)\,du \int_0^{\min(u,s)} \gamma_+(u-r)\gamma_-(r-s)\,dr$$
$$+ \int_\tau^{\tau+s} k(t-u)\,du \int_u^{\tau+s} \gamma_-(u-r)\gamma_+(r-s)\,dr. \tag{8.7}$$

We transform the integral $\chi_2(t, s)$ by replacing the function $\gamma_-(t-r)$ in it by its value from (8.2). Then

$$\chi_2(t, s) = \int_\tau^{\tau+s} \left[k(t-r) + \int_0^\tau k(u+t-r)\gamma_-(-u)\,du \right] \gamma_+(r-s)\,dr$$
$$= \int_s^{\tau+s} k(t-u)\gamma_+(u-s)\,du - \int_s^\tau k(t-u)\gamma_+(u-s)\,du$$
$$+ \int_\tau^{\tau+s} \gamma_+(r-s)\,dr \int_{r-\tau}^r k(t-u)\gamma_-(u-r)\,du.$$

Switching the order of integration in the last integral, we arrive at the equality

$$\int_\tau^{\tau+s} \gamma_+(r-s)\,dr \int_{r-\tau}^r k(t-u)\gamma_-(u-r)\,du$$
$$= \int_0^\tau k(t-u)\,du \int_\tau^{\tau+\min(u,s)} \gamma_-(u-r)\gamma_+(r-s)\,dr$$
$$+ \int_\tau^{\tau+s} k(t-u)\,du \int_u^{\tau+s} \gamma_-(u-r)\gamma_+(r-s)\,dr.$$

Therefore

$$\chi_2(t, s) = \int_0^\tau k(t-s-r)\gamma_+(r)\,dr - \int_s^\tau k(t-u)\gamma_+(u-s)\,du$$
$$+ \int_0^\tau k(t-u)\,du \int_\tau^{\tau+\min(u,s)} \gamma_-(u-r)\gamma_+(r-s)\,dr$$
$$+ \int_\tau^{\tau+s} k(t-u)\,du \int_u^{\tau+s} \gamma_-(u-r)\gamma_+(r-s)\,dr. \tag{8.8}$$

Now substituting for $\chi_1(t, s)$ and $\chi_2(t, s)$ in (8.6) their values (8.7) and (8.8), we obtain

$$\gamma(t, s) = \gamma_+(t-s) - \int_0^\tau k(t-s-r)\gamma_+(r)\,dr - \int_s^\tau k(t-u)\gamma_+(u-s)\,du$$
$$+ \int_0^s k(t-u)\gamma_-(u-s)\,du + \int_0^\tau k(t-u)\,du \int_0^{\min(u,s)} \gamma_+(u-r)\gamma_-(r-s)\,dr$$
$$- \int_0^\tau k(t-u)\,du \int_\tau^{\tau+\min(u,s)} \gamma_-(u-r)\gamma_+(r-s)\,dr.$$

Since the function $\gamma_+(t) \in L_1(0, \tau)$ is the solution of equation (8.1), we have

$$\gamma(t, s) = k(t-s) + \int_0^\tau k(t-u)\Big[\gamma_+(u-s) + \gamma_-(u-s) + \int_0^{\min(u,s)} \gamma_+(u-r)\gamma_-(r-s)\,dr - \int_\tau^{\tau+\min(u,s)} \gamma_-(u-r)\gamma_+(r-s)\,dr\Big]\,du.$$

Thus equality (8.5) holds.

Let us pass to the proof of (8.5) in the case $t < s$. In this case the function $\gamma(t, s)$ is defined by

$$\gamma(t, s) = \gamma_-(t-s) + \int_0^t \gamma_+(t-r)\gamma_-(r-s)\,dr + \int_\tau^{\tau+t} \gamma_-(t-r)\gamma_+(r-s)\,dr.$$

Equality (8.4) implies that

$$\int_0^\tau k(t-u)\gamma(u, s)\,du = \int_s^\tau k(t-u)\gamma_+(u-s)\,du + \int_0^s k(t-u)\gamma_-(u-s)\,du + \omega_1(t, s) - \omega_2(t, s), \tag{8.9}$$

where

$$\omega_1(t, s) = \int_0^\tau k(t-u)\,du \int_0^{\min(u,s)} \gamma_+(u-r)\gamma_-(r-s)\,dr,$$

$$\omega_2(t, s) = \int_0^\tau k(t-u)\,du \int_\tau^{\tau+\min(u,s)} \gamma_-(u-r)\gamma_+(r-s)\,dr.$$

It $u + s - r$ is taken as a new variable of integration in the inner integral for $\omega_1(t, s)$, then

$$\omega_1(t, s) = \int_0^\tau k(t-u)\,du \int_{\max(u,s)}^{u+s} \gamma_-(u-r)\gamma_+(r-s)\,dr.$$

Effecting interchanges of the order of integration in the integrals $\omega_1(t, s)$ and $\omega_2(t, s)$, we obtain

$$\omega_1(t, s) = \int_s^\tau \gamma_+(r-s)\,dr \int_{r-s}^{r} k(t-u)\gamma_-(u-r)\,du + \int_\tau^{\tau+s} \gamma_+(r-s)\,dr \int_{r-s}^{\tau} k(t-u)\gamma_-(u-r)\,du$$

and

$$\omega_2(t, s) = \int_\tau^{\tau+s} \gamma_+(r-s)\,dr \int_{-\tau+r}^{\tau} k(t-u)\gamma_-(u-r)\,du.$$

Therefore the equality

$$\omega_1(t, s) - \omega_2(t, s) = \int_s^\tau \gamma_+(r - s)\, dr \int_{r-s}^{r} k(t - u)\gamma_-(u - r)\, dr$$
$$- \int_\tau^{\tau+s} \gamma_+(r - s)\, dr \int_{-\tau+r}^{r-s} k(t - u)\gamma_-(u - r)\, du$$

holds. After the obvious changes in the variables of integration, this difference can be represented in the form

$$\omega_1(t, s) - \omega_2(t, s) = \int_0^{\tau-s} \gamma_+(r)\, dr \int_0^s k(u + t - s - r)\gamma_-(-u)\, du$$
$$- \int_{\tau-s}^{\tau} \gamma_+(r)\, dr \int_s^\tau k(u + t - s - r)\gamma_-(-u)\, du,$$

or, what is the same thing,

$$\omega_1(t, s) - \omega_2(t, s) = \int_0^{\tau-s} \gamma_+(r)\, dr \int_0^\tau k(u + t - s - r)\gamma_-(-u)\, du$$
$$- \int_s^\tau \gamma_-(-u)\, du \int_0^\tau k(u + t - s - r)\gamma_+(r)\, dr.$$

Since the functions $\gamma_+(t) \in L_1(0, \tau)$ and $\gamma_-(t) \in L_1(-\tau, 0)$ are the solutions of equations (8.1) and (8.2), we have

$$\int_s^\tau \gamma_-(-u)\, du \int_0^\tau k(u + t - s - r)\gamma_+(r)\, dr$$
$$= \int_s^\tau \gamma_+(u + t - s)\gamma_-(-u)\, du - \int_s^\tau k(u+t-s)\gamma_-(-u)\, du \tag{8.10}$$

and

$$\int_0^{\tau-s} \gamma_+(r)\, dr \int_0^\tau k(u + t - s - r)\gamma_-(-u)\, du$$
$$= \int_0^{\tau-s} \gamma_-(t - s - r)\gamma_+(r)\, dr - \int_0^{\tau-s} k(t - s - r)\gamma_+(r)\, dr. \tag{8.11}$$

In addition, the equalities

$$\int_0^{\tau-s} \gamma_-(t - s - r)\gamma_+(r)\, dr$$
$$= \int_0^t \gamma_-(t - s - r)\gamma_+(r)\, dr + \int_t^{\tau-s} \gamma_-(t - s - r)\gamma_+(r)\, dr \tag{8.12}$$
$$= \int_0^t \gamma_+(t - r)\gamma_-(r - s)\, dr + \int_{\tau+s}^{\tau} \gamma_-(t - r)\gamma_+(r - s)\, dr$$

and

$$\int_s^\tau \gamma_+(u + t - s)\,\gamma_-(-u)\,du = \int_{t+s}^{\tau+t} \gamma_-(t - r)\,\gamma_+(r - s)\,dr \tag{8.13}$$

hold. Subtracting (8.10) from (8.11) and taking (8.12) and (8.13) into consideration, we obtain

$$\begin{aligned} &\omega_1(t, s) - \omega_2(t, s) \\ &\quad = \int_s^\tau k\,(u + t - s)\,\gamma_-(-u)\,du - \int_s^\tau k(t - u)\,\gamma_+(u - s)\,du \\ &\quad + \int_0^t \gamma_+(t - r)\,\gamma_-(r - s)\,dr - \int_\tau^{\tau+t} \gamma_-(t - r)\,\gamma_+(r - s)\,dr. \end{aligned}$$

Finally, we obtain

$$\begin{aligned} k(t - s) + \int_0^\tau k(t - u)\,\gamma\,(u, s)\,du &= k(t - s) + \int_0^\tau k(u + t - s)\,\gamma_-(-u)\,du \\ &+ \int_0^t \gamma_+(t - r)\,\gamma_-(r - s)\,dr - \int_\tau^{\tau+t} \gamma_-(t - r)\,\gamma_-(r - s)\,dr \end{aligned}$$

from (8.9) and the latter equality. Considering (8.2), this means

$$\begin{aligned} &k(t - s) + \int_0^\tau k(t - u)\,\gamma\,(u, s)\,du \\ &\quad = \gamma_-(t - s) + \int_0^t \gamma_+(t - r)\,\gamma_-(r - s)\,dr - \int_\tau^{r+t} \gamma_-(t - r)\,\gamma_+(r - s)\,dr. \end{aligned}$$

Therefore (8.5) holds in the case $t < s$ also.

The theorem is proved.

2. The following theorem is the continuous analogue of Theorem 6.3.

THEOREM 8.2. *Let the functions $\gamma_+(t) \in L_1(0, \tau)$ and $\gamma_-(t) \in L_1(-\tau, 0)$ be given. For there to exist a function $k(t) \in L_1(-\tau, \tau)$ for which the equalities*

$$\gamma_+(t) - \int_0^\tau k(t - s)\,\gamma_+(s)\,ds = k(t) \qquad (0 \leqq t \leqq \tau) \tag{8.14}$$

and

$$\gamma_-(-t) - \int_0^\tau k(s - t)\,\gamma_-(-s)\,ds = k(-t) \qquad (0 \leqq t \leqq \tau) \tag{8.15}$$

are fulfilled, it is necessary and sufficient that the operator $\mathscr{P}$ defined in the space $L_1(-\tau, \tau)$ by the equalities

$$(\mathscr{P}\,\omega)(t) = \begin{cases} \omega(t) + \displaystyle\int_t^{\tau+t} \gamma_-(t - s)\,\omega\,(s)\,ds & (-\tau \leqq t \leqq 0), \\ \omega(t) + \displaystyle\int_{t-\tau}^t \gamma_+\,(t - s)\,\omega\,(s)\,ds & (0 \leqq t \leqq \tau) \end{cases}$$

be invertible. If this condition is fulfilled, the function $k(t)$ is uniquely determined.

PROOF. Let us begin with the proof of necessity.

Let us introduce into the discussion the operators $\mathscr{X}$, $\mathscr{V}$, and $\mathscr{H}_+$ defined in the respective L_1 spaces by the equalities

$$(\mathscr{X}\varphi)(t) = \varphi(t) + \int_0^t \gamma_+(t-s)\varphi(s)\,ds,$$

$$(\mathscr{V}\psi)(t) = \int_{-\tau+t}^{0} \gamma_+(t-s)\psi(s)\,ds, \qquad (0 \leqq t \leqq \tau)$$

$$(\mathscr{H}_+\psi)(t) = \psi(t-\tau),$$

where $\varphi(t) \in L_1(0, \tau)$ and $\psi(t) \in L_1(-\tau, 0)$, as well as the operators $\mathscr{Y}$, $\mathscr{U}$ and $\mathscr{H}_-$ prescribed by the equalities

$$(\mathscr{Y}\psi)(t) = \psi(t) + \int_t^0 \gamma_-(t-s)\psi(s)\,ds,$$

$$(\mathscr{U}\varphi)(t) = \int_0^{\tau+t} \gamma_-(t-s)\varphi(s)\,ds, \qquad (-\tau \leqq t \leqq 0)$$

$$(\mathscr{H}_-\varphi)(t) = \varphi(t+\tau),$$

where, as before, $\varphi(t) \in L_1(0, \tau)$ and $\psi(t) \in L_1(-\tau, 0)$.

Considering the space $L_1(-\tau, \tau)$ as the direct sum of the spaces $L_1(-\tau, 0)$ and $L_1(0, \tau)$, the operator $\mathscr{P}$ can be represented in the matrix form

$$\mathscr{P} = \left\| \begin{matrix} \mathscr{Y} & \mathscr{U} \\ \mathscr{V} & \mathscr{X} \end{matrix} \right\|.$$

The operator $\mathscr{X}$ is invertible in $L_1(0, \tau)$. It is not difficult to ascertain the validity of the following equality:

$$\mathscr{P} = \left\| \begin{matrix} I_- & \mathscr{U}\mathscr{X}^{-1} \\ 0 & I_+ \end{matrix} \right\| \cdot \left\| \begin{matrix} \mathscr{Y} - \mathscr{U}\mathscr{X}_{-1}\mathscr{V} & 0 \\ 0 & \mathscr{X} \end{matrix} \right\| \cdot \left\| \begin{matrix} I_- & 0 \\ \mathscr{X}^{-1}\mathscr{V} & I_+ \end{matrix} \right\|,$$

where I_- and I_+ are the identity operators in $L_1(-\tau, 0)$ and $L_1(0, \tau)$.

The end factors in the right side of the latter equality are invertible. Consequently the operator $\mathscr{P}$ is invertible in $L_1(-\tau, \tau)$ if and only if $\mathscr{Y} - \mathscr{U}\mathscr{X}^{-1}\mathscr{V}$ is invertible in $L_1(-\tau, 0)$.

By Theorem 8.1, the operator $I_+ - \mathscr{K}$, where

$$(\mathscr{K}\varphi)(t) = \int_0^{\tau} k(t-s)\varphi(s)\,ds,$$

is invertible in $L_1(0, \tau)$, and

$$(I_+ - \mathscr{K})_{-1} = \mathscr{X}\mathscr{H}_+\mathscr{Y}\mathscr{H}_- - \mathscr{H}_+\mathscr{U}\mathscr{V}\mathscr{H}_-.$$

The equality

$$\mathscr{U}\mathscr{X}^{-1} = \mathscr{H}_-\mathscr{X}^{-1}\mathscr{H}_+\mathscr{U}$$

is verified without difficulty. It follows from the latter relation that

$$\mathscr{Y} - \mathscr{U}\mathscr{X}^{-1}\mathscr{V} = \mathscr{H}_-\mathscr{X}^{-1}(I_+ - \mathscr{K})^{-1}\mathscr{H}_+.$$

Therefore the operator $\mathscr{Y} - \mathscr{U}\mathscr{X}^{-1}\mathscr{V}$ is invertible, and, together with it, so is $\mathscr{P}$.

Let us pass to the proof of the sufficiency of the theorem's conditions.

Let us consider relations (8.14) and (8.15) as equations with the unknown function $k(t) \in L_1(-\tau, \tau)$. After simple transformations these equations take the form

$$k(t) + \int_t^{\tau+t} \gamma_-(t-s)k(s)\,ds = \gamma_-(t) \qquad (-\tau \leqq t \leqq 0),$$

$$k(t) + \int_{-\tau+t}^{t} \gamma_+(t-s)k(s)\,ds = \gamma_+(t) \qquad (0 \leqq t \leqq \tau). \tag{8.16}$$

The function $k(t)$ is uniquely determined from (8.16), by virtue of the invertibility of $\mathscr{P}$.

The theorem is proved.

CHAPTER IV

WIENER-HOPF EQUATIONS WITH DISCONTINUOUS FUNCTIONS

In the present chapter some results of the preceding chapters are generalized to *piecewise continuous* functions of isometric operators operating in *Hilbert* space. The problem of inverting such operators by approximation methods is solved.

The results obtained in abstract form are then applied to Toeplitz matrices composed of the Fourier coefficients of piecewise continuous functions, as well as to their continuous analogues.

We succeed in extending some results even to the case of bounded measurable functions. Finite difference equations are considered in the last section.

§ 1. Discontinuous functions of isometric operators

1. Let V be an isometric operator operating in Hilbert space $\mathfrak{H}$; U, a unitary extension of V which operates in some extended Hilbert space $\tilde{\mathfrak{H}} \supset \mathfrak{H}$; P, the operator of the orthogonal projection of $\tilde{\mathfrak{H}}$ onto $\mathfrak{H}$.

Let us denote by Λ the algebra of all piecewise continuous[1] and left continuous functions on the unit circle. A norm is defined on it by the equality

$$\|a(\zeta)\| = \sup_{|\zeta|=1} |a(\zeta)|.$$

An operator[2] $a(U) \in \mathfrak{A}(\tilde{\mathfrak{H}})$ can be put into correspondence in a natural way with each function $a(\zeta) \in \Lambda$. The set of all such operators is an algebra isomorphic and isometric to the algebra Λ. However, for reasons which will soon be explained, we shall not call the function $a(\zeta)$ the symbol of $a(U)$.

If $a(\zeta) \in \Lambda$, we denote by $a(V)$ the restriction of the operator $Pa(U)P$ to a subspace of $\mathfrak{H}$ $(= P\tilde{\mathfrak{H}})$.[3]

---

[1] A function $a(\zeta)$ is called piecewise continuous if the limits $a(\zeta \pm 0)$ exist at any point $\zeta \in \Gamma$ and the number of points of discontinuity is finite.

[2] The correspondence $a(\zeta) \to a(U)$ is made as follows. Let E_ζ $(|\zeta| = 1)$ be the left-continuous resolution of the identity for the unitary operator U, and $\zeta_0, \cdots, \zeta_n$, the points of discontinuity of the function $a(\zeta)$. The operator $a(u)$ is defined by the equality

$$a(U) = \sum_{j=0}^{n-1} \int_{\zeta_j+0}^{\zeta_{j+1}} a(\zeta)\, dE_\zeta + \int_{\zeta_n+0}^{\zeta_0} a(\zeta)\, dD_\zeta.$$

[3] The function $a(V)$ can also be defined without leaving the basic Hilbert space $\mathfrak{H}$. A representation of the operator V as the orthogonal sum of a unitary operator and a finite, or infinite, number of shift operators can be utilized for this purpose.

LEMMA 1.1. *Let the function* $b(\zeta) \in \Lambda$ *have the form*

$$b(\zeta) = a_-(\zeta)a(\zeta)a_+(\zeta),$$

where $a_+(\zeta)$ *and* $a_-(\zeta)$ *are continuous functions whose Fourier coefficients with nonpositive (respectively nonnegative) indices equal zero, and the function* $a(\zeta) \in \Lambda$. *Then*

$$b(V) = a_-(V)a(V)a_+(V).$$

PROOF. Indeed, the equalities $PUP = UP$ and $PU^{-1}P = PU^{-1}$ imply that

$$Pa_+(U)P = a_+(U)P, \qquad Pa_-(U)P = Pa_-(U).$$

Therefore

$$b(V) = Pa_-(U)a(U)a_+(U)P = Pa_-(U)Pa(U)Pa_+(U)P = a_-(V)a(V)\,a_+(V).$$

2. Let the function $a(\zeta) \in \Lambda$. Let us consider the function

$$a\,(\zeta, \mu) = \mu a(\zeta + 0) + (1 - \mu)a(\zeta) \qquad (|\zeta| = 1;\ 0 \leqq \mu \leqq 1).$$

The set Γ_a of values of the function $a(\zeta, \mu)$ represents a closed curve on the plane. This curve is the union of the set of values of the function $a(\zeta)$ ($|\zeta| = 1$) and the segments

$$\mu a(\zeta_k + 0) + (1 - \mu)a(\zeta_k) \qquad (0 \leqq \mu \leqq 1;\ k = 1,\cdots, n),$$

where $\zeta_1,\cdots, \zeta_n$ are all the points of discontinuity of the function $a(\zeta)$.

The curve Γ_a can be oriented in a natural way. If this curve does not pass through zero, the function $a(\zeta)$ is called *nonsingular*. The number of revolutions of this curve around the point $\zeta = 0$ is called the *index* of the function $a(\zeta)$ and is denoted by ind $a(\zeta)$.

It is easy to verify that *if* $a(\zeta)$ ($\in \Lambda$) *is a nonsingular function and* $b(\zeta)$ *is a continuous function which does not vanish on the circle* Γ, *then the function* $c(\zeta) = a(\zeta)b(\zeta)$ *is nonsingular, and* ind c = ind a + ind b.

We shall call the function $a(\zeta, \mu)$ of the two variables $|\zeta| = 1$, $0 \leqq \mu \leqq 1$ *the symbol of the operator* $a(U)$.

LEMMA 1.2. *Every nonsingular function* $a(\zeta)$ *can be represented in the form*

$$a(\zeta) = r(\zeta)g(\zeta), \tag{1.1}$$

where $r(\zeta)$ *is a polynomial in integral powers of* ζ, $g(\zeta) \in \Lambda$, *and*

$$\sup_{|\zeta|=1} |g(\zeta) - 1| = 1.$$

PROOF. By virtue of its nonsingularity, the function $a(\zeta)$ can be represented in the form

$$a(\zeta) = |a(\zeta)| \exp(i\varphi(\zeta)),$$

where the real function $\varphi(\zeta)$ is such that the relations

$$|\varphi(\zeta_k) - \varphi(\zeta_k + 0)| < \pi - \delta \qquad (\delta > 0)$$

are fulfilled at the points $\zeta_1, \cdots, \zeta_n$ of discontinuity of $a(\zeta)$. A point of continuity ζ_0 of $a(\zeta)$ is taken as the argument's origin of reference. At it the difference $\varphi(\zeta_0 + 0) - \varphi(\zeta_0)$ is a multiple of 2π, and the function $\varphi(\zeta)$ is continuous at all remaining points.

Let us define real functions $b(\zeta)$ and $c(\zeta)$ by setting

$$\begin{aligned} &b(\zeta_k) = \varphi(\zeta_k), \qquad b(\zeta_k + 0) = \varphi(\zeta_k + 0) \qquad (k = 0, \cdots, n),\\ &c(\zeta_0) = \varphi(\zeta_0), \qquad c(\zeta_0 + 0) = \varphi(\zeta_0 + 0),\\ &c(\zeta_k) = \tfrac{1}{2}(\varphi(\zeta_k) + \varphi(\zeta_k + 0)) \qquad (k = 1, \cdots n), \end{aligned}$$

while these functions are defined by linear interpolation along the arc at the remaining points of the circle. It is easy to see that

$$\sup_{|\zeta|=1} |b(\zeta) - c(\zeta)| < \frac{\pi - \delta}{2}. \tag{1.2}$$

Since the function

$$f(\zeta) = \exp[i(\varphi(\zeta) - b(\zeta) + c(\zeta))]$$

is continuous on Γ, there is a polynomial $r_1(\zeta)$ such that

$$f(\zeta) = r_1(\zeta)(1 + m(\zeta)),$$

where

$$\sup_{|\zeta|=1} |m(\zeta)| < \frac{1}{2}, \qquad |\arg(1 + m(\zeta))| < \frac{\delta}{4}. \tag{1.3}$$

Now let us consider the function

$$F(\zeta) = |a(\zeta)|(1 + m(\zeta)) \exp[i(b(\zeta) - c(\zeta))].$$

Since $\inf |F(\zeta)| = 0$ and according to (1.2) and (1.3) the relation $|\arg F(\zeta)| < \pi/2 - \delta/4$ holds, there is a number $\gamma > 0$ such that $\sup |1 - \gamma F(\zeta)| < 1$.

Let us now set $g(\zeta) = \gamma F(\zeta)$ and $r(\zeta) = \gamma^{-1} r_1(\zeta)$. Since $a(\zeta) = r(\zeta)g(\zeta)$, the lemma is proved.

3. THEOREM 1.1. *Let $a(\zeta) \in \Lambda$. The invertibility of an operator $A = a(V)$ is determined by its symbol $a(\zeta, \mu)$. If the function $a(\zeta)$ is nonsingular, the index of A equals $-\kappa$ dim Coker V, where $\kappa =$ ind $a(\zeta)$.*

PROOF. If $a(\zeta)$ is a nonsingular function, the polynomial $r(\zeta)$ in (1.1) obviously does not vanish on Γ, and ind r = ind $a = \kappa$.

Let the equality

$$r(\zeta) = r_-(\zeta)\zeta^\kappa r_+(\zeta) \tag{1.4}$$

give a factorization of the polynomial $r(\zeta)$ (cf. equality (1.24), Chapter I). Then, by virtue of (1.1) and (1.4),

$$a(\zeta) = r_-(\zeta)\, \zeta^\kappa g(\zeta)\, r_+(\zeta).$$

According to Lemma 1.1, the equality

$$A = r_-(V)g(V)\, V^\kappa r_+(V) \tag{1.5}$$

is valid for $\kappa \geqq 0$, and the equality

$$A = r_-(V)V^{(\kappa)}g(V)r_+(V) \tag{1.6}$$

for $\kappa < 0$.

As we have already noted repeatedly, the operators $r_\pm(V)$ are invertible. The operator $g(V)$ is also invertible. This follows from the relations

$$|g(V) - I| = |P(g(U) - I)P| \leqq \sup_{|\zeta|=1} |g(\zeta) - 1| < 1.$$

The formula for the index obviously follows from (1.5) and (1.6).

Thus the sufficiency of the theorem's conditions is established.

Passing to the proof of the necessity,[4] let us assume that the operator A is invertible on at least one side, and that the curve Γ_a passes through the point $\zeta = 0$. Let us consider first the case when the curve Γ_a is a simple, smooth line in some neighborhood S of the point $\zeta = 0$. We assume that this neighborhood of zero is so small that for any point $\lambda \in S$ the operator $A - \lambda I$ is invertible on the same side as A, while its index equals the index of A. If the points $\lambda_1, \lambda_2 \in S$ and lie on different sides of Γ_a, then ind $(a(\zeta) - \lambda_1) \neq$ ind $(a(\zeta) - \lambda_2)$, and the formula for the index leads to a contradiction.

The general case reduces in the usual way to the one already considered: there is a function $b(\zeta) \in \Lambda$ such that the magnitude sup $|a(\zeta) - b(\zeta)|$ is so small that $b(V)$ is invertible on at least one side, and the curve Γ_b passes through the point $\zeta = 0$ and is a simple, smooth line in some neighborhood of this point. The theorem is proved.

Let us note, moreover, that if the function $a(\zeta)$ is singular, then, as the above argument implies, the operator A is neither a Φ_+- nor a Φ_--operator.

§ 2. Applications to discrete and integral equations

1. As in § 7, Chapter I, let l_2 be the Hilbert space of sequences of complex numbers $\{\xi_j\}_1^\infty$, and V the isometric operator defined in l_2 by the equality

[4] For the sake of simplicity it is here assumed that dim Coker $V < \infty$. The proof of the general case can be reduced to that.

$$V\{\xi_j\}_1^\infty = \{0, \xi_1, \xi_2, \cdots\}. \tag{2.1}$$

If $a(\zeta) \in \Lambda$, the operator $a(V)$,as is easy to see, is defined in l_2 by the matrix

$$\|a_{j-k}\|_{j,k=1}^\infty, \tag{2.2}$$

where the a_j ($j = 0, \pm 1, \cdots$) are the Fourier coefficients of the function $a(\zeta)$.

The following theorem is a corollary of Theorem 1.1.

THEOREM 2.1. *Let the function* $a(\zeta) \in \Lambda$, *and let* $\{a_j\}_{-\infty}^\infty$ *be its Fourier coefficients. The invertibility of the operator A prescribed in* l_2 *by matrix* (2.2) *is determined by its symbol* $a(\zeta, \mu)$.

2. Let H_2 be the Hardy space in the upper half-plane, i.e. the Hilbert space of all functions $\Phi(\lambda)$ of the form

$$\Phi(\lambda) = \int_0^\infty \varphi(t) e^{i\lambda t}\, dt \qquad (\operatorname{Im} \lambda \geqq 0),$$

where $\varphi(t) \in L_2(0, \infty)$. The space H_2 can be regarded as a subspace of $L_2(-\infty, \infty)$ by confining the argument to the real axis $(-\infty < \lambda < \infty)$. Let us denote by P the operator of the orthogonal projection of $L_2(-\infty, \infty)$ onto H_2. Let us consider the isometric operator

$$(V\Phi)(\lambda) = \frac{\lambda - i}{\lambda + i}\Phi(\lambda) \qquad (-\infty < \lambda < \infty) \tag{2.3}$$

in the space H_2.

The operator U defined in $L_2(-\infty, \infty)$ by the equality

$$(Uf)(\lambda) = \frac{\lambda - i}{\lambda + i} f(\lambda) \qquad (-\infty < \lambda < \infty)$$

is a unitary extension of the operator V.

If $a(\zeta) \in \Lambda$, it is easy to verify that the operator $a(V)$ operates in H_2 according to the rule

$$a(V)\Phi = Pa\left(\frac{\lambda - i}{\lambda + i}\right)\Phi(\lambda) \qquad (\Phi(\lambda) \in H_2). \tag{2.4}$$

Let us note that the equation

$$Pa\left(\frac{\lambda - i}{\lambda + i}\right)\Phi(\lambda) = F(\lambda)$$

is equivalent to the integral equation

$$\int_0^\infty k(t - s)\varphi(s)\, ds = f(t) \qquad (0 < t < \infty)$$

considered in $L_2(0, \infty)$, where $k(t)$ is in a certain sense a generalized function whose Fourier transform equals $a((\lambda - i)/(\lambda + i))$.

Application of Theorem 1.1 reduces to the following result.

THEOREM 2.2. *Let the function* $a(\zeta) \in \Lambda$. *The invertibility of the operator* $a(V)$ *prescribed in* H_2 *by equality* (2.4) *is determined by its symbol* $a(\zeta, \mu)$.

§ 3. Projection method for piecewise continuous functions of an isometric operator

In this section it is assumed that the isometric operator V satisfies the condition

$$\dim \operatorname{Coker} V < \infty. \tag{3.1}$$

We need the following lemma.

LEMMA 3.1. *Let* $f(\zeta)$ $(|\zeta| = 1)$ *be an arbitrary continuous function. For any operator* $B \in \mathfrak{A}(\mathfrak{H})$ *possessing the property*

$$V^{(-1)}BV = B, \tag{3.2}$$

the operator $f(V)B - Bf(V)$ *is completely continuous.*

PROOF. It is easy to see that it is sufficient to establish the lemma in the case $f(\zeta) = \zeta^j$, $j = 0, \pm 1, \cdots$. From (3.2) it follows that $V^{(-k)}BV^k = B$ for all $k \geq 0$. Multiplying this equality on the left by V^k and then on the right by $V^{(-k)}$, we obtain the relations

$$(I - P_k)BV^k = V^kB, \qquad V^{(-k)}B(I - P_k) = BV^{(-k)}, \tag{3.3}$$

where $P_k = I - V^kV^{(-k)}$. By virtue of (3.1), the orthoprojections P_k are finite-dimensional.

Relations (3.3) imply that the operators $V^{(j)}B - BV^{(j)}$, $j = 0, \pm 1, \cdots$, are finite-dimensional.

The lemma is proved.

Let us suppose that conditions analogous to those applied in § 1, Chapter III, are fulfilled, i.e. let Ω be some unbounded set of positive numbers and let the orthogonal projections P_τ $(\in \mathfrak{A}(\mathfrak{H});\ \tau \subset \Omega)$ converge strongly to the identity operator as $\tau \to \infty$; and let the operators V and P_τ satisfy the following condition:

$$P_\tau V P_\tau = P_\tau V \qquad (\tau \in \Omega). \tag{3.4}$$

THEOREM 3.1. *Let the function* $a(\zeta) \in \Lambda$. *If*

$$\liminf_{\tau \to \infty} |(P_\tau a(V) P_\tau)^{-1}| < \infty, \tag{3.5}$$

then $a(\zeta)$ *is nonsingular, and* ind $a(\zeta) = 0$.

Under fulfillment of these conditions, $a(V) \in \Pi\{P_\tau, P_\tau\}$.

PROOF. If condition (3.5) is fulfilled, then by virtue of Lemma 1.1, Chapter III, the operator $a(V)$ is invertible, and consequently $a(\zeta)$ is a nonsingular function and ind $a(\zeta) = 0$.

Under fulfillment of the latter conditions, the operator $a(V)$ can be represented in the form

$$a(V) = r_-(V)g(V)r_+(V)$$

by virtue of factorization (1.5).

Let us recall that the operator $V^{(-1)}$ coincides with the adjoint operator V^*. Therefore the relation

$$P_\tau V^{(-1)} P_\tau = V^{(-1)} P_\tau \qquad (\tau \in \Omega) \tag{3.6}$$

follows from (3.4).

Let us take an auxiliary operator $B = r_+(V)g(V)r_-(V)$ and show that it is included in the class $\Pi\{P_\tau, P_\tau\}$. Indeed, the operator $g(V) \in \Pi\{P_\tau, P_\tau\}$ by virtue of proposition 1°, § 3, Chapter II. The equalities

$$P_\tau r_+(V)P_\tau = P_\tau r_+(V), \qquad P_\tau r_-(V)P_\tau = r_-(V)P_\tau$$

and analogous equalities for $r_\pm^{-1}(V)$ follow from conditions (3.4) and (3.6). Therefore

$$(P_\tau r_\pm(V)\, P_\tau)^{-1} = P_\tau r_\pm^{-1}\,(V)\, P_\tau$$

and by virtue of the multiplication theorem (Theorem 2.2, Chapter II), the operator $B \in \Pi\{P_\tau, P_\tau\}$.

The operator $a(V)$ is invertible, and, by virtue of Lemma 3.1, $a(V) - B$ is completely continuous; consequently in accordance with Theorem 3.1, Chapter II, $a(V)$ also belongs to $\Pi\{P_\tau, P_\tau\}$.

§ 4. Reduction method for Toeplitz matrices and their continuous analogues

Let us now apply the results of the preceding section to specific classes of equations.

1. Let an isometric operator V be defined in the space l_2 by (2.1), and let the orthogonal projections P_n $(n = 1, 2, \cdots)$ be defined by the equalities

$$P_n\{\xi_j\}_1^\infty = \{\xi_1, \cdots, \xi_n, 0, 0, \cdots\}.$$

The operators V and P_n, $n = 1, 2, \cdots$, satisfy the conditions of Theorem 3.1. If $a(\zeta) \in \Lambda$, the operators $a(V)$ and $P_n a(V) P_n$ are defined by the matrices

$$a(V) = \|a_{j-k}\|_{j,k=1}^\infty, \qquad P_n a(V) P_n = \|a_{j-k}\|_{j,k=1}^n,$$

where $\{a_j\}_{-\infty}^{\infty}$ are the Fourier coefficients of the function $a(\zeta)$. The following theorem is a concretization of Theorem 3.1.

THEOREM 4.1. *Let the function $a(\zeta)$ belong to Λ, and a_j $(j = 0, \pm 1, \cdots)$ its Fourier coefficients. If the relation*

$$\liminf_{n\to\infty} |A_n^{-1}| < \infty$$

is valid for the truncated matrices $A_n = \|a_{j-k}\|_{j,k=1}^{n}$, then the function $a(\zeta)$ is nonsingular, and ind $a(\zeta) = 0$.

Under fulfillment of the latter conditions the system of equations

$$\sum_{k=1}^{n} a_{j-k}\xi_k = \eta_j \qquad (j = 1, \cdots, n),$$

beginning with some n, has the unique solution $\{\xi_j^{(n)}\}_{j=1}^{n}$, and as $n \to \infty$ the vectors

$$\xi^{(n)} = \{\xi_1^{(n)}, \cdots, \xi_n^{(n)}, 0, 0, \cdots\}$$

converge in the norm of l_2 to a solution of the full system

$$\sum_{k=1}^{\infty} a_{j-k}\xi_k = \eta_j \qquad (j = 1, 2, \cdots),$$

whatever the vector $\{\eta_j\}_1^{\infty} \in l_2$ may be.

2. Now let us establish the analogue of Theorem 4.1 for a system of equations which is infinite on both sides. Let us put into correspondence with each function $a(\zeta) \in \Lambda$ an operator $\tilde{A}$ defined in the Hilbert space $\tilde{l}_2$ by a sequence $\{\xi_j\}_{-\infty}^{\infty}$ of matrices $\|a_{j-k}\|_{j,k=-\infty}^{\infty}$, where, as before, $\{a_j\}_{-\infty}^{\infty}$ are the Fourier coefficients of $a(\zeta)$.

Obviously the operator $\tilde{A}$ is invertible if and only if

$$\inf_{|\zeta|=1} |a(\zeta)| > 0.$$

Let us denote by $\tilde{P}_n$, $n = 1, 2, \cdots$, the orthogonal projection defined in $\tilde{l}_2$ by the equality

$$\tilde{P}_n\{\xi_j\}_{-\infty}^{\infty} = \{\cdots, 0, \xi_{-n}, \cdots, \xi_n, 0, \cdots\}.$$

THEOREM 4.2. *Let the function $a(\zeta) \in \Lambda$, and let $\{a_j\}_{-\infty}^{\infty}$ be its Fourier coefficients. If the relation*

$$\liminf_{n\to\infty} |\tilde{A}_n^{-1}| < \infty$$

is valid for the truncated matrices $\tilde{A}_n = \|a_{j-k}\|_{j,k=-n}^{n}$, then $a(\zeta)$ is nonsingular, and ind $a(\zeta) = 0$.

Under fulfillment of these conditions the system of equations

$$\sum_{k=-n}^{n} a_{j-k}\xi_k = \eta_j \qquad (j = 0, \pm 1, \cdots, \pm n),$$

beginning with some n, has the unique solution $\{\xi_j^{(n)}\}_{j=-n}^{n}$, *and as* $n \to \infty$ *the vectors*

$$\xi^{(n)} = \{\cdots, 0, \xi_{-n}^{(n)}, \cdots, \xi_n^{(n)}, 0, \cdots\}$$

converge in the norm of $\tilde{l}_2$ *to a solution of the full system*

$$\sum_{k=-\infty}^{\infty} a_{j-k}\xi_k = \eta_j \qquad (j = 0, \pm 1, \cdots),$$

whatever the vector $\{\eta_j\}_{-\infty}^{\infty} \in l_2$ *may be.*

This theorem is a corollary of the preceding one and Theorem 2.1, Chapter II, since the matrix $\tilde{A}_n$ coincides with the matrix A_{2n+1} of Theorem 4.1.

3. Let us now consider in the space H_2 the isometric operator V defined by equality (2.3) and the projections $\tilde{P}_\tau = FP_\tau F^{-1}$ $(0 < \tau < \infty)$, where F is the Fourier transform and the projections P_τ are defined in $L_2(0, \infty)$ by the equalities

$$(P_\tau \varphi)(t) = \begin{cases} \varphi(t), & 0 < t < \tau, \\ 0, & \tau < t < \infty. \end{cases}$$

It is not difficult to verify that the operator V and the system of projections $\tilde{P}_\tau$ satisfy the conditions of Theorem 3.1, which reduces to a criterion for the applicability of the projection method $(\tilde{P}_\tau, \tilde{P}_\tau)$ to the operator $a(V)$ defined by equality (2.4).

Let us consider the system of functions

$$\Psi_0(\lambda) = \frac{1}{\sqrt{\pi}} \frac{1}{\lambda + i}$$

$$\Psi_k(\lambda) = (V^k \Psi_0)(\lambda) = \frac{1}{\sqrt{\pi}}\left(\frac{\lambda - i}{\lambda + i}\right)^k \frac{1}{\lambda + i} \qquad (k = 1, 2, \cdots) \tag{4.1}$$

and the operators

$$\mathscr{P}_n \Phi = \sum_{k=0}^{n-1} (\Phi, \Psi_k)\Psi_k \qquad (\Phi(\lambda) \in H_2;\ n = 1, 2, \cdots).$$

System (4.1) is an orthonormal basis in H_2; consequently the operators $\mathscr{P}_n$ are orthogonal projections. They converge strongly to the identity operator as $n \to \infty$.

It is easy to see that the operator V and the projections $\mathscr{P}_n$ satisfy the conditions of Theorem 3.1, while the equation

$$\mathscr{P}_n a(V) \mathscr{P}_n \Phi = \mathscr{P}_n \mathscr{B} \qquad (\mathscr{B}(\lambda) \in H_2)$$

is equivalent to the algebraic system

$$\sum_{k=0}^{n-1} a_{j-k}\xi_k = b_j \qquad (j = 0,\cdots, n-1), \tag{4.2}$$

where $\{a_j\}_{j=-\infty}^{\infty}$ and $\{b_j\}_0^\infty$ are the Fourier coefficients of the functions $a(\zeta)$ and $\mathscr{B}(i(1+\zeta)/(1-\zeta))$.

The following theorem is yet another realization of Theorem 3.1.

THEOREM 4.3. *Let the function* $a(\zeta) \in \Lambda$ *and the operator* $a(V)$ *be defined by* (2.4). *If*

$$\liminf_{n\to\infty} \left|(\mathscr{P}_n a(V)\mathscr{P}_n)^{-1}\right| < \infty,$$

then $a(\zeta)$ *is nonsingular, and* ind $a(\zeta) = 0$.

Under fulfillment of these conditions the truncated system of equations (4.2), *beginning with some n, has the unique solution* $\{\xi_j^{(n)}\}_0^{n-1}$, *and as* $n \to \infty$ *the functions*

$$\Phi_n(\lambda) = \sum_{j=0}^{n-1} \xi_j^{(n)} \Psi_j(\lambda)$$

converge in the norm of H_2 *to a solution of the full equation*

$$a(V)\Phi = \mathscr{B},$$

whatever the function $\mathscr{B}(\lambda) \in H_2$ *may be.*

We shall not formulate here the corresponding theorem for integral equations on the entire axis.

§ 5. Toeplitz matrices composed of the Fourier coefficients of measurable functions

Let us denote by M the algebra of all bounded measurable functions on the unit circle Γ, and by M_R the set of all functions $a(\zeta) \in M$ representable in the form

$$a(\zeta) = r(\zeta)(1 + m(\zeta)), \tag{5.1}$$

where $r(\zeta)$ is a polynomial which does not vanish on Γ and the function $m(\zeta) \in M$ satisfies the condition

$$\sup_{|\zeta|=1} |m(\zeta)| < 1.$$

By virtue of Lemma 1.2, every nonsingular function $a(\zeta)$ of Λ belongs to M_R also.

If a function $a(\zeta)$ of M_R admits two representations of the form (5.1)

$$a(\zeta) = r_1(\zeta)(1 + m_1(\zeta)) = r_2(\zeta)(1 + m_2(\zeta)),$$

then, as is easy to see,

$$\operatorname{ind} r_1(\zeta) = \operatorname{ind} r_2(\zeta).$$

This permits the concept of *index* to be introduced for any function $a(\zeta) \in M_R$:

$$\operatorname{ind} a(\zeta) = \operatorname{ind} r(\zeta),$$

where $r(\zeta)$ is the polynomial in representation (5.1) of the function $a(\zeta)$.

Let us put into correspondence with each function $a(\zeta) \in M$ (and in particular with each function $a(\zeta) \in M_R$) a bounded linear operator $A = a(V)$ generated in l_2 by the matrix $\|a_{j-k}\|_{j,k=1}^{\infty}$, where the a_j, $j = 0, \pm 1, \cdots$, are the Fourier coefficients of the functions $a(\zeta)$.

This definition of the operator $a(V)$ is a natural extension to a broader class of functions[5] of the concept introduced in § 2 of a function of the isometric operator V.

THEOREM 5.1. *Let $a(\zeta) \in M_R$. Then the invertibility of the operator $a(V)$ is consistent with the index of the function $a(\zeta)$.*

Let us note that Lemma 1.1 remains valid for the functions $b(\zeta)$, $a(\zeta) \in M_R$, and the proof of Theorem 5.1 is carried out exactly as the proof of sufficiency in Theorem 1.1.

THEOREM 5.2. *Let the function $a(\zeta) \in M_R$, and let $\{a_j\}_{-\infty}^{\infty}$ be its Fourier coefficients. If* ind $a(\zeta) = 0$, *then the truncated system of equations*

$$\sum_{k=1}^{n} a_{j-k}\xi_k = \eta_j \qquad (j = 1, \cdots, n),$$

beginning with some n, has the unique solution $\{\xi_j^{(n)}\}_{j=1}^{n}$, and as $n \to \infty$ the vectors

$$\xi^{(n)} = \{\xi_1^{(n)}, \cdots, \xi_n^{(n)}, 0, 0, \cdots\}$$

converge in the norm of l_2 to a solution of the full system

$$\sum_{k=1}^{\infty} a_{j-k}\xi_k = \eta_j \qquad (j = 1, 2, \cdots),$$

whatever the vector $\{\eta_j\}_1^{\infty} \in l_2$ may be.

The proof of this theorem is carried out exactly as the proof of the second assertion of Theorem 3.1.

THEOREM 5.3. *Let the function $a(\zeta) \in M_R$, and let $\{a_j\}_{-\infty}^{\infty}$ be its Fourier coefficients. If* ind $a(\zeta) = 0$, *then the truncated system of equations*

[5] The question of the proper definition of the symbol remains open in this case.

$$\sum_{k=-n}^{n} a_{j-k}\xi_k = \eta_j \qquad (j = 0, \pm 1, \cdots, \pm n),$$

beginning with some n, has the unique solution $\{\xi_j^{(n)}\}_{j=-n}^{n}$, *and as* $n \to \infty$ *the vectors*

$$\xi^{(n)} = \{\cdots, 0, \xi_{-n}^{(n)}, \cdots, \xi_n^{(n)}, 0, \cdots\}$$

converge in the norm of $\tilde{l}_2$ *to a solution of the full system*

$$\sum_{k=-\infty}^{\infty} a_{j-k}\xi_k = \eta_j \qquad (j = 0, \pm 1, \cdots),$$

whatever the vector $\{\eta_j\}_{-\infty}^{\infty} \in \tilde{l}_2$ *may be.*

This theorem is a corollary of the preceding one and Theorem 2.1, Chapter II.

§ 6. Finite difference equations

1. As in § 10, Chapter I, let an isometric operator V be defined in Hilbert space $L_2(0, \infty)$ by the equality

$$(Vf)(t) = \begin{cases} f(t-1), & t > 1, \\ 0, & t < 1. \end{cases}$$

In this case the operator $a(V)$ can be defined for any bounded measurable function $a(\zeta)$ ($|\zeta| = 1$). Let us consider a unitary extension U of the operator V which operates in $\tilde{L}_2 = L_2(-\infty, \infty)$ according to the rule $(Uf)(t) = f(t-1)$, and the unitary operator $U_1 = FUF^{-1}$, where F is the Fourier transform.

It is easy to see that U_1 is an operator of multiplication by the function $e^{i\lambda}$. The operator $a(U_1)$ is defined as an operator of multiplication by the function $a(e^{i\lambda})$, and the operator $a(U)$, by the equality $a(U) = F^{-1}a(U_1)F$.

If $a(\zeta) \in \Lambda$ (i.e. $a(\zeta)$ is a piecewise continuous function), then, as is easy to see, this definition of $a(U)$ coincides with the definition in § 1.

If $a(\zeta)$ expands into an absolutely convergent Fourier series, then obviously

$$(a(U)f)(t) = \sum_{j=-\infty}^{\infty} a_j f(t-j) \qquad (-\infty < t < \infty), \tag{6.1}$$

where a_j ($j = 0, \pm 1, \cdots$) are the Fourier coefficients of the function $a(\zeta)$. Let us note that the series (6.1) converges in the norm of L_2 on any finite interval for any $a(\zeta) \in L_2(\Gamma)$. Indeed, this follows from the estimates

$$\int_{-N}^{N} \Big| \sum_{j=n}^{m} a_j f(t-j) \Big|^2 dt \leq \int_{-N}^{N} \sum_{j=n}^{m} |a_j|^2 \sum_{j=n}^{m} |f(t-j)|^2 dt$$
$$= \sum_{j=n}^{m} |a_j|^2 \sum_{r=-N}^{N-1} \sum_{j=n}^{m} \int_{r}^{r+1} |f(t-j)|^2 dt \leq 2N \sum_{j=n}^{m} |a_j|^2 |f|_{\tilde{L}_2}^2$$

and analogous estimates for the sum

$$\sum_{j=-m}^{-n} a_j f(t-j).$$

It is easy to verify that equality (6.1) remains valid as well for any bounded measurable function $a(\zeta)$.

Now, in accordance with the definition in § 1, let us define the operator $a(V)$ for any bounded measurable function $a(\zeta)$ by setting

$$(a(V)f)(t) = \sum_{j=-\infty}^{\infty} a_j f(t-j) \qquad (0 < t < \infty), \tag{6.2}$$

where, as previously, we set $f(t-j) = 0$ for $t < j$.

Here the following theorem is a realization of Theorem 1.1 (concerning invertibility).

THEOREM 6.1. *Let the function $a(\zeta) \in \Lambda$, and let $a_j, j = 0, \pm 1, \cdots$, be its Fourier coefficients. The invertibility of the operator $a(V)$ prescribed in $L_2(0, \infty)$ by equality (6.2) is determined by its symbol $a(\zeta, \mu)$. If $a(\zeta)$ is nonsingular, then*

$$\begin{aligned} \dim \operatorname{Coker} a(V) &= \infty \qquad (\operatorname{ind} a(\zeta) > 0). \\ \dim \operatorname{Ker} a(V) &= \infty \qquad (\operatorname{ind} a(\zeta) < 0). \end{aligned}$$

2. Now let us consider the projection method for finite difference equations. Let us denote by P_n, $n = 1, 2, \cdots$, the projector defined in $L_2(0, \infty)$ by the equality

$$(P_n f)(t) = \begin{cases} f(t), & t < n, \\ 0, & t > n. \end{cases}$$

THEOREM 6.2. *Let the function $a(\zeta) \in \Lambda$, and let a_j, $j = 0, \pm 1, \cdots$, be its Fourier coefficients. If*

$$\liminf_{n \to \infty} \left| (P_n a(V) P)_n^{-1} \right| < \infty, \tag{6.3}$$

then $a(\zeta)$ is nonsingular, and ind $a(\zeta) = 0$.

Under fulfillment of these conditions the equation

$$\sum_{t-n<j<t} a_j \varphi(t-j) = f(t) \qquad (0 < t < n), \tag{6.4}$$

beginning with some n, has the unique solution $\varphi_n(t) \in L_2(0, n)$, and as $n \to \infty$ the functions

$$\tilde{\varphi}_n(t) = \begin{cases} \varphi_n(t), & t < n, \\ 0, & t > n \end{cases} \tag{6.5}$$

converge in the norm of $L_2(0, \infty)$ to a solution of the equation

$$\sum_{j=-\infty}^{\infty} a_j \varphi(t-j) = f(t) \qquad (0 < t < \infty), \tag{6.6}$$

whatever the function $f(t) \in L_2(0, \infty)$ *may be.*

This theorem is not a realization of Theorem 3.1, since the operator V does not satisfy condition (3.1) for the finite-dimensionality of the cokernel. We need the following simple lemma to prove this theorem. Let us denote by $(L_p)_n$ the space of n-dimensional vector-functions $f = \{f_1, \cdots, f_n\}$ with components in $L_p(0, 1)$, on which a norm is defined by the equality

$$|f| = \left(\sum_{j=1}^{n} |f_j|^p\right)^{1/p}.$$

LEMMA 6.1. *Let A be an operator generated in the n-dimensional space l_p by the matrix $\|a_{jk}\|_1^{\infty}$ with complex elements, and $\tilde{A}$ the operator generated by the same matrix in the space $(L_p)_n$. Then $|A|_p = |\tilde{A}|_p$.*

PROOF. The inequality $|A|_p \leqq |\tilde{A}|_p$ is trivial. Let $f(x)$ be a continuous vector-function of $(L_p)_n$. We have

$$|(\tilde{A}f)(x)|_p^p = \sum_{j=1}^{n} \left| \sum_{k=1}^{n} a_{jk} f_k(x) \right|^p \leqq |A|_p^p \sum_{k=1}^{n} |f_k(x)|^p$$

for each fixed x. Integrating with respect to x from 0 to 1, we obtain

$$|\tilde{A}f|_p^p \leqq |A|_p^p |f|_p^p,$$

which implies $|\tilde{A}|_p \leqq |A|_p$.

The lemma is proved.

PROOF OF THEOREM 6.2. If condition (6.3) is fulfilled, then by virtue of Lemma 1.1, Chapter III, the operator $a(V)$ is invertible; consequently $a(\zeta)$ is a nonsingular function, and ind $a(\zeta) = 0$. To prove the second assertion of the theorem, let us observe that equation (6.4), or, what is the same thing, the equation $P_n a(V) P_n \varphi = P_n f$, is equivalent to the system

$$\sum_{k=1}^{n} a_{j-k} \varphi_k(t) = f_j(t) \qquad (j = 1, \cdots, n),$$

where the functions $\varphi_j(t)$ and $f_j(t)$ are defined by

$$\varphi_j(t) = \varphi(t + j - 1), \qquad f_j(t) = f(t + j - 1) \qquad (0 \leqq t \leqq 1; j = 1, \cdots, n).$$

They belong to the space $L_2(0, 1)$. By virtue of this equivalence and Lemma 6.1,

$$|(P_n a(V) P_n)^{-1}| = |A_n^{-1}|,$$

where $A_n = \|a_{j-k}\|_{j,k=1}^{n}$, and the norm A_n^{-1} is understood to be in n-dimensional l_2. Now applying Theorem 4.1 of the present chapter and Theorem 2.1 of Chapter II, we obtain $a(V) \in \Pi\{P_n, P_n\}$.

The theorem is proved.

REMARK. Theorem 6.2 and its proof remain valid in the space $L_p(0, \infty)$ ($p \geqq 1$) if the function $a(\zeta)$ belongs to the corresponding algebra $\mathfrak{R}(\zeta)$ (cf. § 2, Chapter I).

The following theorem is a corollary of Theorems 5.1 and 5.2 and Lemma 6.1.

THEOREM 6.3. *Let the function* $a(\zeta) \in M_R$, *and let* a_j, $j = 0, \pm 1, \cdots$, *be its Fourier coefficients. The operator* $a(V)$ *is invertible, only left invertible, or only right invertible, depending on whether the number* ind $a(\zeta)$ *is equal to zero, positive, or negative.*

If ind $a(\zeta) = 0$, *equation* (6.4), *beginning with some* n, *has the unique solution* $\varphi_n(t) \in L_2(0, n)$, *and as* $n \to \infty$ *the functions* (6.5) *converge in the norm of* $L_2(0, \infty)$ *to a solution of equation* (6.6), *whatever the function* $f(t) \in L_2(0, \infty)$ *may be.*

CHAPTER V
PAIR EQUATIONS

Continuous and discrete pair equations in convolutions and their transposes are examined in this chapter. In particular, singular integral equations on the circle and boundary value problem equations for functions of a complex variable fall into this class.

The basic propositions of this chapter are obtained as corollaries of the results of the first and fourth chapters and some simple general propositions set forth in the first section.

§ 1. General propositions

Let $\mathfrak{M} \subset \mathfrak{A}(\mathfrak{B})$ be an algebra of linear operators acting in the Banach space $\mathfrak{B}$, $P \in \mathfrak{A}(\mathfrak{B})$ a projection in the same space, and $Q = I - P$.

Let A and B be two operators of $\mathfrak{M}$. An operator of the form $PA + QB$ is called a *pair operator*. An operator of the form $AP + BQ$ is called the *transpose* of the pair.

If no ambiguity arises, we shall speak of both of these operators as *pair* operators.

Some simple relationships between the pair operators $AP + BQ$ and $PA + QB$ (A, $B \in \mathfrak{M}$) and operators of the form PXP or QXQ ($X \in \mathfrak{M}$), considered respectively in the subspaces $P\mathfrak{B}$ and $Q\mathfrak{B}$, are explained in this section. We shall assume that the algebra $\mathfrak{M}$ possesses the following two properties:

1) *The set of operators invertible in* $\mathfrak{M}$ *forms a complete* (*in the sense of the operator norm*) *set in* $\mathfrak{M}$.

2) *If at least one of the restrictions* $PXP\,|P\mathfrak{B}$ *or* $QXQ|Q\mathfrak{B}$ ($X \in \mathfrak{M}$) *is invertible on either side, the operator* X *is invertible in* $\mathfrak{M}$.

EXAMPLE. Let us regard as $\mathfrak{M}$ the algebra of all operators defined in the space $\tilde{l}_2$ by the Toeplitz matrices $A = \|a_{j-k}\|_{j,k=-\infty}^{\infty}$ composed of the Fourier coefficients of continuous functions. Let us take as P, as usual, the projection annihilating all components with negative indices.

Obviously the algebra $\mathfrak{M}$ possesses property 1), since $|A| = \max|a(\zeta)|$, and the operators of $\mathfrak{M}$ which correspond to functions which nowhere vanish form an everywhere dense set in $\mathfrak{M}$.

The algebra $\mathfrak{M}$ also possesses property 2), since the unilateral invertibility of the discrete Wiener-Hopf operator for A, i.e. the operator $PAP|P\bar{l}_2$, implies the nonvanishing of the corresponding function $a(\zeta)$ and therefore also the invertibility of A itself.

It is clear that analogous examples can be constructed with integral, finite-difference, and other operators.

The following theorem is valid for such *specific* algebras.

THEOREM 1.1. *Let the operators $A, B \in \mathfrak{M}$. For the operator $AP + BQ$ to be invertible on at least one side, it is necessary and sufficient that the operators A and B be invertible in $\mathfrak{M}$ and the operator $PB^{-1}AP|P\mathfrak{B}$ be invertible on the same side as $AP + BQ$.*

The theorem remains valid under replacement of the operator $PB^{-1}AP|P\mathfrak{B}$ by the operator $QA^{-1}BQ|Q\mathfrak{B}$.

PROOF. Let A and B be invertible operators of $\mathfrak{M}$; then the operator $AP + BQ$ can be represented in the form

$$AP + BQ = B(PB^{-1}AP + Q)(I + QB^{-1}AP). \tag{1.1}$$

Since $(QB^{-1}AP)^2 = 0$, the operator $I + QB^{-1}AP$ is invertible, where

$$(I + QB^{-1}AP)^{-1} = I - QB^{-1}AP.$$

Also valid is the equality

$$AP + BQ = A(P + QA^{-1}BQ)(I + PA^{-1}BQ), \tag{1.2}$$

in which the operator $I + PA^{-1}BQ$ is invertible:

$$(I + PA^{-1}BQ)^{-1} = I - PA^{-1}BQ.$$

Representations (1.1) and (1.2) for the operator $AP + BQ$ imply that if at least one of the operators $AP + BQ$, $PB^{-1}AP|P\mathfrak{B}$ or $QA^{-1}BQ|Q\mathfrak{B}$ is invertible on either side, the remaining operators are invertible on the same side.

To complete the theorem's proof it remains for us to show that the invertibility on either side of the operator $AP + BQ$ implies the bilateral invertibility of the operators A and B ($\in \mathfrak{M}$).

Let us assume first that only one of the operators A or B is invertible. For the sake of definiteness, let that one be B. Then equality (1.1) holds. This equality implies that the operator $PB^{-1}AP|P\mathfrak{B}$ is invertible on the same side as the operator $AP + BQ$. According to condition 2), this implies that the operator $B^{-1}A$ is invertible, and therefore A is also invertible, which contradicts the hypothesis.

Now let us assume that both operators A and B are noninvertible, while $AP + BQ$ is invertible on some side. By virtue of property 1), B can be approximated with any degree of accuracy by invertible operators of $\mathfrak{M}$. An invertible operator B' of $\mathfrak{M}$ can be chosen so close (in norm) to B that the operator $AP + B'Q$ will be invertible on the same side as $AP + BQ$. By virtue of what has been proved, this implies that A is invertible. The latter contradicts the hypothesis.

The theorem is proved.

Let us note that if A and B are invertible, formulas (1.1) and (1.2) imply the equalities

$$\begin{aligned}\dim \operatorname{Ker}(AP + BQ) &= \dim \operatorname{Ker}(PB^{-1}AP|P\mathfrak{B}) \\ &= \dim \operatorname{Ker}(QA^{-1}BQ|Q\mathfrak{B}) \end{aligned} \tag{1.3}$$

and

$$\begin{aligned}\dim \operatorname{Coker}(AP + BQ) &= \dim \operatorname{Coker}(PB^{-1}AP|P\mathfrak{B}) \\ &= \dim \operatorname{Coker}(QA^{-1}BQ|Q\mathfrak{B}). \end{aligned} \tag{1.4}$$

From (1.1) and (1.2) also follows the rule for constructing an operator which is inverse to $AP + BQ$ on one side if the corresponding inverse of at least one of the operators $PB^{-1}AP|P\mathfrak{B}$ or $QA^{-1}BQ|Q\mathfrak{B}$ is known.

Let us note, moreover, that if condition 2) on the algebra $\mathfrak{M}$ is replaced by the stronger condition

2′) *if even one of the operators $PXP|P\mathfrak{B}$ or $QXQ|Q\mathfrak{B}$ ($X \in \mathfrak{M}$) is a Φ_+- or Φ_--operator, the operator X is invertible,*[1)]

then the following sharpening of Theorem 1.1 is valid.

THEOREM 1.2. *For the operator $AP + BQ$ ($A, B \in \mathfrak{M}$) to be a Φ_+-operator it is necessary and sufficient that the operators A and B be invertible in $\mathfrak{M}$ and that the operator $PB^{-1}AP|P\mathfrak{B}$ ($QA^{-1}BQ|Q\mathfrak{B}$) be a Φ_+-operator.*

Equalities (1.3) and (1.4) remain valid in this case.

The proof of this theorem is essentially no different from the proof of the preceding one.

The theorem obtained from Theorem 1.2 by substituting Φ_- for Φ_+ is also valid.

In concluding the section let us note that everything said concerning operators of the form $AP + BQ$ can be reformulated in a natural way for operators of the form $PA + QB$. In this connection the equalities

$$PA + QB = (I + PAB^{-1}Q)(PAB^{-1}P + Q)B$$

and

$$PA + QB = (I + QBA^{-1}P)(P + QBA^{-1}Q)A$$

assume the role of the basic equalities (1.1) and (1.2).

§ 2. Unilateral invertibility tests for one class of pair operators

Let the invertible operator $U \in \mathfrak{A}(\mathfrak{B})$ and the projection $P \in \mathfrak{A}(\mathfrak{B})$ satisfy the conditions of § 3.2, Chapter I, i.e.:

---

1)The examples cited above possess this stronger property.

1) the spectral radii of the operators U and U^{-1} equal unity, and

2) $$UP = PUP,\ UP \neq PU,\ PU^{-1} = PU^{-1}P. \tag{2.1}$$

As in subsection 2, let us introduce the algebras $\mathfrak{K}(U)$, $\mathfrak{R}(U)$ and $\mathfrak{R}^{\pm}(U)$. The equalities

$$A_+P = PA_+P, \qquad A_-Q = QA_-Q \qquad (Q = I - P,\ A_\pm \in \mathfrak{R}_\pm(U))$$

are easily derived from relations (2.1). In the sequel we shall be interested in pair operators of the form $A_1P + A_2Q$ (or $PA_1 + QA_2$), where A_1 and A_2 range over the algebra $\mathfrak{R}(U)$. If $A_1, A_2 \in \mathfrak{R}(U)$ and $A_\pm \in \mathfrak{R}_\pm(U)$, then the following multiplication rule holds:

$$(A_1P + A_2Q)(A_+P + A_-Q) = A_1A_+P + A_2A_-Q. \tag{2.2}$$

Equality (2.2) will be applied often to expand an operator on the right into the factors on the left.

Let us note in addition the analogous rule

$$PA_1A_- + QA_2A_+ = (PA_- + QA_+)(PA_1 + QA_2). \tag{2.2'}$$

In the sequel the algebra $\mathfrak{R}(U)$ will assume the role of the algebra $\mathfrak{M}$. Indeed, it was shown in the second scheme of § 3, Chapter I, that every algebra $\mathfrak{R}(U)$ possesses property 2) and even the strengthened property 2′) (cf. § 11, Chapter I). Let us show that $\mathfrak{R}(U)$ possesses property 1) also, i.e. that the set of invertible operators of $\mathfrak{R}(U)$ is dense in it. As a matter of fact, every operator $X \in \mathfrak{R}(U)$ can be approximated with any degree of accuracy by operators of $\mathfrak{K}(U)$, i.e. by operators of the form

$$Y = U^l \prod_{k=1}^{m} (U - \alpha_k I),$$

where l is an integer, m is a natural number, and the α_k are complex numbers. If none of the numbers α_k lies on the unit circle, the operator Y is the desired invertible operator. If, however, some of the numbers α_k do lie on the unit circle, we obtain the desired operator by replacing the α_k with sufficiently close numbers not on the circle.

1. The following theorem is basic in this section.

THEOREM 2.1. *Let $U \in \mathfrak{A}$ be an invertible operator satisfying conditions* 1) *and* 2) *of this section.*

For the operator $AP + BQ$, where $A, B \in \mathfrak{R}(U)$, to be invertible on at least one side, it is necessary and sufficient that

$$A(\zeta) \neq 0, \quad B(\zeta) \neq 0 \qquad (|\zeta| = 1). \tag{2.3}$$

If conditions(2.3) *are fulfilled, the invertibility of the operator $AP + BQ$ is consistent with the index of the function $A(\zeta)/B(\zeta)$:*

$$\kappa = \text{ind}\,[A(\zeta)/B(\zeta)] = (1/2\pi)\,[\arg\,(A(e^{i\varphi})/B(e^{i\varphi}))]_{\varphi=0}^{2\pi}.$$

The theorem remains valid under replacement of the operators of the form $AP + BQ$ by operators of the form $PA + QB$.

By analogy with the definition of the symbol for the individual operator A, let us call the pair of individual symbols $(A(\zeta), B(\zeta))$ *the symbol of the pair operators* $AP + BQ$ or $PA + QB$ $(A, B \in \mathfrak{R}(U))$. Let us take as the *index* of the pair $(A(\zeta), B(\zeta))$ the difference between the indices of $A(\zeta)$ and $B(\zeta)$, or, what is the same thing, the index of the quotient $A(\zeta)/B(\zeta)$:

$$\kappa = \text{ind}\,(A(\zeta),B(\zeta)) = \text{ind}\,A\,(\zeta) - \text{ind}\,B\,(\zeta) = \text{ind}\,[A(\zeta)/B(\zeta)].$$

Naturally it is assumed in this connection that neither of the functions $A(\zeta)$ or $B(\zeta)$ vanishes anywhere on the unit circle $|\zeta| = 1$.

By analogy with the case of an individual operator, let us also say that the invertibility of a pair operator *is determined by its symbol* if and only if both components of its symbol nowhere vanish and under fulfillment of this condition the invertibility of the pair operator is consistent with the index of its symbol.

We can now reformulate Theorem 2.1 as follows.

Let $U \in \mathfrak{A}$ be an invertible operator satisfying conditions 1) *and* 2) *of this section. Then the invertibility of the pair operator $AP + BQ$ with the coefficients A and B of $\mathfrak{R}(U)$ is determined by its symbol.*

Proof. If we take $\mathfrak{M} = \mathfrak{R}(U)$, the theorem just formulated will follow immediately from Theorem 1.1 of the present chapter and Theorem 3.2, Chapter I.

Nevertheless, we shall go into a direct proof here of the sufficiency of the theorem's conditions, since formulas emerge in this connection which are suitable for inverting operators of the form $AP + BQ$.

Let conditions (2.3) be fulfilled. Then the operators A and B are invertible. Let us denote by R some operator of $\mathfrak{K}(U)$ for which

$$|CR - I| < \varepsilon \quad (< 1), \tag{2.4}$$

where $C = B^{-1}A$. From (2.4) it follows that

$$\max_{|\zeta|=1} |C(\zeta)R(\zeta) - 1| < 1,$$

and therefore $R(\zeta) \neq 0$ $(|\zeta| = 1)$ and $\text{ind}\,R(\zeta) = -\kappa$.

The polynomial $R(\zeta)$ can be factored as $R(\zeta) = R_-(\zeta)\zeta^{-\kappa}R_+(\zeta)$ (cf. equality (1.24), Chapter I), the respective operators $R_\pm$ $(\in \mathfrak{K}^\pm(U))$ being invertible and $R_\pm^{-1} \in \mathfrak{R}^\pm(U)$.

Taking into account the equality $C = R_-^{-1}U^\kappa R_+^{-1}(I + S)$, where $S = CR - I$ $(|S| < \varepsilon)$, we obtain

$$AP + BQ = B\,[R_-^{-1}U^\kappa R_+^{-1}(I + S)P + Q].$$

Utilizing (2.2), we obtain from this the expansion into the factors

$$AP + BQ = BR_-^{-1}(U^\kappa(I + S)P + Q)(R_+^{-1}P + R_-Q). \tag{2.5}$$

In turn the operator $U^\kappa(I + S)P + Q$ for $\kappa \geqq 0$ can be expanded in the same way into the factors

$$U^\kappa(I + S)P + Q = ((I + S)P + Q)(U^\kappa P + Q) = (I + SP)(U^\kappa P + Q).$$

Thus for $\kappa \geqq 0$

$$AP + BQ = BR_-^{-1}(I + SP)(U^\kappa P + Q)(R_+^{-1}P + R_-Q). \tag{2.6}$$

Obviously, for sufficiently small ε all of the factors in (2.6) are invertible, with the exception perhaps of the factor $U^\kappa P + Q$. Therefore the operator $AP + BQ$ is invertible on some side if and only if the operator $U^\kappa P + Q$ is invertible on the same side. For $\kappa = 0$ the operator $U^\kappa P + Q$ equals I; hence the operator $AP + BQ$ is invertible. For $\kappa > 0$ the operator $U^\kappa P + Q$ is left invertible only. Indeed, by virtue of the same rule (2.2),

$$(U^{-\kappa}P + Q)(U^\kappa P + Q) = I.$$

Let us multiply these operators in a different order:

$$(U^\kappa P + Q)(U^{-\kappa}P + Q) = U^\kappa PU^{-\kappa} + QU^{-\kappa}P + Q.$$

If it is assumed that $U^\kappa PU^{-\kappa} + QU^\kappa P + Q = I$, then $U^\kappa PU^{-\kappa} = P$ or $PU^\kappa P = PU^\kappa$, which contradicts condition (2.1).

Now let $\kappa < 0$. By virtue of what has been proved, the auxiliary operator $U^{-\kappa}AP + BQ$ is invertible. Factoring it in accordance with (2.2), we obtain

$$U^{-\kappa}AP + BQ = (AP + BQ)(U^{-\kappa}P + Q),$$

which implies that the operator $AP + BQ$ is right invertible. If it were simply invertible, the operator $U^{-\kappa}P + Q$ would also be invertible, which contradicts what has already been proved. Thus the sufficiency of the theorem's conditions is established.

Equality (2.6) permits a left inverse operator to be constructed for $AP + BQ$ for $\kappa \geqq 0$.

In addition, it is easy to obtain the equality

$$AP + BQ = BR_{-\kappa}^{-1}(U^\kappa P + Q)[I + (U^{-\kappa}P + Q)U^\kappa SP](R_+^{-1}P + R_-Q) \tag{2.7}$$

from equality (2.5) for $\kappa < 0$.

By means of this equality a right inverse of $AP + BQ$ can be constructed for $\kappa < 0$.

The direct proof of sufficiency is completed.

Although formulas (2.6) and (2.7) just obtained permit unilateral inverses to be constructed for $AP + BQ$, such a method proves to be inefficient. More effective formulas can be constructed if the operator $B^{-1}A$ admits the factorization

$$B^{-1}A = C_-U^\kappa C_+ \qquad (C_\pm \in \mathfrak{R}^\pm(U),\ C_\pm^{-1} \in \mathfrak{R}^\pm(U)).$$

(We are assuming that A and B are invertible.)

Then the appropriate inverse of the operator $AP + BQ$ is given (for any κ) by

$$(AP + BQ)^{(-1)} = (C_+^{-1}P + C_-Q)(U^{-\kappa}P + Q)C_-^{-1}B^{-1}.$$

The formula

$$(PA + QB)^{(-1)} = B^{-1}C_+^{-1}(PU^{-\kappa} + Q)(PC_-^{-1} + QC_+)$$

also holds under the same conditions.

Factorization of the operator $B^{-1}A$ is easily obtained if the function $A(\zeta)/B(\zeta)$ admits factorization and the algebra $\mathfrak{R}(U)$ does not have a radical.

Let us note in addition that if even one of the conditions (2.3) is not fulfilled, then, by virtue of Theorem 1.2 of the present chapter and Theorem 3.2 of Chapter I, the operator $AP + BQ$ $(PA + QB)$ is neither a Φ_+- nor a Φ_--operator.

2. Let us explain how the index of a pair operator is canceled. Let us consider first the case of the operator $C = AP + BQ$. The method of canceling the index here depends on the sign of $\kappa = \kappa_1 - \kappa_2$, where $\kappa_1 = \operatorname{ind} A(\zeta)$ and $\kappa_2 = \operatorname{ind} B(\zeta)$.

For $\kappa_1 \geqq \kappa_2$ multiplication rule (2.2) gives the equality

$$C = (AU^{-\kappa}P + BQ)(U^\kappa P + Q) = (A_1P + B_1Q)(U^\kappa P + Q), \tag{2.8}$$

where $A_1 = AU^{-\kappa}$ and $B_1 = B$.

It is easy to see that $\operatorname{ind} A_1(\zeta) = \operatorname{ind} B_1(\zeta) = \kappa_2$, which implies that the operator $A_1P + B_1Q$ is invertible. Equality (2.8) permits us to apply proposition 5°, § 1, Chapter I. Let us observe that the indices of *both* operators A and B can be canceled by means of the expansion

$$C = U^{\kappa_2}(AU^{-\kappa_1}P + BU^{-\kappa_2}Q)(U^{\kappa_1-\kappa_2}P + Q). \tag{2.9}$$

Such cancellation is convenient for the application of projection methods. In the case $\kappa_1 \leqq \kappa_2$ cancellation is carried out by means of the equality

$$C(U^{-\kappa}P + Q) = AU^{-\kappa}P + BQ = A_1P + B_1Q, \tag{2.10}$$

where $A_1 = AU^{-\kappa}$, $B_1 = B$ and

$$U^{-\kappa_2}C(U^{\kappa_2-\kappa_1}P + Q) = AU^{-\kappa_1}P + BU^{-\kappa_2}Q. \tag{2.11}$$

These equalities permit proposition 8° of § 1, Chapter I, to be applied to the case under consideration.

Let us formulate the respective concretizations of propositions 5° and 6°. Let us introduce the projection

$$\mathscr{P} = I - (U^{|\kappa|}P + Q)(U^{-|\kappa|}P + Q). \tag{2.12}$$

For $\kappa > 0$ the equation

$$(AP + BQ)\varphi = f \tag{2.13}$$

is solvable if and only if the condition

$$\mathscr{P}\psi = 0 \tag{2.14}$$

is fulfilled for the solution ψ *of the equation*

$$(AU^{-\kappa}P + BQ)\psi = f. \tag{2.15}$$

Under fulfillment of this condition, a solution φ *of equation* (2.13) *is given by the formula*

$$\varphi = (U^{-\kappa}P + Q)\psi. \tag{2.16}$$

For $\kappa < 0$ *the vector*

$$\varphi = (U^{-\kappa}P + Q)\psi, \tag{2.17}$$

where ψ *is a solution of the equation*

$$(AU^{-\kappa}P + BQ)\psi = f, \tag{2.18}$$

is always one of the solutions of the original equation. The general solution of the homogeneous equation

$$(AP + BQ)\chi = 0 \tag{2.19}$$

is given by the formula

$$\chi = [I - (U^{-\kappa}P + Q)(AU^{-\kappa}P + BQ)^{-1}(AP + BQ)]\mathscr{P}h, \tag{2.20}$$

where h is an arbitrary vector.

Formula (2.20) *is equivalent to the following*:

$$\chi = (U^{-\kappa}P + Q)g + \mathscr{P}h, \tag{2.21}$$

where h is an arbitrary vector and g is the solution of the equation

$$(AU^{-\kappa}P + BQ)g = -(AP + BQ)\mathscr{P}h. \tag{2.22}$$

Analogous results can be formulated for the operator $C = PA + QB$. In this case the relations

$$(PU^{-\kappa} + Q)C = PAU^{-\kappa} + QB \tag{2.23}$$

for $\kappa > 0$ and

$$C = (PU^{\kappa} + Q)(PAU^{-\kappa} + QB) \tag{2.24}$$

for $\kappa < 0$ are utilized.

Thus propositions 6° and 7° of § 1, Chapter I, can be applied in this case.

3. The results of § 1, together with the results of § 9, Chapter I, permit the following theorem to be obtained concerning algebras generated by a one-parameter group of operators.

Let $U(t)$ $(-\infty < t < \infty)$ be a strongly continuous group of isometric operators in the Banach space $\mathfrak{B}$, and let P be a projection with the following properties:

$$PU_tP = U_tP, \quad U_tP \neq PU_t \qquad (0 < t < \infty),$$
$$PU_tP = PU_t \quad (-\infty < t < 0).$$

Let us denote by H the generating operator of the group U_t, and by $\mathfrak{W}$ ($\neq 0$) the number $\dim [(H - iI) P\mathfrak{B}/P\mathfrak{B}]$.

THEOREM 2.4. *For the operator $A_1P + A_2Q$ to be invertible on some side, where*

$$A_j = \alpha_j I + \int_{-\infty}^{\infty} k_j(t) U_t \, dt, \qquad Q = I - P,$$

the α_j are complex numbers, and $k_j(t) \in L_1(-\infty, \infty)$ $(j = 1, 2)$, it is necessary and sufficient that the conditions

$$\mathscr{A}_j(\lambda) \overset{\text{def}}{=} \alpha_j + \int_{-\infty}^{\infty} k_j(t) e^{i\lambda t} \, dt \neq 0 \qquad (-\infty < \lambda < \infty;\ i = 1,2)$$

be fulfilled.

If these conditions are fulfilled, the invertibility of the operator $A_1P + A_2Q$ is consistent with the index

$$\kappa = \operatorname{ind} (A_1(\lambda)/A_2(\lambda)).$$

If $\kappa > 0$,

$$\dim \operatorname{Coker} (A_1P + A_2Q) = \mathfrak{M}\kappa,$$

and if $\kappa < 0$,

$$\dim \operatorname{Ker} (A_1P + A_2Q) = -\mathfrak{M}\kappa.$$

We shall not here go into the formulations of the analogues of the other propositions of this section.

§ 3. General theorems concerning discrete systems

Let us denote by U an operator defined in the Hilbert space $\bar{l}_2$ of bilateral sequences $\{\xi_j\}_{-\infty}^{\infty}$ by the equality $U\{\xi_j\} = \{\xi_{j-1}\}$. The operator U is invertible, and its inverse operates according to the rule $U^{-1}\{\xi_j\} = \{\xi_{j+1}\}$.

The operator U is unitary, and its spectrum consists of all the points of the unit circle. The algebra $\mathfrak{R}(U)$ does not contain a radical and is isomorphic to the algebra of all functions continuous on the unit circle.

Let us denote by P the projection defined in $\bar{l}_2$ by the equalities

$$P\{\xi_j\} = \{\eta_j\}, \qquad \eta_j = \begin{cases} \xi_j, & j = 0,1,\cdots, \\ 0, & j = -1, -2,\cdots. \end{cases}$$

The operators U, U^{-1} and P obviously satisfy conditions (2.1).

Let $a(\zeta)$ be an arbitrary function continuous on the unit circle and a_j, $j = 0, \pm 1,\cdots$, its Fourier coefficients. As in § 7, Chapter I, it can be shown that an operator A of $\mathfrak{R}(U)$ with symbol $a(\zeta)$ is defined by the matrix $\|a_{j-k}\|_{j,k=-\infty}^{\infty}$.

Applying Theorem 2.1 to the case under consideration, we obtain the following proposition.

THEOREM 3.1. *Let $a(\zeta)$ and $b(\zeta)$ be arbitrary functions continuous on the unit circle, and $\{a_j\}_{-\infty}^{\infty}$ and $\{b_j\}_{-\infty}^{\infty}$ their Fourier coefficients. The invertibility of the pair operator $AP + BQ$ prescribed in $\tilde{l}_2$ by the system*

$$\sum_{k=0}^{\infty} a_{j-k}\xi_k + \sum_{k=-\infty}^{-1} b_{j-k}\xi_k = \eta_j \qquad (j = 0, \pm 1, \cdots) \tag{3.1}$$

is determined by its symbol $(a(\zeta), b(\zeta))$. In other words, for this operator to be invertible on at least one side, it is necessary and sufficient that

$$a(\zeta) \neq 0, \quad b(\zeta) \neq 0 \qquad (|\zeta| = 1). \tag{3.2}$$

If conditions (3.2) *are fulfilled, the operator $AP + BQ$ is invertible, only left invertible, or only right invertible, depending on whether the number*

$$\kappa = (1/2\pi)\,[\arg a(e^{i\varphi})/b(e^{i\varphi})]_{\varphi=0}^{2\pi}$$

is equal to zero, positive, or negative.

The theorem is also valid for the operator $PA + QB$ defined by the system

$$\begin{aligned} &\sum_{k=-\infty}^{\infty} a_{j-k}\xi_k = \eta_j \qquad (j = 0, 1, \cdots), \\ &\sum_{k=-\infty}^{\infty} b_{j-k}\xi_k = \eta_j \qquad (j = -1, -2, \cdots). \end{aligned} \tag{3.3}$$

System (3.3) is the pair, while system (3.1) is the *transpose* of the pair.

If conditions (3.2) are fulfilled, the index of system (3.1) or system (3.3) can be canceled. The following two theorems are concretizations of the propositions of § 2.2.

THEOREM 3.2. *Let conditions* (3.2) *be fulfilled for functions $a(\zeta)$ and $b(\zeta)$ which are continuous on the unit circle.*

If $\kappa > 0$, system (3.1) *is solvable if and only if*

$$\psi_0 = \psi_1 = \cdots = \psi_{\kappa-1} = 0, \tag{3.4}$$

where $\psi = \{\psi_j\}_{-\infty}^{\infty}$ is the solution of the system

$$\sum_{k=-\infty}^{-1} b_{j-k}\psi_k + \sum_{k=0}^{\infty} a_{j-k+\kappa}\psi_k = \eta_j \qquad (j = 0, \pm 1, \cdots). \tag{3.5}$$

Under fulfillment of conditions (3.4) *the vector $\xi = \{\xi_j\}_{-\infty}^{\infty}$,*

$$\xi_j = \begin{cases} \psi_j, & j = -1, -2, \cdots, \\ \psi_{j+\kappa}, & j = 0, 1, \cdots, \end{cases}$$

is a solution of system (3.1).

If, however, $\kappa < 0$, *the vector* $\xi = \{\xi_j\}_{-\infty}^{\infty}$,

$$\xi_j = \begin{cases} \psi_j, & j = -1, -2, \cdots, \\ 0, & j = -0, 1, \cdots, -\kappa - 1, \\ \psi_{j+\kappa}, & j = -\kappa, -\kappa + 1, \cdots, \end{cases}$$

where $\{\psi_j\}_{-\infty}^{\infty}$ *is the solution of system* (3.5), *is one of the solutions of system* (3.1), *while the general solution of the homogeneous system*

$$\sum_{k=0}^{\infty} a_{j-k}\xi_k + \sum_{k=-\infty}^{-1} b_{j-k}\xi_k = 0$$

is defined by the equalities

$$\xi = \{\xi_j\}_{-\infty}^{\infty}, \qquad \xi_j = \begin{cases} \psi_j, & j = -1, -2, \cdots, \\ \alpha_j, & j = 0, 1, \cdots, -\kappa - 1, \\ \psi_{j+\kappa}, & j = -\kappa, -\kappa + 1, \cdots, \end{cases}$$

where the α_j, $j = 0, 1, \cdots, -\kappa - 1$, *are arbitrary numbers, and* $\{\psi_j\}_{-\infty}^{\infty}$ *is the solution of the system*

$$\sum_{k=-\infty}^{-1} b_{j-k}\psi_k + \sum_{k=0}^{\infty} a_{j-k+\kappa}\psi_k = -\sum_{k=0}^{-\kappa-1} a_{j-k}x_k \qquad (j = 0, \pm 1, \cdots).$$

THEOREM 3.3. *Let conditions* (3.2) *be fulfilled for functions* $a(\zeta)$ *and* $b(\zeta)$ *which are continuous on the unit circle.*

If $\kappa > 0$, *the system*

$$\sum_{k=-\infty}^{\infty} b_{j-k}\xi_k = \eta_j \qquad (j = -1, -2, \cdots),$$

$$\sum_{k=-\infty}^{\infty} a_{j-k}\xi_k = \eta_j \qquad (j = \kappa, \kappa + 1, \cdots)$$

has the unique solution $\xi = \{\xi_j\}_{-\infty}^{\infty}$, *which is a solution of system* (3.3) *if and only if the conditions*

$$\sum_{k=-\infty}^{\infty} a_{j-k}\xi_k = \eta_j \qquad (j = 0, 1, \cdots, \kappa - 1)$$

are fulfilled. If, however, $\kappa < 0$, *the unique solution of the system*

$$\sum_{k=-\infty}^{\infty} b_{j-k}\xi_k = \eta_j \qquad (j = -1, -2, \cdots),$$

$$\sum_{k=-\infty}^{\infty} a_{j-k+\kappa}\xi_k = 0 \qquad (j = 0, 1, \cdots, -\kappa - 1),$$

$$\sum_{k=-\infty}^{\infty} a_{j-k+\kappa}\xi_k = \eta_{j+\kappa} \qquad (j = -\kappa, -\kappa + 1, \cdots)$$

is one of the solutions of system (3.3), *while the general solution of the homogeneous system*

$$\sum_{k=-\infty}^{\infty} a_{j-k}\xi_k = 0 \qquad (j = 0, 1, \cdots),$$

$$\sum_{k=-\infty}^{\infty} b_{j-k}\xi_k = 0 \qquad (j = -1, -2, \cdots)$$

is the totality of all solutions of the system

$$\sum_{k=-\infty}^{\infty} b_{j+k}\xi_k = 0 \qquad (j = -1, -2, \cdots),$$

$$\sum_{\kappa=-\infty}^{\infty} a_{j-k+\kappa}\xi_k = \alpha_j \qquad (j = 0, 1, \cdots, -\kappa - 1),$$

$$\sum_{k=-\infty}^{\infty} a_{j-k+\kappa}\xi = 0 \qquad (j = -\kappa, -\kappa + 1, \cdots),$$

where the $\alpha_j, j = -0, 1, \cdots, -\kappa - 1$, *are arbitrary numbers.*

Let us note that the theorems of this section are also valid for the spaces $\tilde{l}_p$ $(1 \leqq p \leqq \infty)$ and $\tilde{c}_0$ only if the continuous functions $a(\zeta)$ and $b(\zeta)$ belong to the algebra $\mathfrak{R}(\zeta)$. Let us recall that for any of these spaces the algebra $\mathfrak{R}(\zeta)$ contains all absolutely convergent Fourier series.

§ 4. Pair integral equations

Let $\tilde{E}$ be one of the Banach spaces introduced in § 8, Chapter I, and let the operator U be defined in $\tilde{E}$ by the equality

$$(Uf)(t) = f(t) - 2\int_{-\infty}^{t} e^{s-t}f(s)\,ds.$$

U is invertible in $\tilde{E}$, and its inverse is defined by

$$(U^{-1}f)(t) = f(t) - 2\int_{t}^{\infty} e^{t-s}f(s)\,ds.$$

As already noted, the spectral radii of the operators U and U^{-1} equal unity, where the operators U and U^{-1} and the projection P defined in $\tilde{E}$ by the equality

$$(Pf)(t) = \begin{cases} f(t), & 0 < t < \infty, \\ 0, & -\infty < t < 0, \end{cases} \tag{4.1}$$

satisfy condition (2.1).

The algebra $\mathfrak{R}(U)$ contains all operators of the form

$$(Af)(t) = f(t) - \int_{-\infty}^{\infty} k(t-s)f(s)\,ds,$$

where $k(t) \in \tilde{L}_1$ and the function $A(\zeta)$ $(|\zeta| = 1)$ corresponding to the operator A is defined by the equalities

$$A(\zeta) = \mathscr{A}\left(i\frac{1+\xi}{1-\xi}\right), \qquad \mathscr{A}(\lambda) = 1 - \int_{-\infty}^{\infty} k(t)\, e^{i\lambda t}\, dt$$
$$(-\infty \leqq \lambda \leqq \infty)$$

(cf. § 8, Chapter I).

If operators A and B have the form

$$Af = f(t) - \int_{-\infty}^{\infty} k_1(t-s)f(s)\, ds, \qquad Bf = f(t) - \int_{-\infty}^{\infty} k_2(t-s)f(s)\, ds,$$

where $k_j(t) \in \tilde{L}_1$ $(j = 1, 2)$, then

$$(AP + BQ)f = f(t) - \int_{0}^{\infty} k_1(t-s)f(s)ds - \int_{-\infty}^{0} k_2(t-s)f(s)\, ds, \tag{4.2}$$

and

$$(PA + QB)f = \begin{cases} f(t) - \int_{-\infty}^{\infty} k_1(t-s)f(s)ds, & 0 < t < \infty, \\ f(t) - \int_{-\infty}^{\infty} k_2(t-s)f(s)ds, & -\infty < t < 0. \end{cases} \tag{4.3}$$

Now let us apply Theorem 2.1 to the case under consideration.

THEOREM 4.1. *Let* $k_j(t) \in \tilde{L}_1$ $(j = 1, 2)$. *The invertibility of the operator* $AP + BQ$ $(PA + QB)$ *prescribed in* $\tilde{E}$ *by equality* (4.2) (*equality* (4.3)) *is determined by its symbol* $(\mathscr{A}(\lambda), \mathscr{B}(\lambda))$, *where*

$$\mathscr{A}(\lambda) = 1 - \int_{-\infty}^{\infty} k_1(t)e^{i\lambda t}\, dt, \qquad \mathscr{B}(\lambda) = 1 - \int_{-\infty}^{\infty} k_2(t)e^{i\lambda t}\, dt,$$
$$\text{ind}\,[\mathscr{A}(\lambda)/\mathscr{B}(\lambda)] = (1/2\pi)\,[\arg \mathscr{A}(\lambda)/\mathscr{B}(\lambda)]_{\lambda=-\infty}^{\infty}. \tag{4.4}$$

This theorem is also a corollary of Theorem 2.4. In this connection a group of operators U_t must be defined in $\tilde{E}$ by the equality

$$(U_t f)(s) = f(s-t),$$

and a projection P, by equailty (4.1).

The equation $(PA + QB)\varphi = f$ is a *pair*, and the equation $(AP + BQ)\varphi = f$ is the transpose of the pair. Under fulfillment of conditions (4.4), as in the discrete case, it is possible to cancel the index of these equations and to obtain continuous analogues of Theorem 3.2 and 3.3.

§ 5. General unilateral invertibility test for pair operators (case of discontinuous functions)

1. Let U be a unitary operator operating in Hilbert space $\mathfrak{H}$, and P an orthoprojection related to U, as previously, by the conditions

$$PUP = UP, \qquad UP \neq PU, \qquad PU^{-1}P = PU^{-1}$$

The following theorem is valid.

THEOREM 5.1. *Let $a(\zeta)$ and $b(\zeta)$ ($|\zeta| = 1$) be piecewise continuous and left continuous functions. The operator $c(U)P + b(U)Q$ ($Pa(U) + Qb(U)$) is invertible on some side if and only if the condition*

$$a(\zeta)\, b(\zeta + 0)\mu + a(\zeta + 0)b(\zeta)\,(1 - \mu) \neq 0$$
$$(|\zeta| = 1,\ 0 \leqq \mu \leqq 1) \tag{5.1}$$

is fulfilled.

If this condition is fulfilled, the invertibility of the operator $a(U)P + b(U)Q$ ($Pa(U) + Qb(U)$) is consistent with the number $\kappa =$ ind $(a(\zeta)/b(\zeta))$.[2)]

PROOF. If $c(\zeta) \in \Lambda$, the operator $Pc(U)P\,|P\mathfrak{H}$, in accordance with the definition of § 1, Chapter IV, equals $c(V)$, the function $c(\zeta)$ of the isometric operator $V = U|P\mathfrak{H}$.

Let us denote by $\mathfrak{M}$ the algebra of all operators $c(U)$, where $c(\zeta) \in \Lambda$. The operator $c(U) \in \mathfrak{M}$ is invertible if and only if the conditions $c(\zeta) \neq 0$ and $c(\zeta + 0) \neq 0$ ($|\zeta| = 1$) are fulfilled. Moreover, $[c(U)]^{-1} = c^{-1}(U)$. By taking into account in addition that

$$|c(U)| = \sup_{|\zeta|=1} |c(\zeta)|,$$

it is easily concluded that the set of operators invertible in $\mathfrak{M}$ is dense in $\mathfrak{M}$ in the operator norm.

Theorem 1.1 of the present chapter and Theorem 1.1 of Chapter IV imply that the operator $A(U)P + B(U)Q$ is invertible on some side if and only if the following conditions are fulfilled:

$$a(\zeta) \neq 0, \quad a(\zeta + 0) \neq 0, \quad b(\zeta) \neq 0, \quad b(\zeta + 0) \neq 0 \qquad (|\zeta| = 1)$$

and

$$\frac{a(\zeta)}{b(\zeta)}\,\mu + \frac{a(\zeta + 0)}{b(\zeta + 0)}\,(1 - \mu) \neq 0 \qquad (|\zeta| = 1, 0 \leqq \mu \leqq 1).$$

Obviously these conditions are equivalent to (5.1).

The last assertion of the theorem follows from Theorem 1.1, Chapter IV, because the operator $a(U)P + b(U)Q$ is invertible on the same side as $Pb^{-1}(U)a(U)P|P\mathfrak{H}$.

The theorem is proved.

---

2) For the definition of index in the case of a piecewise continuous function, cf. § 1, Chapter IV. The reader has presumably already taken note of the fact that we are not using the symbol concept here. This is explained by the fact that the regular symbol concept is more complex here. That is, the matrix

$$\begin{Vmatrix} a(\zeta)\,(1 - \mu) + a(\zeta + 0)\mu & \sqrt{\mu(1 - \mu)}\,(a(\zeta + 0) - a(\zeta)) \\ \sqrt{\mu(1 - \mu)}\,(b(\zeta + 0) - b(\zeta)) & b(\zeta)\mu + b(\zeta + 0)\,(1 - \mu) \end{Vmatrix}$$

is the symbol here. (Cf. I. C. Gohberg and N. Ja. Krupnik [5].)

The remaining theorems of §2 also carry over to the case of discontinuous functions under consideration.

2. Our results apply in a natural way to various discrete and continuous pair Winer-Hopf equations. As an example, let us cite the theorem for a discrete pair system of Wiener-Hopf equations.

THEOREM 5.2. *Let* $a(\zeta)$ *and* $b(\zeta)$ $(|\zeta| = 1)$ *be piecewise continuous and left continuous functions, and let* a_j *and* b_j, $j = 0, \pm 1, \cdots$, *be their Fourier coefficients. For an operator defined in the space* $\bar{l}_2$ *by the pair system of equalities*

$$\sum_{k=-\infty}^{\infty} a_{j-k}\xi_k = \eta_j \qquad (j = 0, 1, \cdots),$$

$$\sum_{k=-\infty}^{\infty} b_{j-k}\xi_k = \eta_j \qquad (j = -1, -2, \cdots)$$

to be invertible on some side, it is necessary and sufficient that condition (5.1) *be fulfilled. If condition* (5.1) *is fulfilled, the invertibility of this operator is consistent with the number* $\kappa = \text{ind}\,(a(\zeta)/b(\zeta))$.

The theorem just formulated follows from the preceding one if the operators U and P are defined in $\bar{l}_2$ the same way as in § 3.

§ 6. Singular integral equations and boundary value problems

1. In the space $L_p(\Gamma)$ $(1 < p < \infty)$, where Γ is the unit circle, let us consider the singular integral equation

$$a(\zeta)\,\varphi(\zeta) + \frac{b(\zeta)}{\pi i}\int_\Gamma \frac{\varphi(z)}{z - \zeta}\,dz = f(\zeta) \qquad (|\zeta| = 1), \tag{6.1}$$

where $a(\zeta)$ and $b(\zeta)$ are arbitrary functions continuous on Γ.

As is well known, the singular operator

$$(S\varphi)(\zeta) = \frac{1}{\pi i}\int_\Gamma \frac{\varphi(z)}{z - \zeta}\,dz$$

is bounded in the space $L_p(\Gamma)$, and $S^2 = I$. Now let us show that equation (6.1) is comprised in the scheme presented in § 2. Let us denote by U the operator defined in $L_p(\Gamma)$ by the equality

$$(Uf)(\zeta) = \zeta f(\zeta).$$

The operator U is obviously bounded and invertible in the space $L_p(\Gamma)$, and the spectra of U and U^{-1} coincide with the unit circle. The operator $R \in \mathfrak{R}(U)$ is obviously defined by the equality $(Rf)(\zeta) = R(\zeta)f(\zeta)$, where $R(\zeta)$ is the polynomial corresponding to R.

It is easy to see that

$$|R| = \max_{|\zeta|=1} |R(\zeta)|. \tag{6.2}$$

It is not difficult to infer from relation (6.2) that the function algebra $\mathfrak{R}(\zeta)$ coincides with the algebra of *all* continuous functions of Γ and the operator $A \in \mathfrak{R}(U)$ is defined by the equality $(Af)(\zeta) = A(\zeta)f(\zeta)$, where $A(\zeta) \in \mathfrak{R}(\zeta)$ is the function corresponding to the operator A.

As is well known, the projection $P = \frac{1}{2}(I + S)$ projects the space $L_p(\Gamma)$ onto the subspace H_p consisting of all functions $\varphi(\zeta) \in L_p(\Gamma)$ which have Fourier coefficients with negative indices equal to zero. Functions of H_p admit holomorphic continuations into the interior of the unit disk.

It is not difficult to verify that the operators U, U^{-1} and P satisfy the conditions of §2. Equation (6.1) can be written in the form

$$c(\zeta)(P\varphi)(\zeta) + d(\zeta)(Q\varphi)(\zeta) = f(\zeta) \qquad (|\zeta| = 1), \tag{6.3}$$

where

$$c(\zeta) = a(\zeta) + b(\zeta), \qquad d(\zeta) = a(\zeta) - b(\zeta), \qquad Q = I - P.$$

Corresponding to the general definition, the pair of functions $(c(\zeta), d(\zeta))$ is called the *symbol* of the singular integral operator (6.1) or (6.3).

Let us note that equation (6.3) is the equation of the function theoretic boundary value problem written in the form

$$c(\zeta)\Phi_+(\zeta) + d(\zeta)\Phi_-(\zeta) = f(\zeta) \qquad (|\zeta| = 1),$$

where $\Phi_+(\zeta) \in H_p$ and $\Phi_-(\zeta)$ belongs to the subspace consisting of all functions $\varphi(\zeta) \in L_p(\Gamma)$ which have Fourier coefficients with nonnegative indices equal to zero.

Since the operators of multiplication by the functions $c(\zeta)$ and $d(\zeta)$ belong to the algebra $\mathfrak{R}(U)$, all the results of § 2 are applicable to equation (6.3). In particular, Theorem 2.1 implies the following well-known theorem in the theory of singular integral equations.

THEOREM 6.1. *Let $a(\zeta)$ and $b(\zeta)$ be arbitrary functions continuous on Γ. The invertibility of the operator A prescribed in the space $L_p(\Gamma)$ by the equality*

$$(A\varphi)(\zeta) = a(\zeta)\,\varphi(\zeta) + \frac{b(\zeta)}{\pi i}\int_\Gamma \frac{\varphi(z)}{z-\zeta}\,dz \qquad (|\zeta| = 1) \tag{6.4}$$

is determined by its symbol $(c(\zeta), d(\zeta))$.

2. Let the functions $a(\zeta), b(\zeta) \in \Lambda$. In the space $L_p(\Gamma)$ let us consider the operator A defined by (6.4). Theorem 5.1 is applicable to an operator defined by equality (6.4). The following theorem is a corollary of it.

THEOREM 6.2. *Let the functions $a(\zeta)$, $b(\zeta) \in \Lambda$. For an operator A defined in $L_2(\Gamma)$ by* (6.4) *to be invertible on at least one side, it is necessary and sufficient that*

$$[a(\zeta) + b(\zeta)]\,[a(\zeta + 0) - b(\zeta + 0)]\mu + [a(\zeta + 0) + b(\zeta + 0)]\,[a(\zeta) - b(\zeta)]\,(1 - \mu) \neq 0$$
$$(|\zeta| = 1,\ 0 \leqq \mu \leqq 1).$$

Under fulfillment of this condition the invertibility of A is consistent with the number

$$\operatorname{ind} \frac{a(\zeta) + b(\zeta)}{a(\zeta) - b(\zeta)}.$$

§ 7. Finite difference pair equations

Equations of the form

$$\left.\begin{aligned} \sum_{j=-\infty}^{\infty} a_j f(t - j) &= g(t) \qquad (0 < t < \infty), \\ \sum_{j=-\infty}^{\infty} b_j f(t - j) &= g(t) \qquad (-\infty < t < 0) \end{aligned}\right\} \tag{7.1}$$

or of the form

$$\sum_{j<t} a_j f(t - j) + \sum_{j>t} b_j f(t - j) = g(t) \qquad (-\infty < t < \infty), \tag{7.2}$$

where the a_j and $b_j, j = 0, \pm 1, \cdots$, are given numbers, $g(t) \in L_2(-\infty, \infty)$ is a known function, and $f(t) \in L_2(-\infty, \infty)$ is the function sought, are considered in this section.

Let us consider in $L_2(-\infty, \infty)$ a unitary operator U defined by the equality

$$(Uf)(t) = f(t - 1),$$

and the orthoprojection p:

$$(Pf)(t) = \begin{cases} f(t), & 0 < t < \infty, \\ 0, & -\infty < t < 0. \end{cases}$$

Let a_j and b_j be the Fourier coefficients of the bounded measurable functions $a(\zeta)$ and $b(\zeta)$. Then, just as in § 6, Chapter IV, it can be shown that the series appearing in equalities (7.1) and (7.2) converge in the norm of the space L_2 on some finite interval, the equalities (7.1) and (7.2) defining respectively the operators $Pa(U) + Qb(U)$ and $a(U)P + b(U)Q$ in $L_2(-\infty, \infty)$.

Since all the necessary properties are fulfilled for the operators U and P, Theorem 5.1, which implies the following theorem, is applicable to equations (7.1) and (7.2).

THEOREM 7.1. *Let $a(\zeta)$ and $b(\zeta)$ be left continuous and piecewise continuous functions, and a_j and b_j their Fourier coefficients.*

For an operator A defined in $L_2(-\infty, \infty)$ *by* (7.1) *or* (7.2) *to be invertible on at least one side, it is necessary and sufficient that the condition*

$$a(\zeta)b(\zeta + 0)\mu + a(\zeta + 0)b(\zeta)(1 - \mu) \neq 0$$
$$(|\zeta| = 1,\ 0 \leqq \mu \leqq 1)$$

be fulfilled. If this condition is fulfilled, the invertibility of A is consistent with the number $\operatorname{ind}(a(\zeta)/b(\zeta))$.

CHAPTER VI

PROJECTION METHODS FOR SOLVING PAIR EQUATIONS

Some projection methods for pair convolution equations (discrete and continuous) are set forth in the present chapter.

As in the preceding chapters, the basic results are established first in the abstract case and then applied to various specific classes of equations.

§ 1. Projection method for approximate inversion of one class of pair operators

Throughout this entire section we shall assume that the operators $U, U^{-1}, P \in \mathfrak{A}(\mathfrak{B})$ satisfy, as in § 2 of Chapter V, the following conditions:

1) The spectral radii of the operators U and U^{-1} equal unity.

2) $UP = PUP$, $UP \neq PU$ and $PU^{-1} = PU^{-1}P$.

Let us assume in addition that

3) $\dim \operatorname{Coker} (U|P\mathfrak{B}) < \infty$.

The following lemma holds under conditions 1), 2) and 3).

LEMMA 1.1. *If the operator F belongs to $\mathfrak{R}(U)$, the operator $PF - FP$ is completely continuous. The operators PFQ and QFP are also completely continuous.*

Here, as usual, $Q = I - P$.

PROOF. It is obviously sufficient to show that the operator $PF - FP$ is completely continuous.

For $j > 0$ the operator

$$PU^j - U^jP = (P - U^jPU^{-j})U^j$$

is finite-dimensional, since the operator $P - U^jPU^{-j}$ is a projection projecting the space $\mathfrak{B}$ onto some complement $\mathfrak{N}_j$ of the subspace $U^jP\mathfrak{B}$ up to the subspace $P\mathfrak{B}$. By virtue of condition 3), this complement is finite-dimensional.

The direct sum decomposition of the subspaces

$$U^{-1}Q\mathfrak{B} \dotplus U^{-1}P\mathfrak{B} = \mathfrak{B}, \quad P\mathfrak{B} = UP\mathfrak{B} \dotplus \mathfrak{N}_1,$$

where $\mathfrak{N}_1$ is finite-dimensional, implies that $\dim \operatorname{Coker} (U^{-1}|Q\mathfrak{B}) < \infty$. This implies that for $j < 0$ the operator

$$PU^j - U^jP = -(Q - U^jQU^{-j})U^j$$

is finite-dimensional, since the projection $Q - U^jQU^{-j}$ projects $\mathfrak{B}$ onto some complement of the subspace $U^jQ\mathfrak{B}$ up to the subspace $Q\mathfrak{B}$. From what has been proved it follows that every operator of the form $PR - RP$, where $R \in \mathfrak{R}(U)$, is finite-dimensional. Now let F be an arbitrary operator of $\mathfrak{R}(U)$, and let the sequence $R_n \in \mathfrak{R}(U)$, $n = 1, 2, \cdots$, converge uniformly to F. By what was proved above, the operators $T_n = PR_n - R_nP$ are finite-dimensional. By passing to the limit in the last equality we obtain the complete continuity of the operator $PF - FP$.

The lemma is proved.

Let Ω be some unbounded set of positive numbers, and $P_\tau \in \mathfrak{A}(\mathfrak{B})$ $(\tau \in \Omega)$ an arbitrary family of projections which converges strongly to the identity operator as $\tau \to \infty$.

Let us suppose that the operators U, U^{-1}, P and the family P_τ are related also by the following two conditions:

4) The projections P_τ $(\tau \in \Omega)$ and P are commutative.

5) The subspaces Im $P_\tau P$, Im $Q_\tau P$, Im $P_\tau Q$ and Im $Q_\tau Q$ $(\tau \in \Omega)$ are invariant relative to the operators PU^{-1}, UP, QU and $U^{-1}Q$, respectively.

Here, naturally, $Q_\tau = I - P_\tau$.

The following theorem is valid under fulfillment of conditions 1)—5).

THEOREM 1.1. *Let the operators A and B of $\mathfrak{R}(U)$ be such that*

$$A(\zeta) \neq 0, \quad B(\zeta) \neq 0 \quad (|\zeta| = 1), \qquad \text{ind } A(\zeta) = \text{ind } B(\zeta) = 0.$$

Then $AP + BQ \in \Pi\{P_\tau, P_\tau\}$ and $PA + QB \in \Pi\{P_\tau, P_\tau\}$.

PROOF. The operator $AP + BQ$ is representable in the form

$$AP + BQ = PAP + QBQ + T, \tag{1.1}$$

where the operator $T = QAP + PBQ$ is completely continuous by virtue of Lemma 1.1. Let us denote by A_1 and $P_{1\tau}$ the restrictions of the operators PAP and P_τ to the subspace $P\mathfrak{B}$, and by A_2 and $P_{2\tau}$ the restrictions of the operators QBQ and P_τ to the subspace $Q\mathfrak{B}$.

By virtue of the remark to Theorem 1.1 of Chapter III, the operator A_j belongs to $\Pi\{P_{j\tau}, P_{j\tau}\}$ $(j = 1, 2)$, and by virtue of the proposition concerning the block applicability of the projection method (proposition 2°, § 2, Chapter II), the operator $PAP + QBQ$ belongs to $\Pi\{P_\tau, P_\tau\}$.

Since the operator $AP + BQ$ is invertible, by virtue of equality (1.1) and the theorem concerning perturbation by a completely continuous operator (Theorem 3.1, Chapter II) it belongs to the class $\Pi\{P_\tau, P_\tau\}$ The theorem is proved analogously for the operator $PA + QB$.

THEOREM 1.2. *Let the operators* $A, B \in \mathfrak{R}(U)$, *let the operators* $P_\tau(AP + BQ)P_\tau - P_\tau$ $(\tau \in \Omega)$ *be completely continuous, and let at least one of the following two conditions be fulfilled*:

1) *As* $\tau \to \infty$, *the operators* P_τ^* *converge strongly to the identity operator in* $\mathfrak{B}^*$.

2) *The space* $\mathfrak{B}$ *and the operators* A, B, P *and* P_τ *possess preadjoints*, $^*P_\tau$ *converging strongly to the identity operator in* $^*\mathfrak{B}$ *as* $\tau \to \infty$.

If

$$\liminf_{\tau\to\infty} \left|[P_\tau(AP + BQ]P_\tau^{-1}\right| < \infty, \tag{1.2}$$

then

$$A(\zeta) \neq 0, \quad B(\zeta) \neq 0 \quad (|\zeta| = 1), \qquad \operatorname{ind} A(\zeta) = \operatorname{ind} B(\zeta) = 0.$$

PROOF. [1]The lemmas concerning invertibility of a strong limit (Lemmas 1.1 and 1.2, Chapter III) imply that the operator $AP + BQ$ is invertible. Therefore by virtue of Theorem 2.1, Chapter V, the relations

$$A(\zeta) \neq 0, \qquad B(\zeta) \neq 0 \qquad (|\zeta| = 1),$$
$$\operatorname{ind} A(\zeta) = \operatorname{ind} B(\zeta) \ (= \kappa)$$

are fulfilled.

It remains to show that $\kappa = 0$.

The invertibility of the operator $AP + BQ$ and condition (1.2) imply, by virtue of Theorem 2.1, Chapter II, that $AP + BQ \in \Pi\{P_{\tau_n}, P_{\tau_n}\}$, where τ_n $(\tau_n \in \Omega, n = 1,2,\cdots)$ is an unboundedly increasing sequence of numbers.

Let us consider the operator

$$S = (PAU^{-\kappa}P + QBU^{-\kappa}Q)U^{\kappa}.$$

By virtue of Theorem 3.2, Chapter I, S is invertible. In addition, the operator $S - (AP + BQ)$ is completely continuous. This follows from Lemma 1.1. Therefore, by virtue of the theorem on perturbation by a completely continuous operator (Theorem 3.1, Chapter II), $S \in \Pi\{P_{\tau_n}, P_{\tau_n}\}$.

For the sake of definiteness, let $\kappa \geqq 0$ (the proof is analogous in the case $\kappa \leqq 0$). Then the space $PP_{\tau_n}\mathfrak{B}$ $(\subset P_{\tau_n}\mathfrak{B})$ is obviously invariant under the operator $P_{\tau_n}SP_{\tau_n}$. The operator $P_{\tau_n}SP_{\tau_n}$ is invertible in the subspace $P_{\tau_n}\mathfrak{B}$ (beginning with some n), and since it differs from P_{τ_n} by a completely continuous term, the restriction of this operator to the subspace $PP_{\tau_n}\mathfrak{B}$ is also an invertible operator. Moreover,

$$(PP_{\tau_n}SPP_{\tau_n})^{-1} = (P_{\tau_n}SP_{\tau_n})^{-1}PP_{\tau_n},$$

and consequently

$$\sup_n \left|(PP_{\tau_n}SPP_{\tau_n})^{-1}\right| < \infty. \tag{1.3}$$

---

[1]The idea of the proof of this theorem becomes intelligible if it is compared with the proof of the concrete Theorem 2.1.

It is easy to see that $SP = PAP$ for $\kappa \geqq 0$. Therefore relation (1.3) is equivalent to the following:

$$\sup_n |(\hat{P}_{\tau_n}\hat{A}\hat{P}_{\tau_n})^{-1}| < \infty,$$

where $\hat{P}_{\tau_n}$ and $\hat{A}$ are the restrictions of P_{τ_n} and PAP to the subspace $P\mathfrak{B}$. The invertibility of $\hat{A}$ is derived from the last relation, just as in Lemma 1.1, Chapter III. Hence by virtue of Theorem 3.2, Chapter I, it follows that $\kappa = 0$.

The theorem is proved.

An analogous theorem holds for the operator $PA + QB$.

The theorem just proved permits the projection methods for the inversion of the pair operators $AP + BQ$ and $PA + QB (A,B \in \mathfrak{R}(U))$ to be applied in the "basic" case, i.e. when the pair of indices (κ_1, κ_2) is zero. If, however, $(\kappa_1, \kappa_2) \neq (0, 0)$, then, as earlier (cf. § 1, Chapter III), the pair operator must be revised; i.e. its indices must be "canceled." For the sake of definiteness, let us consider $AP + BQ$. The operator $C = U^{-\kappa_1}AP + U^{-\kappa_2}BQ$ is the operator with canceled indices here.

By virtue of the theorem just proved, the operators $P_\tau CP_\tau$ are invertible on the subspace $P_\tau\mathfrak{B}$; beginning with some τ, their inverse operators converge strongly to the operator C^{-1}. Therefore the operators

$$(U^{\kappa_2-\kappa_1}P + Q)(P_\tau CP_\tau)^{-1}U^{-\kappa_2}$$

converge strongly to the appropriate (left or right) inverse of $AP + BQ$.

Let us explain that the equality

$$AP + BQ = U^{\kappa_2}(U^{-\kappa_1}AP + U^{-\kappa_2}BQ)(U^{\kappa}P + Q)$$

is valid for $\kappa_1 \geqq \kappa_2$; consequently $(U^{-\kappa}P + Q)C^{-1}U^{-\kappa_2}$ is a left inverse of the operator $AP + BQ$.

The equality

$$(AP + BQ)(U^{-\kappa}P + Q) = U^{\kappa_2}(U^{-\kappa_1}AP + U^{-\kappa_2}BQ)$$

is valid for $\kappa_1 \leqq \kappa_2$, which implies that $(U^{-\kappa}P + Q)C^{-1}U^{-\kappa_2}$ is a right inverse of the operator $AP + BQ$.

An analogous method is also applicable to the operator $PA + QB$.

In conclusion let us consider the case $A = B$; then $AP + BQ = A$.

The results of this section imply in particular that if the conditions

$$A(\zeta) \neq 0 \quad (|\zeta| = 1), \qquad \text{ind}\, A(\zeta) = 0 \tag{1.4}$$

are fulfilled for an operator $A \in \mathfrak{R}(U)$, then $A \in \Pi\{P_\tau, P_\tau\}$. Under certain natural additional restrictions (cf. Theorem 1.2) conditions (1.4) are also necessary for $A \in \Pi\{P_\tau, P_\tau\}$.

Let us note in addition that if $\kappa = \text{ind}\, A(\zeta) \neq 0$, the projection method (P_τ, P_τ) is not applicable to the equation $A\varphi = f$. However, this method is certainly applicable to the equation with canceled index $U^{-\kappa}A\varphi = U^{-\kappa}f$, which is equivalent to it. This method is also applicable to the equation $AU^{-\kappa}\phi = f$.

§ 2. Discrete equations

As in § 3, Chapter V, let us introduce a unitary shift operator U and an orthogonal projection$P$. Let us denote by P_n the symmetric truncation projection in $\tilde{l}_2$:

$$P_n\{\xi_j\} = \{\eta_j\}, \qquad \eta_j = \begin{cases} \xi_j & |j| \leqq n, \\ 0, & |j| > n. \end{cases}$$

It is not difficult to verify that the operators U, U^{-1} and the projections P, P_n ($n = 1, 2, \cdots$) satisfy all of the conditions of § 1 (cf. Theorem 1.1).

THEOREM 2.1. *Let $a(\zeta)$ and $b(\zeta)$ be arbitrary functions continuous on the unit circle, and a_j and b_j, $j = 0, \pm 1, \cdots$, their Fourier coefficients. If* $\liminf_{n\to\infty} |C_n^{-1}| < \infty$, *where*

$$C_n = \| c_{jk} \|_{j,k=-n}^{n},$$

$$c_{jk} = \begin{cases} a_{j-k}, & k = 0, 1, \cdots, n, \\ b_{j-k}, & k = -1, -2, \cdots, -n \end{cases} \qquad (j = -n, -n+1, \cdots, n),$$

then $a(\zeta)$, $b(\zeta) \neq 0$ ($|\zeta| = 1$) and $\operatorname{ind} a(\zeta) = \operatorname{ind} b(\zeta) = 0$. *Under fulfillment of these conditions, the truncated system of equations, beginning with some n,*

$$\sum_{k=0}^{n} a_{j-k}\xi_k + \sum_{k=-n}^{-1} b_{j-k}\xi_k = \eta_j \qquad (j = 0, \pm 1, \cdots, \pm n)$$

has the unique solution $\{\xi_j^{(n)}\}_{j=-n}^{n}$, and as $n \to \infty$ the vectors

$$\xi^{(n)} = \{\cdots, 0, \xi_{-n}^{(n)}, \cdots, \xi_n^{(n)}, 0, \cdots\}$$

converge in the norm of $\tilde{l}_2$ to a solution of the full system

$$\sum_{k=0}^{\infty} a_{j-k}\xi_k + \sum_{k=-\infty}^{-1} b_{j-k}\xi_k = \eta_j \qquad (j = 0, \pm 1, \cdots), \tag{2.1}$$

whatever the vector $\{\eta_j\}_{-\infty}^{\infty} \in \tilde{l}_2$ may be.

PROOF. The second assertion of the theorem follows from Theorem 1.1, and the first, from Theorem 1.2. Let us cite another, more intuitive, proof of the theorem's first assertion.

The lemma concerning the invertibility of a strong limit (Lemma 1.1, Chapter III) implies that the operator $AP + BQ$ defined in $\tilde{l}_2$ by system (2.1) is invertible. Therefore by virtue of Theorem 3.1, Chapter V, $a(\zeta)$, $b(\zeta) \neq 0$ ($|\zeta| = 1$) and $\operatorname{ind} a(\zeta) = \operatorname{ind} b(\zeta)$.

Let us assume that $\kappa = \operatorname{ind} a(\zeta) \neq 0$. For the sake of definiteness let $\kappa > 0$. Let us denote by n_k, $k = 1, 2, \cdots$, a sequence of natural numbers such that $\sup_k |C_{n_k}^{-1}| < \infty$. Then $AP + BQ \in \Pi\{P_{n_k}, P_{n_k}\}$.

Let $\| c_{jk} \|_{j,k=-\infty}^{\infty}$ be the matrix corresponding to the operator $AP + BQ$, i.e. the matrix of the full system (2.1.) Let us consider an operator S defined in $\bar{l}_2$ by the matrix $\| s_{jk} \|_{j,k=-\infty}^{\infty}$, where $s_{jk} = 0$ for all pairs j, k lying in one of the following angles:

1) $-\infty < j < \kappa,\ 0 \leqq k < \infty$,

2) $\kappa \leqq j < \infty,\ -\infty < k < 0$,

and $s_{jk} = c_{jk}$ for all the remaining pairs j, k. The matrix $\| s_{jk} \|_{j,k=-\infty}^{\infty}$ has the form

$$\left\| \begin{array}{cc:cc} \vdots & \vdots & & \\ \vdots & \vdots & & 0 \\ \cdots\cdots b_1 & b_0 & & \\ \hdashline b_2 & b_1 & & \\ \cdots\cdots & & & 0 \\ \cdots\cdots b_{\kappa+1} & b_\kappa & & \\ 0 & & a_\kappa & a_{\kappa-1}\cdots\cdots \\ & & a_{\kappa+1} & a_\kappa \\ & & \cdots & \cdots \quad \cdots\cdots \end{array} \right\|$$

It is not difficult to verify that operator S is invertible and differs from the operator $AP + BQ$ by a completely continuous term. By virtue of Theorem 3.1, Chapter II, $S \in \Pi\{P_{n_k}, P_{n_k}\}$. Therefore, beginning with some k, the operators $P_{n_k} S P_{n_k}$ are invertible in $P_{n_k}\bar{l}_2$. But the matrix of the operator $P_{n_k} S P_{n_k}$, which has the form

$$\left\| \begin{array}{ccc:ccc} b_0 & \cdots & b_{-n_k+1} & & & \\ \cdot & \cdots & \cdot & & & \\ \cdot & \cdots & \cdot & & 0 & \\ \cdot & \cdots & \cdot & & & \\ b_{n_k-1} & \cdots & b_0 & & & \\ \hdashline b_n & \cdots & b_1 & & & \\ \cdot & \cdots & \cdot & & & \\ \cdot & \cdots & \cdot & & 0 & \\ \cdot & \cdots & \cdot & & & \\ b_{n_k-1+\kappa} & \cdots & b_\kappa & & & \\ & & & a_\kappa & \cdots & a_{\kappa-n_k} \\ & & & \cdot & \cdots & \cdot \\ & 0 & & \cdot & \cdots & \cdot \\ & & & \cdot & \cdots & \cdot \\ & & & a_{n_k} & \cdots & a_0 \end{array} \right\|$$

is obviously singular. The contradiction just obtained proves the theorem.

An analogous theorem holds for the pair system of equations

$$\left. \begin{array}{ll} \sum\limits_{k=-\infty}^{\infty} a_{j-k}\xi_k = \eta_j & (j = 0, 1, \cdots), \\ \sum\limits_{k=-\infty}^{\infty} b_{j-k}\xi_k = \eta_j & (j = -1, -2, \cdots). \end{array} \right\} \tag{2.2}$$

REMARK. The projection method for solving pair discrete equations which was cited in Theorem 2.1 can be modified so that it becomes applicable also in the case when ind $a(\zeta)$ and ind $b(\zeta)$ do not equal zero. This case can be reduced to the one under consideration by passing to an equation with canceled indices (cf. the remark after the proof of Theorem 1.2 and Theorems 3.2, 3.3, Chapter V). Formulation of the corresponding propositions is left to the reader (cf. the authors' paper [4]).

In concluding this section let us consider the case when $a(\zeta) = b(\zeta)$. In this case system (2.1), like system (2.2), is converted into the system

$$\sum_{k=-\infty}^{\infty} a_{j-k}\xi_k = \eta_j \qquad (j = 0, \pm 1, \cdots), \tag{2.3}$$

a solution of which is the sequence of Fourier coefficients of a quotient of two functions.

The results of the present section imply the following theorem, which is closely related to Theorem 0.2 formulated in the Introduction.

THEOREM 2.2. *Let $a(\zeta)$ be an arbitrary function continuous on the unit circle, and $a_j, j = 0, \pm 1, \cdots$, its Fourier coefficients. If the relation*

$$\liminf_{n\to\infty} |A_n^{-1}| < \infty \tag{2.4}$$

is valid for the matrix $A_n = \| a_{j-k} \|_{j,k=-n}^{n}$, then

$$a(\zeta) \neq 0 \quad (|\zeta| = 1) \qquad \text{and} \qquad k = \operatorname{ind} a(\zeta) = 0. \tag{2.5}$$

Under fulfillment of only the first of the conditions (2.5) *the truncated system*

$$\sum_{k=-n}^{n} a_{j-k}\xi_k = \eta_j \qquad (j = -n + \kappa, \cdots, n + \kappa),$$

beginning with some n, has the unique solution $\{\xi_j^{(n)}\}_{j=-n}^{n}$, and as $n \to \infty$ the vectors $\xi^{(n)} = \{\bar{\xi}_j^{(n)}\}_{j=-\infty}^{\infty}$, where

$$\bar{\xi}_j^{(n)} = \begin{cases} \xi_j^{(n)}, & |j| \leqq n, \\ 0, & |j| > n, \end{cases}$$

converge in the norm of $\bar{l}_2$ to a solution of the full system (2.3), *whatever the vector $\{\eta_j\}_{-\infty}^{\infty} \in l_2$ may be.*

Let us explain the relationship of this theorem to Theorem 0.2. We are here restricting ourselves to the $\bar{l}_2$ case. In Theorem 2.2 the function $a(\zeta)$ is merely continuous, while in Theorem 0.2 it expands into an absolutely convergent Fourier series. Requirement (2.4) is a relaxation of the requirement of the existence and convergence in the norm of $\bar{l}_p$ of the sequence of vectors $\xi^{(n)}$; i.e. the first part (the "necessity") of Theorem 2.2 is a true strengthening of the first part of Theorem 0.2 for the $\bar{l}_2$ case.

The second part of Theorem 2.2 (the "sufficiency") coincides for $\kappa = 0$ in the $\bar{l}_2$

case with the second part of Theorem 0.2, while for $\kappa \neq 0$ it represents a new, additional assertion.

Let us note that all of the theorems of the present section are also valid for the spaces l_p $(p \geqq 1)$ and $\tilde{c}_0$ under appropriate restrictions on the functions $a(\zeta)$ and $b(\zeta)$. Theorem 2.2 generalizes, in addition, to the case of a piecewise continuous function $a(\zeta)$ (cf. Theorem 4.2, Chapter IV, where this was done for the case $\kappa = 0$).

§ 3. Integral equations

1. As before, let us denote by P_τ $(0 < \tau < \infty)$ the symmetric truncation projection defined in the space $\tilde{L}_p$ $(p \geqq 1)$ by the equality

$$(P_\tau f)(t) = \begin{cases} f(t), & |t| < \tau, \\ 0, & |t| > \tau. \end{cases} \tag{3.1}$$

The projections P_τ and the operators U, U^{-1} and P introduced in § 4, Chapter V, satisfy all of the conditions of § 1. We obtain the following theorem as a realization of Theorems 1.1 and 1.2.

THEOREM 3.1. *Let $k_j(t) \in \tilde{L}_1$ $(j = 1,2)$, and let the operator $AP + BQ$ be defined in L_p $(p \geqq 1)$ by the equation*

$$\varphi(t) - \int_0^\infty k_1(t-s)\varphi(s)\,ds - \int_{-\infty}^0 k_2(t-s)\varphi(s)\,ds = f(t) \tag{3.2}$$
$$(-\infty < t < \infty).$$

If

$$\liminf_{\tau\to\infty} \left|[P_\tau(AP + BQ)P_\tau]^{-1}\right| < \infty, \tag{3.3}$$

then

$$\mathscr{A}(\lambda) \overset{\text{def}}{=} 1 - \int_{-\infty}^{\infty} k_1(t)e^{i\lambda t}\,dt \neq 0, \qquad (-\infty < \lambda < \infty)$$
$$\mathscr{B}(\lambda) \overset{\text{def}}{=} 1 - \int_{-\infty}^{\infty} k_2(t)e^{i\lambda t}\,dt \neq 0,$$
$$\operatorname{ind} \mathscr{A}(\lambda) = \operatorname{ind} \mathscr{B}(\lambda) = 0. \tag{3.4}$$

Under fulfillment of these conditions the equation

$$\varphi(t) - \int_0^\tau k_1(t-s)\varphi(s)\,ds - \int_{-\tau}^0 k_2(t-s)\varphi(s)\,ds = f(t)$$
$$(-\tau < t < \tau),$$

beginning with some τ, has the unique solution $\varphi_\tau(t)$ whatever the function $f(t) \in \tilde{L}_p$ may be, and as $\tau \to \infty$ the functions

$$\tilde{\varphi}_\tau(t) = \begin{cases} \varphi(t), & |t| < \tau, \\ 0, & |t| > \tau, \end{cases}$$

converge in the norm of the space $\tilde{L}_p$ to a solution of equation (3.2).

An analogous theorem holds for the pair equation

$$\begin{aligned} \varphi(t) - \int_{-\infty}^{\infty} k_1(t-s)\varphi(s)\,ds &= f(t) \qquad (0 < t < \infty), \\ \varphi(t) - \int_{-\infty}^{\infty} k_2(t-s)\varphi(s)\,ds &= f(t) \qquad (-\infty < t < 0). \end{aligned} \tag{3.5}$$

Let us note that Theorem 1.3 concerning the cancellation of indices permits an approximation method of solving equation (3.2) or (3.5) to be justified also for $\mathscr{A}(\lambda) \neq 0$, ind $\mathscr{B}(\lambda) \neq 0$.

2. Let us establish for the case of Hilbert space $\tilde{L}_2$ the applicability to a pair equation and to the transpose equation of the Galerkin method relative to one special system of functions.

This system is defined on the entire axis by the following equalities:

$$\begin{aligned} \psi_0(t) &= \begin{cases} \sqrt{2}\,e^{-t}, & t > 0, \\ 0, & t < 0, \end{cases} \\ \psi_j(t) &= (U^j\psi_0)(t) \qquad (-\infty < t < \infty;\ j = \pm 1, \pm 2, \cdots), \end{aligned} \tag{3.6}$$

which are analogous to equalities (3.3) of Chapter III for the case of the semiaxis.

It is easy to see that for $j \geqq 0$ the functions $\psi_j(t)$ coincide with functions (3.3) of Chapter III when continued as zero on the negative semiaxis, while for $j < 0$

$$\psi_j(t) = \psi_{-j-1}(-t), \qquad j = -1, -2, \cdots.$$

Just as in § 3, Chapter III, it can be established that (3.6) is a basis in $\tilde{L}_2$ and that equations (3.2) and (3.5) are equivalent respectively to the systems

$$\sum_{k=0}^{\infty} a_{j-k}\xi_k + \sum_{k=-\infty}^{-1} b_{j-k}\xi_k = \eta_j \qquad (j = 0, \pm 1, \cdots) \tag{3.7}$$

and

$$\left.\begin{aligned} \sum_{k=-\infty}^{\infty} a_{j-k}\xi_k &= \eta_j \qquad (j = 0, 1, \cdots), \\ \sum_{k=-\infty}^{\infty} b_{j-k}\xi_k &= \eta_j \qquad (j = -1, -2, \cdots) \end{aligned}\right\} \tag{3.8}$$

considered in $\tilde{l}_2$. Here a_j, b_j and η_j are the Fourier coefficients of the functions

$$1 - K_1\Big(i\,\frac{1+\zeta}{1-\zeta}\Big), \qquad 1 - K_2\Big(i\,\frac{1+\zeta}{1-\zeta}\Big), \qquad F\Big(i\,\frac{1+\zeta}{1-\zeta}\Big) \qquad (|\zeta| = 1)$$

and the functions $K_1(\lambda)$, $K_2(\lambda)$ and $F(\lambda)$ are the Fourier transforms of the functions

$k_1(t)$, $k_2(t)$ and $f(t)$. In addition, if the vector $\{\xi_j\}_{-\infty}^{\infty}$ is a solution, for example, of system (3.7), the function

$$\varphi(t) = \sum_{-\infty}^{\infty} \xi_j \psi_j(t)$$

is a solution of equation (3.2), and conversely.

Let us note, moreover, that the system

$$\sum_{k=0}^{n} a_{j-k}\xi_k + \sum_{k=-n}^{-1} b_{j-k}\xi_k = \eta_j \qquad (j = 0, \pm 1, \cdots, \pm n)$$

coincides with the system of the Galerkin method for equation (3.2), i.e. with the system

$$\sum_{k=0}^{n} (A\psi_k, \psi_j)c_k + \sum_{k=-n}^{-1} (B\psi_k, \psi_j)c_k = (f, \psi_j) \tag{3.9}$$

$$(j = 0, \pm 1, \cdots, \pm n),$$

where A and B are defined by

$$\left.\begin{aligned} (A\varphi)(t) &= \varphi(t) - \int_{-\infty}^{\infty} k_1(t-s)\varphi(s)\,ds, \\ (B\varphi)(t) &= \varphi(t) - \int_{-\infty}^{\infty} k_2(t-s)\varphi(s)\,ds. \end{aligned}\right\} \tag{3.10}$$

Now applying the results of § 2, we obtain the following propositions.

THEOREM 3.2. *Let* $k_j(t) \in \tilde{L}_1$, $j = 1, 2$, *and let* A *and* B *be defined by* (3.10). *If* $\liminf_{n\to\infty} |D_n^{-1}| < \infty$, *where* D_n *is the matrix of system* (3.9), *then conditions* (3.4) *are fulfilled. Under fulfillment of these conditions, system* (3.9), *beginning with some* n, *has the unique solution* $\{c_j^{(n)}\}_{j=-n}^{n}$, *and as* $n \to \infty$ *the functions*

$$\varphi_n(t) = \sum_{-n}^{n} c_j^{(n)} \psi_j(t)$$

converge in the norm of $\tilde{L}_2$ *to a solution of equation* (3.2), *whatever the function* $f(t) \in \tilde{L}_2$ *may be.*

An analogous theorem holds for pair equation (3.5).

A remark analogous to the remark to Theorem 2.2 can be made for Theorem 3.2.

The above results imply a criterion for the applicability to the operator A defined by the equation

$$\varphi(t) - \int_{-\infty}^{\infty} k(t-s)\varphi(s)\,ds = f(t) \qquad (-\infty < t < \infty;\, k(t) \in \tilde{L}_1) \tag{3.11}$$

of the projection method (P_τ, P_τ) and the Galerkin method relative to the system of functions (3.6).

Let us, for example, formulate a theorem concerning the applicability of the Galerkin method in $\tilde{L}_2$.

Let us denote by D_n the matrix of the system

$$\sum_{k=-n}^{n} (A\psi_k, \psi_j) c_k = (f, \psi_j) \qquad (j = 0, \pm 1, \cdots, \pm n).$$

THEOREM 3.3. *Let $k(t) \in \tilde{L}_1$, and let the operator A be defined by* (3.11). *If $\liminf_{n\to\infty} |D_n^{-1}| < \infty$, then*

$$\mathscr{A}(\lambda) \overset{\text{def}}{=} 1 - \int_{-\infty}^{\infty} k(t) e^{i\lambda t}\, dt \neq 0 \qquad (-\infty < \lambda < \infty),$$

$$\kappa \overset{\text{def}}{=} \operatorname{ind} \mathscr{A}(\lambda) = 0. \tag{3.12}$$

Under fulfillment of the first of conditions (3.12) *the system*

$$\sum_{k=-n}^{n} (A\psi_k, \psi_j) c_k = (f, \psi_j) \qquad (j = -n + \kappa, \cdots, n + \kappa),$$

beginning with some n, has the unique solution $\{c_j^{(n)}\}_{j=-n}^{n}$, and as $n \to \infty$ the functions

$$\varphi_n(t) = \sum_{-n}^{n} c_j^{(n)} \psi_j(t)$$

converge in the norm of $\tilde{L}_2$ to a solution of equation (3.11), *whatever the function $f(t) \in \tilde{L}_2$ may be.*

Let us note in addition that analogous criteria for the applicability of projection methods hold also for the case of the piecewise continuous function $\mathscr{A}(\lambda)$ (cf. §§ 2 and 4 of Chapter IV).

§ 4. Singular integral equations

We shall use the notation and results of § 6, Chapter V, in the present section. Let us introduce in addition the projections P_n, $n = 1, 2, \cdots$, defined in the space $L_p(\Gamma)$ $(1 < p < \infty)$ by the equalities

$$(P_n g)(\zeta) = \sum_{-n}^{n} g_j \zeta^j,$$

where g_j, $j = 0, \pm 1, \cdots$, are the Fourier coefficients of the function $g(\zeta)$. The projections P_n converge strongly to the identity operator as $n \to \infty$ (cf. S. Karlin [**1**]), and together with the operators U and P satisfy all of the conditions of § 1.

The following theorem is a criterion for the applicability of a projection method to the singular integral equation

$$a(\zeta)\varphi(\zeta) + \frac{b(\zeta)}{\pi i}\int_{\Gamma}\frac{\varphi(z)}{z-\zeta}\,dz = f(\zeta) \qquad (\zeta \in \Gamma). \tag{4.1}$$

Here $a(\zeta)$ and $b(\zeta)$ are arbitrary functions continuous on Γ, and $f(\zeta) \in L_p(\Gamma)$; as before, let $c(\zeta) = a(\zeta) + b(\zeta)$ and $d(\zeta) = a(\zeta) - b(\zeta)$, and let c_j, d_j and f_j, $j = 0, \pm 1, \cdots$, be the Fourier coefficients of the functions $c(\zeta)$, $d(\zeta)$ and $f(\zeta)$.

THEOREM 4.1. *For the system of equations*

$$\sum_{k=0}^{n} c_{j-k}\xi_k + \sum_{k=-n}^{-1} d_{j-k}\xi_k = f_j \qquad (j = 0, \pm 1, \cdots, \pm n) \tag{4.2}$$

beginning with some n, to have the unique solution $\{\xi_j^{(n)}\}_{j=-n}^{n}$, *and for the functions*

$$\varphi_n(\zeta) = \sum_{-n}^{n} \xi_j^{(n)}\zeta^j$$

to converge in the norm of $L_p(\Gamma)$ *to a solution of equation* (4.1) *as* $n \to \infty$ (*whatever may be the function* $f(\zeta) \in L_p(\Gamma)$), *it is necessary and sufficient that the conditions*

$$\begin{gathered} c(\zeta) \neq 0, \quad d(\zeta) \neq 0 \qquad (|\zeta| = 1), \\ \operatorname{ind} c(\zeta) = \operatorname{ind} d(\zeta) = 0 \end{gathered} \tag{4.3}$$

be fulfilled.

PROOF. The necessity of conditions (4.3) follows from Theorem 1.2, and their sufficiency, from Theorem 1.1.

The first part of Theorem 4.1 (the "necessity") can be sharpened and formulated analogously to Theorem 2.1.

We shall not cite here the theorems concerning the approximate solution of equation (4.1) in the case of nonzero indices for the functions $c(\zeta)$ and $d(\zeta)$ (cf. the remark on p.146).

Let us note in addition that system (4.2) coincides with the system of the Galerkin method for equation (4.1) relative to the functions ζ^j, $j = 0, \pm 1, \cdots$.

§ 5. Finite difference equations

The applicability of a projection method to equations of the forms

$$\left.\begin{aligned} \sum_{j=-\infty}^{\infty} a_j f(t-j) &= g(t) \qquad (0 < t < \infty), \\ \sum_{j=-\infty}^{\infty} b_j f(t-j) &= g(t) \qquad (-\infty < t < 0) \end{aligned}\right\} \tag{5.1}$$

and

$$\sum_{j<t} a_j f(t-j) + \sum_{j>t} b_j f(t-j) = g(t) \qquad (-\infty < t < \infty) \tag{5.2}$$

is established in this section.

These equations are considered in the space $\tilde{L}_2$. The theory of such equations is set forth in § 7, Chapter V. Let $a(\zeta)$ and $b(\zeta)$ be piecewise continuous functions, and let a_j and b_j, $j = 0, \pm 1, \cdots$, be their Fourier coefficients. Then equalities (5.1) and (5.2) define the pair operators $Pa(U) + Qb(U)$ and $a(U)P + b(U)Q$ in the space $\tilde{L}_2$.

Let us denote by P_n, $n = 1, 2, \cdots$, the projection defined in $\tilde{L}_2$ by the equality

$$(P_n\varphi)(t) = \begin{cases} \varphi(t), & |t| < n, \\ 0, & |t| > n. \end{cases}$$

THEOREM 5.1. *Let $a(\zeta)$ and $b(\zeta)$ be arbitrary functions continuous on the unit circle, and let a_j and b_j, $j = 0, \pm 1, \cdots$, be their Fourier coefficients. If*

$$\liminf_{n\to\infty} |[P_n(a(U)P + b(U)Q)P_n]^{-1}| < \infty,$$

then $a(\zeta), b(\zeta) \neq 0$ $(|\zeta| = 1)$ *and* $\operatorname{ind} a(\zeta) = \operatorname{ind} b(\zeta) = 0$. *Under fulfillment of these conditions the equation*

$$\sum_{t-n<j<t} a_j f(t-j) + \sum_{t<j<t+n} b_j f(t-j) = g(t) \qquad (-n < t < n), \tag{5.3}$$

beginning with some n, has the unique solution $f_n(t) \in L_2(-n, n)$, and as $n \to \infty$ the functions

$$\tilde{f}_n(t) = \begin{cases} f_n(t), & |t| < n, \\ 0, & |t| > n \end{cases} \tag{5.4}$$

converge in the norm of $\tilde{L}_2$ to a solution of equation (5.2), *whatever the function $g(t) \in \tilde{L}_2$ may be.*

PROOF. (5.3), or what is the same, the equation $P_n(a(U)P + b(U)Q)P_nf = P_ng$, is equivalent to the system

$$\sum_{k=1}^{n} a_{j-k} f_k(t) + \sum_{k=-n+1}^{0} b_{j-k} f_k(t) = g_j(t) \qquad (j = -n+1, \cdots, n), \tag{5.5}$$

where the functions

$$f_j(t) = f(t + j - 1), \quad g_j(t) = g(t + j - 1)$$
$$(0 \leqq t \leqq 1;\ j = -n+1, \cdots, n)$$

belong to the space $L_2(0, 1)$. By virtue of this equivalence and Lemma 6.1 of Chapter IV,

$$|[P_n(a(U)P + b(U)Q)P_n]^{-1}| = |A_n^{-1}|,$$

where A_n is the matrix of system (5.5). To complete the proof it is sufficient to apply Theorem 2.1 of this chapter and Theorem 2.1 of Chapter II.

Theorem 5.1 and its proof remain valid in the space $\tilde{L}_p$ $(1 \leqq p < \infty)$ also if the continuous functions $a(\zeta)$ and $b(\zeta)$ belong to the corresponding algebra $\mathfrak{R}(\zeta)$.

Analogous results are valid for equation (5.1).

In conclusion let us consider the case when $a(\zeta) = b(\zeta)$.

THEOREM 5.2. *Let the function* $a(\zeta) \in \Lambda$, *and let* $a_j, j = 0, \pm 1, \cdots$, *be its Fourier coefficients. If*

$$\liminf_{n\to\infty} |(P_n a(U) P_n)^{-1}| < \infty,$$

then $a(\zeta)$ *is nonsingular, and* ind $a(\zeta) = 0$.

Under fulfillment of these conditions the equation

$$\sum_{t-n<j<t+n} a_j f(t-j) = g(t) \qquad (-n < t < n),$$

beginning with some n, has the unique solution $f_n(t) \in L_2(-n, n)$, *and as* $n \to \infty$ *the functions* (5.4) *converge in the norm of* $\tilde{L}_2$ *to a solution of the equation*

$$\sum_{j=-\infty}^{\infty} a_j f(t-j) = g(t) \qquad (-\infty < t < \infty),$$

whatever the function $g(t) \in \tilde{L}_2$ *may be.*

This theorem is proved analogously to the preceding one. It is a corollary of Theorem 4.2 and Lemma 6.1, Chapter IV.

§ 6. Galerkin method for an operator of multiplication by a function

Let A be a bounded linear operator of multiplication by a function in $L_2(0, 1)$:

$$(A\varphi)(t) = a(t)\varphi(t) \qquad (0 \leqq t \leqq 1), \tag{6.1}$$

where $a(t)$ $(0 \leqq t \leqq 1)$ is a bounded measurable function. The conditions under which the Galerkin method relative to the orthogonal system of trigonometric functions $\{e^{i2\pi jt}\}, j = 0, \pm 1, \cdots$, is applicable to A are explained in this section.

The conditions obtained below are not only sufficient but also necessary in the case of a continuous or piecewise continuous function $a(t)$.

1. First let $a(t)$ be a *continuous* function which does not vanish on the segment $0 \leqq t \leqq 1$. The continuous function arg $a(t)$ is not uniquely defined. However, the difference arg $a(t_2)$ − arg $a(t_1)$, $0 \leqq t_1, t_2 \leqq 1$, obviously does not depend on the choice of arg $a(t)$.

THEOREM 6.1. *Let* $a(t)$, $0 \leqq t \leqq 1$, *be a continuous function, and let*

$$\varphi_j(t) = e^{i2\pi jt}, \qquad j = 0, \pm 1, \cdots. \tag{6.2}$$

For the Galerkin[2] *method* (φ_j, φ_j) *to be applicable to the operator A defined by*

[2] For determination of applicability of the Galerkin method see § 1, Chapter II.

(6.1), *it is necessary and sufficient that the following conditions be fulfilled*:

a) $a(t) \neq 0$ $(0 \leqq t \leqq 1)$,

b) $|\arg a(1) - \arg a(0)| < \pi$.

Let us note that if the continuous function $a(t)$ takes equal values at the ends of the segment [0, 1], condition b) takes the form $\arg a(1) - \arg a(0) = 0$.

This theorem follows from the more general Theorem 6.2 for piecewise continuous $a(t)$. Here we make a few simple observations.

Let us introduce the integer

$$\kappa = (1/2\pi)[\arg a(1) - \arg a(0) + \arg a_0(0) - \arg a_0(1)], \tag{6.3}$$

where

$$a_0(\mu) = a(0)\mu + a_1(1)(1-\mu) \neq 0 \qquad (0 \leqq \mu \leqq 1).$$

The number κ equals the index[3] of the function $\tilde{a}(\zeta)$ $(|\zeta| = 1)$ defined by the equality $\tilde{a}(e^{i2\pi t}) = a(t)$, $0 \leqq t \leqq 1$. Condition b) simply means that $\kappa = 0$. The case of $\kappa \neq 0$ is reduced to the case $\kappa = 0$ by cancellation of the index, i.e. by replacing the original operator by the operator

$$e^{-i2\pi\kappa t}A = e^{-i2\pi\kappa t}a(t)I.$$

In other words, under fulfillment of condition a) and the condition $a_0(\mu) \neq 0$ $(0 \leqq \mu \leqq 1)$ the Galerkin method (φ_j, ψ_j), where

$$\varphi_j(t) = e^{i2\pi jt} \qquad \text{and} \qquad \psi_j(t) = e^{i2\pi(j+\kappa)t} \qquad (j = 0, \pm 1, \cdots),$$

is applicable to the operator A defined by (6.1).

If only condition a) of Theorem 6.1 is fulfilled,[4] there is a *complex* number α such that the Galerkin method relative to the system of functions (6.2) is applicable to the operator $e^{i2\pi\alpha t}$.

Indeed, $a(0)/a(1)$ can be represented in the form $e^{i2\pi\alpha}$, where α is a complex number.

It is easy to see that the function $a(t)e^{i2\pi\alpha t}$, $0 \leqq t \leqq 1$, is continuous and takes identical values at the ends of the segment $0 \leqq t \leqq 1$. In addition, the number α can be so chosen that

$$[\arg (a(t)e^{i2\pi\alpha t})]_{t=0}^{1} = 0.$$

What has been said means also that under condition a) the Galerkin method (φ_j, ψ_j), where

$$\varphi_j(t) = e^{i2\pi jt} \qquad \text{and} \qquad \psi_j(t) = e^{i2\pi(j-\alpha)t} \qquad (j = 0, \pm 1, \cdots),$$

is applicable to A itself.

---

[3]For determination of the index of discontinuous functions cf. § 1, Chapter IV.

[4]Condition a) signifies the invertibility of the operator A, and it is necessary for the applicability of the Galerkin method relative to any system of functions.

2. Now let us pass to the case of piecewise continuous functions.

Let $a(t)$, $0 \leqq t \leqq 1$, be a piecewise continuous function which is continuous on the left, and let $0 \leqq t_1 < t_2 < \cdots < t_n < 1$ be all its points of discontinuity.

Let us fill in the discontinuities of the graph of the function $a(t)$ by joining as segments the points $a(t_j)$ and $a(t_j + 0)$ $(j = 1,\cdots, n)$, as well as the points $a(0)$ and $a(1)$.

Let us denote by Γ_a the continuous curve obtained. This curve can be oriented in a natural way. If it does not pass through the coordinate origin, then by $\kappa = \operatorname{ind} \Gamma_a$ we denote the number of revolutions of this curve about the origin.

THEOREM 6.2. *For the Galerkin method relative to the system of functions* (6.2) *to be applicable to the operator $A = a(t)I$ in $L_2(0, 1)$, it is necessary and sufficient that the following conditions be fulfilled*:

a) *The curve Γ_a does not pass through zero.*

b) $\kappa = \operatorname{ind} \Gamma_a = 0$.

PROOF. It is easy to see that in the basis (6.2) the Toeplitz matrix $\| a_{j-k} \|_{j,k=-\infty}^{\infty}$ corresponds to the operator $A = a(t)I$, where a_j, $j = 0, \pm 1 \cdots$, are the Fourier coefficients of the function $a(t)$ relative to system (6.2). The Galerkin method relative to system (6.2) for this operator coincides with the reduction method, under which an equation with the infinite matrix $\| a_{j-k} \|_{j,k=-\infty}^{\infty}$ is replaced by a finite system of equations with the matrix $\| a_{j-k} \|_{j,k=-n}^{n}$.

It remains to make use of Theorem 4.2, Chapter IV, in order to complete the proof. Let us note that the same remarks can be made on Theorem 6.2 as on the preceding one.

3. Let us cite one more theorem following from Theorem 5.3, Chapter IV.

THEOREM 6.3. *Let the function $f(\zeta) \in M_R$,*[5] *let* $\operatorname{ind} f(\zeta) = 0$ *and let* $a(t) = f(e^{2\pi it})$. *Then the Galerkin method relative to the system of trigonometric functions* (6.2) *is applicable to the operator $A = a(t)I$.*

The following proposition holds for a real bounded measurable function $a(t)$.

For the Galerkin method relative to the trigonometric system (6.2) *to be applicable to the operator $A = a(t)I$, it is necessary and sufficient that one of the two conditions*

$$\operatorname*{ess\,inf}_{0 \leqq t \leqq 1} a(t) > 0, \qquad \operatorname*{ess\,sup}_{0 \leqq t \leqq 1} a(t) < 0$$

be fulfilled.

This proposition is easily proved with the help of a theorem of H. Widom [**1**] concerning the connectedness of the spectrum of a discrete Wiener-Hopf operator.

---

[5] For the definition of M_R cf. § 5 of Chapter IV.

CHAPTER VII

WIENER-HOPF INTEGRAL-DIFFERENCE EQUATIONS

The theory of integral-difference equations of the form

$$\sum_{j=-\infty}^{\infty} a_j\varphi(t-\delta_j) + \int_0^{\infty} k(t-s)\varphi(s)ds = f(t) \qquad (0<t<\infty) \tag{1}$$

and of equations similar to them is set forth in the present chapter. In this equation the $a_j, j = 0, \pm 1, \cdots$, are complex numbers, the δ_j are arbitrary real numbers, $k(t)$ $(-\infty < t < \infty)$ and $f(t)$ $(0 < t < \infty)$ are given functions, while $\varphi(t)$ $(0 < t < \infty)$ is the function sought. We assume in addition that for $\delta_j > 0$ the function $\varphi(t-\delta_j)$ vanishes on the segment $0 \leqq t \leqq \delta_j$.

In the simplest case equation (1) is considered in the space $L_p(0, \infty)$ $(1 \leqq p \leqq \infty)$ under the following restrictions:

$$\sum_{j=-\infty}^{\infty} |a_j| < \infty, \qquad \int_{-\infty}^{\infty} |k(t)|\,dt < \infty. \tag{2}$$

Under these restrictions equation (1) can be written in the form

$$\int_0^{\infty} \varphi(s)d\omega(t-s) = f(t) \qquad (0 < t < \infty),$$

where $\omega(t)$ $(-\infty < t < \infty)$ is a function of bounded variation without singular component (cf. Gel'fand, Raĭkov and Šilov [1]). Under the restrictions (2) the symbol

$$\mathscr{A}(\lambda) = \sum_{j=-\infty}^{\infty} a_j e^{i\delta_j\lambda} + \int_{-\infty}^{\infty} k(t)e^{i\lambda t}dt \qquad (-\infty < \lambda < \infty) \tag{3}$$

is associated with equation (1).

It turns out that the operator A defined by equation (1) is invertible on at least one side if and only if

$$\inf_{-\infty<\lambda<\infty} |\mathscr{A}(\lambda)| > 0.$$

Under fulfillment of this condition the two numbers

$$\nu = \lim_{l\to\infty} \frac{1}{2l} [\arg a(\lambda)]_{-l}^{l}, \qquad n = \frac{1}{2\pi} [\arg(1 + a^{-1}(\lambda)K(\lambda))]_{-\infty}^{\infty},$$

where

$$a(\lambda) = \sum_{j=-\infty}^{\infty} a_j e^{i\delta_j\lambda}, \qquad K(\lambda) = \int_{-\infty}^{\infty} k(t)e^{i\lambda t}dt,$$

can be defined.

The nature of A's invertibility is determined from the signs of the numbers (indices) ν and n as follows:

$n \backslash \nu$	$\nu>0$	$\nu=0$	$\nu<0$
$n>0$	L	L	R
$n=0$	L	B	R
$n<0$	L	R	R

Here L denotes invertibility only on the left, R, invertibility only on the right, and B, bilateral invertibility.

It is evident from this table that the index ν is "dominant": for $\nu > 0$ invertibility is only on the left; for $\nu < 0$, only on the right; and for $\nu = 0$ invertibility is consistent with the index n.

These results are set forth in § 2. The theorem concerning factorization of functions of the form (3) which is established in § 1 plays a vital role in their proof.

In § 3, these results are obtained under weaker restrictions on the symbol $\mathscr{A}(\lambda)$. In the case of the space L_2, in particular, they are obtained for an arbitrary continuous symbol. Ring-theoretic methods are essentially utilized here.

Next some abstract analogues of equation (1) are considered, and their invertibility is analyzed. Then the criterion of applicability of one projection method to equation (1) is established. The last section is devoted to pair integral-difference equations and their transposes.

§ 1. Theorem concerning factorization

Let us denote by $\mathfrak{G}$ the Banach algebra of all functions of the form

$$\mathscr{A}(\lambda) = \sum_{j=-\infty}^{\infty} a_j e^{i\delta_j\lambda} + \int_{-\infty}^{\infty} k(t)e^{i\lambda t}dt \qquad (-\infty < \lambda < \infty), \tag{1.1}$$

where the δ_j are arbitrary distinct real numbers, the a_j are arbitrary complex numbers the series of which is absolutely convergent, and $k(t) \in L_1(-\infty, \infty)$. The norm of the element $\mathscr{A}(\lambda)$ in $\mathfrak{G}$ is defined by the equality

$$\| \mathscr{A}(\lambda) \| = \sum_{j=-\infty}^{\infty} |a_j| + \int_{-\infty}^{\infty} |k(t)|\, dt,$$

and the operations are defined as usual.[1)]

The algebra $\mathfrak{G}$ obviously contains the algebra $\mathfrak{P}$ of almost-periodic functions expanding into absolutely convergent Fourier series, as well as the algebra $\mathfrak{L}_0$ of the Fourier transforms of the functions of $L_1(-\infty, \infty)$. The algebra $\mathfrak{G}$ is the direct sum of its subalgebras $\mathfrak{P}$ and $\mathfrak{L}_0$. That $\mathfrak{L}_0$ is an ideal of $\mathfrak{G}$ is established by immediate verification. It is also easy to establish that for any pair of functions $a(\lambda) \in \mathfrak{P}$ and $K(\lambda) \in \mathfrak{L}_0$ the relations

$$\inf_{-\infty<\lambda<\infty} |a(\lambda) + K(\lambda)| \leqq \inf_{-\infty<\lambda<\infty} |a(\lambda)| \tag{1.2}$$

and

$$\sup_{-\infty<\lambda<\infty} |a(\lambda) + K(\lambda)| \geqq \sup_{-\infty<\lambda<\infty} |a(\lambda)| \tag{1.3}$$

are fulfilled. To prove these relations it is sufficient to make use of the definition of an almost-periodic function and of the fact that $K(\lambda) \to 0$ as $\lambda \to \pm\infty$. The fact is that the points at which the greatest lower, or least upper, bound of $|a(\lambda)|$ is almost reached occur arbitrarily far out.

Let us call a function $\mathscr{A}(\lambda) \in \mathfrak{G}$ *nondegenerate* if

$$\inf_{-\infty<\lambda<\infty} |\mathscr{A}(\lambda)| > 0.$$

Relation (1.2) implies that if the function

$$\mathscr{A}(\lambda) = a(\lambda) + K(\lambda) \qquad (a(\lambda) \in \mathfrak{P},\ K(\lambda) \in \mathfrak{L}_0)$$

is nondegenerate, its almost-periodic component $a(\lambda)$ is also nondegenerate.

Let us associate with each nondegenerate function $\mathscr{A}(\lambda) = a(\lambda) + K(\lambda) \in \mathfrak{G}$ the two numbers $\nu(\mathscr{A})$ and $n(\mathscr{A})$ defined by the equalities

$$\nu(\mathscr{A}) = \lim_{l\to\infty} \frac{1}{2l} [\arg a(\lambda)]_{-l}^{l}, \qquad n(\mathscr{A}) = \frac{1}{2\pi} [\arg(1 + a^{-1}(\lambda) K(\lambda)]_{\lambda=-\infty}^{\infty}.$$

The limit in the first equality exists by virtue of a well-known property of almost-periodic functions (cf. B. M. Levitan [**1**], Theorem 2.7.1). Let us call the real number $\nu(\mathscr{A})$ and the integer $n(\mathscr{A})$ the *indices* of the function $\mathscr{A}(\lambda)$. Let us note that a function's indices do not vary under small (in the norm of the algebra $\mathfrak{G}$) perturbations.

For functions of $\mathfrak{G}$ of the form $1 + \omega(\lambda)$, where $\|\omega(\lambda)\|$ is small, the indices ν and n equal 0.

Let us denote by $\mathfrak{G}^+$ ($\mathfrak{G}^-$) the subalgebra of $\mathfrak{G}$ which consists of all functions of the form (1.1) for which the numbers δ_j are nonnegative (nonpositive) and the function $k(t)$ vanishes on the negative (positive) semiaxis. Obviously all functions of $\mathfrak{G}^+$ ($\mathfrak{G}^-$) admit holomorphic continuation into the upper (lower) half-plane.

[1)]Verification of the fact that $\mathfrak{G}$ actually is an algebra is left to the reader.

It is easy to see that $\mathfrak{G}^+ \cap \mathfrak{L}_0$ ($\mathfrak{G}^- \cap \mathfrak{L}_0$) is an ideal of $\mathfrak{G}^+$ ($\mathfrak{G}^-$).
This section is devoted to the proof of the following proposition.

THEOREM 1.1. *Every nondegenerate function $\mathscr{A}(\lambda) \in \mathfrak{G}$ admits factorization of the form*

$$\mathscr{A}(\lambda) = \mathscr{A}_-(\lambda)e^{i\nu\lambda}\left(\frac{\lambda-i}{\lambda+i}\right)^n \mathscr{A}_+(\lambda) \qquad (-\infty < \lambda < \infty), \tag{1.4}$$

where $\mathscr{A}_+^{\pm 1} \in \mathfrak{G}^+$, $\mathscr{A}_-^{\pm 1} \in \mathfrak{G}^-$; $\nu = \nu(\mathscr{A})$ and $n = n(\mathscr{A})$.

PROOF. Let $\mathscr{A}(\lambda) = a(\lambda) + K(\lambda) \in \mathfrak{G}$ be some nondegenerate function. Then, as already noted, its almost-periodic component $a(\lambda)$ is also nondegenerate. It is easy to see that the index $\nu(b)$ of the almost-periodic function $b(\lambda) = a(\lambda)e^{-i\nu\lambda}$ equals zero. According to a well-known generalization of the Wiener-Levy theorem (cf. Gel'fand, Raĭkov and Šilov [1]), the function $c(\lambda) = \log b(\lambda)$ belongs to the algebra $\mathfrak{P}$, i.e. the function $c(\lambda)$ has the form

$$c(\lambda) = \sum_{j=-\infty}^{\infty} c_j e^{i\gamma_j\lambda} \qquad \left(\sum_{-\infty}^{\infty} |c_j| < \infty\right),$$

where the γ_j are distinct real numbers.
Let us set

$$c_+(\lambda) = \sum_{\gamma_j \geqq 0} c_j e^{i\gamma_j\lambda} \qquad \text{and} \qquad c_-(\lambda) = \sum_{\gamma_j < 0} c_j e^{i\gamma_j\lambda}.$$

Then obviously

$$a(\lambda) = a_-(\lambda)e^{i\nu\lambda}a_+(\lambda), \tag{1.5}$$

where $a_\pm(\lambda) = \exp c_\pm(\lambda)$, and therefore

$$a_+^{\pm 1}(\lambda) \in \mathfrak{G}^+ \cap \mathfrak{P}, \qquad a_-^{\pm 1}(\lambda) \in \mathfrak{G}^- \cap \mathfrak{P}.$$

Since $\mathfrak{L}^0$ is an ideal of the algebra $\mathfrak{G}$, the function $a^{-1}(\lambda)K(\lambda)$ also belongs to $\mathfrak{L}_0$, and

$$1 + a^{-1}(\lambda)K(\lambda) \neq 0 \qquad (-\infty < \lambda < \infty).$$

By virtue of Theorem 8.3, Chapter I, the function $1 + a^{-1}(\lambda)K(\lambda)$ can be represented in the form

$$1 + a^{-1}(\lambda)K(\lambda) = G_-(\lambda)\left(\frac{\lambda-i}{\lambda+i}\right)^n G_+(\lambda), \tag{1.6}$$

where the functions $G_\pm(\lambda)$ and $G_\pm^{-1}(\lambda)$ belong to $\mathfrak{G}^\pm \cap \mathfrak{L}$. Here $\mathfrak{L}$ denotes the algebra obtained from $\mathfrak{L}_0$ by adjoining the identity element.

Multiplying equalities (1.5) and (1.6) and combining factors with one and the same sign in the index, we obtain factorization (1.4). The theorem is proved.

Let us note that the functions $\mathscr{A}_\pm(\lambda)$ in factorization (1.4) are uniquely determined to within a numerical factor.

§ 2. Integral-difference operators with absolutely convergent symbols

1. *Integral-difference opterators and their symbols.* Let us denote by $\hat{\mathfrak{G}}$ the set of all operators A operating in the space $L_p(0, \infty)$ in accordance with the formula

$$(A\varphi)(t) = \sum_{j=-\infty}^{\infty} a_j\varphi(t - \delta_j) + \int_0^{\infty} k(t - s)\varphi(s)ds, \tag{2.1}$$

where the δ_j are distinct real numbers, the a_j are complex numbers the series of which is absolutely convergent, and $k(t) \in L_1(-\infty, \infty)$. Here, as previously, $\varphi(t - \delta_j) = 0$ for $t < \delta_j$. It is easy to see that each operator $A \in \hat{\mathfrak{G}}$ is a bounded linear operator, where

$$| A |_{L_p} \leqq \sum_{j=-\infty}^{\infty} | a_j | + \int_{-\infty}^{\infty} | k(t) | \, dt.$$

As will be shown below, the equality sign is attained in the last relation for $p = 1$.

Let us associate with an operator $A \in \hat{\mathfrak{G}}$ defined by equality (2.1) a function

$$\mathscr{A}(\lambda) = \sum_{j=-\infty}^{\infty} a_j e^{i\delta_j\lambda} + \int_{-\infty}^{\infty} k(t)e^{i\lambda t}\, dt \qquad (-\infty < \lambda < \infty),$$

which we shall call the *symbol* of A.

A one-to-one correspondence exists between the operators A of the set $\hat{\mathfrak{G}}$ and their symbols $\mathscr{A}(\lambda) \in \mathfrak{G}$. It is obviously linear but not multiplicative (what is more, a product of operators of $\hat{\mathfrak{G}}$ need not even be an operator of that set as, for example, the projection $U_\nu U_{-\nu}$ of subsection 2, below). However, this correspondence is partially multiplicative in the following sense: If operators A_1 and A_2 of $\hat{\mathfrak{G}}$ are such that their symbols belong to $\mathfrak{G}^-$ and $\mathfrak{G}^+$, respectively ($\mathscr{A}_1(\lambda) \in \mathfrak{G}^-$; $\mathscr{A}_2(\lambda) \in \mathfrak{G}^+$), then the operator $B = A_1AA_2$ (for an arbitrary A of $\hat{\mathfrak{G}}$) also belongs to $\hat{\mathfrak{G}}$, and the equality $\mathscr{B}(\lambda) = \mathscr{A}_1(\lambda)\mathscr{A}(\lambda)\mathscr{A}_2(\lambda)$ holds for its symbol. The latter assertion is easily verified, and we shall not go into its proof.

What has been said implies in particular that the set $\hat{\mathfrak{G}}^+$ ($\hat{\mathfrak{G}}^-$) consisting of all operators $A \in \hat{\mathfrak{G}}$ with symbols $\mathscr{A}(\lambda) \in \mathfrak{G}^+$ ($\mathfrak{G}^-$) forms a commutative algebra. In particular, if the operator $A \in \hat{\mathfrak{G}}^+$ ($\hat{\mathfrak{G}}^-$) and if its symbol $\mathscr{A}(\lambda)$ is such that $\mathscr{A}^{-1}(\lambda) \in \mathfrak{G}^+(\mathfrak{G}^-)$, then A is invertible, and the operator A^{-1} with symbol $\mathscr{A}^{-1}(\lambda)$ is its inverse.

2. *Reduction of the general case to an elementary one.* Let us denote by U_ν ($-\infty < \nu < \infty$) the bounded linear operator defined in $L_p(0, \infty)$ by the equality

$$(U_\nu\varphi)(t) = \begin{cases} \varphi(t - \nu), & \max(\nu, 0) < t < \infty, \\ 0, & 0 \leqq t \leqq \max(\nu, 0). \end{cases}$$

Obviously $U_{-\nu}U_\nu = I$ for all $\nu \geqq 0$, and the difference $I - U_\nu U_{-\nu}$ is the following projection:

$$(I - U_\nu U_{-\nu})\varphi(t) = \begin{cases} \varphi(t), & 0 < t < \nu, \\ 0, & \nu < t < \infty. \end{cases}$$

For all real ν the operator $U_\nu \in \widehat{\mathfrak{G}}$, and the function $e^{i\nu\lambda}$ is obviously its symbol.

Let us denote by $V^{(n)}$, $n = 0, \pm 1, \cdots$, the operator of $\widehat{\mathfrak{G}}$ whose symbol equals $(\lambda - i)^n / (\lambda + i)^n$. By virtue of the results of § 8, Chapter I, $V^{(n)} = V^n$, $n \geqq 0$, and $V^{(n)} = (V^{(-1)})^{-n}$, $n < 0$, where the operators V and $V^{(-1)}$ are defined by the equalities

$$(V\varphi)(t) = \varphi(t) - 2\int_0^t e^{s-t}\varphi(s)\,ds, \qquad (V^{(-1)}\varphi) = \varphi(t) - 2\int_t^\infty e^{t-s}\varphi(s)\,ds.$$

Let us recall, moreover, that for all natural n the operator V^n is left invertible: $V^{(-n)}V^n = I$. The image Im V^n of the operator V^n, $n>0$, coincides with the closed linear span $\mathfrak{E}_{p,n}$ of the functions $e^{-t}\Lambda_j(2t)$, $j = n, n+1, \cdots$, in the space $L_p(0, \infty)$, while the kernel Ker $V^{(-n)}$ of the operator $V^{(-n)}$ $(n > 0)$ is the linear span of the functions $e^{-t}\Lambda_j(2t)$, $j = 0,1,\cdots, n-1$, or, what is the same, of the functions $t^j e^{-t}$ for the same j, and does not depend on p. Here the Λ_j are Laguerre polynomials. The equation $V^n\varphi = g$ $(n > 0)$ is solvable in $L_p(0, \infty)$ if and only if the conditions

$$\int_0^\infty g(t)t^j e^{-t}\,dt = 0 \qquad (j = 0, 1, \cdots, n-1)$$

are fulfilled.

The difference $P_n = I - V^{(n)}V^{(-n)}$ $(n > 0)$ is a projection projecting $L_p(0, \infty)$ onto the subspace Ker $V^{(-n)}$ parallel to $\mathfrak{E}_{p,n}$ The projection P_n is orthogonal in $L_2(0, \infty)$.

For all ν $(0 \leqq \nu < \infty)$ and $n = 1, 2, \cdots$ the operators $U_\nu V^n$ $(= V^n U_\nu)$, $U_{-\nu}V^{(-n)}$ $(= V^{(-n)}U_{-\nu})$, $U_{-\nu}V^n$ and $V^{(-n)}U_\nu$ belong[2] to the set $\widehat{\mathfrak{G}}$. Obviously the functions

$$e^{i\nu\lambda}\left(\frac{\lambda-i}{\lambda+i}\right)^n, \qquad e^{-i\nu\lambda}\left(\frac{\lambda-i}{\lambda+i}\right)^{-n}, \qquad e^{-i\nu\lambda}\left(\frac{\lambda-i}{\lambda+i}\right)^n, \qquad e^{i\nu\lambda}\left(\frac{\lambda-i}{\lambda+i}\right)^{-n}$$

are respectively the symbols of these elementary integral-difference operators.

Let the symbol $\mathscr{A}(\lambda)$ of an operator $A \in \widehat{\mathfrak{G}}$ be nondegenerate, and let equality (1.4) give the factorization of the function $\mathscr{A}(\lambda)$. Then, by virtue of what was said at the end of subsection 1, A can be represented in the form

$$A = A_- U_\nu V^{(n)} A_+ \tag{2.2}$$

for $\nu \leqq 0$, and in the form

$$A = A_- V^{(n)} U_\nu A_+ \tag{2.3}$$

for $\nu > 0$, where $A_\pm$ are operators of $\widehat{\mathfrak{G}}^\pm$ with symbols equal respectively to $\mathscr{A}_\pm(\lambda)$; $\nu = \nu(\mathscr{A})$ and $n = n(\mathscr{A})$.

[2]Let us note that $U_\nu V^{(-n)}$ and $V^n U_{-\nu}$ do not belong to $\widehat{\mathfrak{G}}$.

Since the operators $A_\pm$ are invertible and the operators $A_\pm^{-1} \in \widehat{\mathfrak{G}}^\pm$ with symbols $\mathscr{A}_\pm^{-1}(\lambda)$ are their inverses, the question of the nature of A's invertibility reduces to analysis of the elementary operators $U_\nu V^{(n)}$ for $\nu \leqq 0$ and $V^{(n)} U_\nu$ for $\nu > 0$ $(n = 0, \pm 1, \cdots)$.

3. *Case* $\nu > 0$. The following are assumed in this subsection and in the two subsequent ones:

1) $A \in \widehat{\mathfrak{G}}$ is an operator with nondegenerate symbol $\mathscr{A}(\lambda)$.

2) Equality (1.4) gives a factorization of the symbol $\mathscr{A}(\lambda)$.

3) $A_\pm$ are operators of $\widehat{\mathfrak{G}}^\pm$ with symbols $\mathscr{A}_\pm(\lambda)$. The numbers $\nu(\mathscr{A})$ and $n(\mathscr{A})$ are denoted by ν and n for the sake of brevity.

THEOREM 2.1. *If* $\nu > 0$, *the operator* A *is left invertible. For the equation*

$$A\varphi = g \tag{2.4}$$

to be solvable, it is necessary and sufficient that

a) *for* $n \geqq 0$ *the function* $A_-^{-1} g$ *vanish on the segment* $[0, \nu]$ *and the condition*

$$\int_0^\infty (A_-^{-1}g)(t)t^k e^{-t}\, dt = 0 \qquad (k = 0, 1, \cdots, n-1) \tag{2.5}$$

be fulfilled;

b) *for* $n < 0$ *the function* $V^{(-n)} A_-^{-1} g$ *coincide on the segment* $[0, \nu]$ *with some linear combination of the functions* $t^j e^{-t}$, $j = 0,1,\cdots, -n-1$.

PROOF. As already noted in subsection 2, analysis of A's invertibility reduces to analysis of the operator $V^{(n)} U_\nu$. If $n \geqq 0$, the operator $U_{-\nu} V^{(-n)}$ is a left inverse of the operator $V^{(n)} U_\nu$. For $n < 0$ and $\nu > 0$ we have

$$U_{-\nu} V^{-n} V^{(n)} U_\nu = I - U_{-\nu} P_{-n} U_\nu, \tag{2.6}$$

where $P_m = I - V^m V^{(-m)}$ $(m = 1,2,\cdots)$ is a finite-dimensional projection projecting L_p onto a subspace with basis $t^j e^{-t}, j = 0,1,\cdots, m-1$ (cf. subsection 2). The operator P_m is an orthogonal projection in L^2.

Since the sets of values of the operators U_ν and P_{-n} intersect only at zero, in the case of L_2 the operator $U_{-\nu} P_{-n} U_\nu$ is of norm less than unity; consequently the operator $I - U_{-\nu} P_{-n} U_\nu$ is invertible, while

$$(I - U_{-\nu} P_{-n} U_\nu)^{-1} = \sum_{j=0}^\infty (U_{-\nu} P_{-n} U_\nu)^j \tag{2.7}$$

Since the operator $I - U_{-\nu} P_{-n} U_\nu$ in any space L_p $(1 \leqq p \leqq \infty)$ can annihilate only linear combinations of the function $t^j e^{-t}$ $(j = 0,1,\cdots, -n-1)$ belonging to L_2, it operates in one-to-one fashion in all L_p spaces. By virtue of the finite-dimensionality of the projection P_{-n}, the operator $I - U_{-\nu} P_{-n} U_\nu$ is invertible in all L_p spaces. It is easy to see that the series in the right side of (2.7) converges in any L_p space; therefore equality (2.7) remains valid in all spaces L_p, $1 \leqq p \leqq \infty$.

Taking equality (2.6) into account, we obtain the fact that $V^{(n)}U_\nu$ is invertible only on the left, and a left inverse of it has the form

$$(I - U_{-\nu}P_{-n}U_\nu)^{-1}U_{-\nu}V^{-n}.$$

Now let us pass to determining conditions for the solvability of equation (2.4), which can be written in the form

$$V^{(n)}U_\nu A_+\varphi = A_-^{-1}g. \tag{2.8}$$

The operators V^n and U_ν commute for $n \geqq 0$. Therefore it is necessary for the solvability of equation (2.8) that the function $A_-^{-1}g$ belong to the intersection of the sets of values of the operators V^n and U_ν. The latter is equivalent to fulfilling conditions a).

Conversely, let the function $A_-^{-1}g$ belong to the sets of values of the operators V^n and U_ν, and let $\psi = U_{-\nu}A_-^{-1}g$ be the unique solution of the equation $U_\nu\psi = A_-^{-1}g$. Let us show that the function $\psi(t)$ belongs to the set of values of the operator V^n. Indeed, denoting the function $(A_-^{-1}g)(t)$ by $\chi(t)$, we obtain

$$\begin{aligned}\int_0^\infty \psi(t)t^je^{-t}dt &= \int_0^\infty (U_{-\nu}\chi)(t)t^je^{-t}dt = \int_0^\infty \chi(t+\nu)t^je^{-t}\,dt \\ &= \int_\nu^\infty \chi(t)(t-\nu)^je^{-t+\nu}dt = \int_0^\infty \chi(t)(t-\nu)^je^{-t+\nu}\,dt = 0.\end{aligned}$$

Thus the equation $U_\nu V^n\varphi = A_-^{-1}g$ is solvable; consequently equation (2.8) is also solvable.

Now let $n < 0$. If the function $\varphi(t) \in L_p(0, \infty)$ is a solution of equation (2.8), then

$$U_\nu A_+\varphi = V^{-n}A_-^{-1}g - f, \tag{2.9}$$

where $f = (V^{-n}V^{(n)} - I)U_\nu A_+\varphi$.

It is easy to show that the function f has the form

$$f = \sum_{j=0}^{-n-1} c_jt^je^{-t}, \tag{2.10}$$

where the coefficients c_j are such that

$$(V^{-n}A_-^{-1}g)(t) = f(t) \qquad (0 \leqq t \leqq \nu). \tag{2.11}$$

Conversely, let condition b) be fulfilled, and let the numbers c_j be so chosen that (2.11) is fulfilled for the function (2.10). Then equation (2.9) is solvable, and its solution is obviously also a solution of equation (2.8).

The theorem is proved.

In the case of $\nu > 0$ under consideration here, the operator A admits the representation $A = A_- V^{(n)} U_\nu A_+$.

An operator $A^{(-1)}$ inverse to A on the left is prescribed by the formula

$$A^{(-1)} = A_+^{-1} V^{(n)} U_{-\nu} A_-^{-1} \qquad \text{for } n \geqq 0,$$

and by the formula

$$A^{(-1)} = A_+^{-1} (I - U_{-\nu} P_{-n} U_\nu)^{-1} U_{-\nu} V^n A_-^{-1} \qquad \text{for} \quad n < 0.$$

The operator $(I - U_{-\nu} P_{-n} U_\nu)^{-1}$ can be obtained from the formula

$$(I - U_{-\nu} P_{-n} U_\nu)^{-1} = \sum_{j=0}^{\infty} (U_{-\nu} P_{-n} U_\nu)^j,$$

where the series converges in the operator norm.[3)]

4. *Case* $\nu < 0$.

THEOREM 2.2. *If* $\nu < 0$, *the operator* A *is right invertible. Every solution* $\varphi \in L_p$ *of the homogeneous equation*

$$A\varphi = 0 \tag{2.12}$$

a) *for* $n \geqq 0$ *has the form*

$$\varphi = A_+^{-1} V^{(-n)} g, \tag{2.13}$$

where $g(t)$ *is an arbitrary function of* $L_p(0, \infty)$ *which satisfies the conditions*

$$g(t) = 0 \qquad (t > -\nu),$$
$$\int_0^\infty g(t) t^j e^{-t} dt = 0 \qquad (j = 0, 1, \cdots, n-1); \tag{2.14}$$

b) *for* $n < 0$ *has the form*

$$\varphi = A_+^{-1}\Big(g(t) + \sum_{j=0}^{-n-1} c_j t^j e^{-t}\Big),$$

where $g(t)$ *is an arbitrary function of* $L_p(0, \infty)$ *which equals zero for* $-\nu < t < \infty$, *and the* c^j *are arbitrary complex numbers.*

PROOF. Let us consider the operator $A^{(-1)}$ defined by the equality

$$A^{(-1)} = A_+^{-1} V^{-n} U_{-\nu} A_-^{-1} \qquad \text{for} \quad n \leqq 0,$$

and by the equality

$$A^{(-1)} = A_+^{-1} V^{(-n)} U_{-\nu} (I - U_\nu P_n U_{-\nu})^{-1} A_-^{-1} \qquad \text{for} \quad n > 0.$$

This operator is a right inverse of A. This is obvious for $n \leqq 0$, while for $n > 0$ it follows from equality (2.6).

[3)]The operator $(I - U_{-\nu} P_{-n} U_\nu)^{-1}$ can also be found as an operator inverse to a Fredholm operator with degenerate kernel.

Equality (2.12) is equivalent to the following:

$$U_\nu V^{(n)} A_+ \varphi = 0. \tag{2.15}$$

In the case $n \geqq 0$ it is obvious that every function of the form (2.13) is a solution of equation (2.15).

Conversely, let $\varphi(t)$ be a solution of (2.15) for $n \geqq 0$. Then the function $g(t) = V^{(n)} A_+ \varphi$ satisfies conditions (2.14), and $\varphi(t)$ is expressed in terms of it by the equality $\varphi = A_+^{-1} V^{(-n)} g$.

In the case $n < 0$ the operators U_ν and $V^{(n)}$ commute; therefore every function of the form $\varphi = A_+^{-1} \psi$, where ψ belongs to the direct sum of the kernels of the operators U_ν and $V^{(n)}$, is a solution of equation (2.15).

Let $\varphi(t)$ be a solution of (2.15) for $n < 0$. Then the function $\psi = U_\nu A_+ \varphi$ is a linear combination of the functions $t^j e^{-t}$, $j = 0, 1, \cdots, -n-1$; consequently $A_+ \varphi = U_{-\nu} \psi + b(t)$, where $b(t) = 0$ for $t > -\nu$. It is easy to see in addition that the function $U_{-\nu}\psi$ is representable as the sum of two functions, one of which is a linear combination of the functions $t^j e^{-t}$, while the second equals zero for $t > -\nu$.

This completes the proof of the theorem.

5. *Case* $\nu = 0$.

THEOREM 2.3. *If* $\nu = 0$, *then the following assertions are true.*

a) *For* $n = 0$ *the operator* A *is invertible, where* $A^{-1} = A_+^{-1} A_-^{-1}$.

b) *For* $n > 0$ *the operator* $A^{(-1)} = A_+^{-1} V^{(-n)} A_-^{-1}$ *is a left inverse of* A; *equation* (2.4) *is solvable if and only if condition* (2.5) *is fulfilled.*

c) *For* $n < 0$ *the operator* $A^{(-1)} = A_+^{-1} V^{-n} A_-^{-1}$ *is a right inverse of* A; *the general solution of the homogeneous equation* (2.12) *is given by the equality*

$$\varphi = A_+^{-1}\left(\sum_{j=0}^{-n-1} c_j t^j e^{-t}\right),$$

where the c_j *are arbitrary complex numbers.*

The proofs of the preceding theorems remain essentially valid for the case of $\nu = 0$ under consideration here.

6. *Necessity of the nondegeneracy condition.*

THEOREM 2.4. *For an operator* $A \in \mathfrak{G}$ *to be invertible in the space* $L_p(0, \infty)$, $1 \leqq p \leqq \infty$, *on at least one side, it is necessary and sufficient that its symbol* $\mathscr{A}(\lambda)$ *be nondegenerate.*

If the symbol $\mathscr{A}(\lambda)$ *degenerates,* A *is neither a* Φ_+- *nor a* Φ_-*-operator.*

PROOF. The sufficiency of the conditions has already been established in the preceding theorems.

Let the symbol $\mathscr{A}(\lambda)$ of an operator $A \in \mathfrak{G}$ degenerate. It is easy to see that in any neighborhood of A there is an operator $A_1 \in \mathfrak{G}$ whose symbol has the form $\mathscr{A}_1(\lambda) = a_1(\lambda) + K_1(\lambda)$, where $a_1(\lambda) \in \mathfrak{P}$ is the almost-periodic polynomial $\sum_1^m c_j e^{i\delta_j\lambda}$, and $K_1(\lambda) \in \mathfrak{L}_0$ is a rational function, while $\mathscr{A}_1(\lambda_0) = 0$ at some point λ_0 of the real axis.

Let us consider the function

$$\mathscr{A}_2^{\pm}(\lambda) = \mathscr{A}_1(\lambda)\frac{\lambda \pm i}{\lambda - \lambda_0}. \tag{2.16}$$

It can be represented in the form

$$\mathscr{A}_2^{\pm}(\lambda) = \mathscr{A}_1(\lambda) + (\lambda_0 \pm i)\left[\frac{a_1(\lambda) - a_1(\lambda_0)}{\lambda - \lambda_0} + \frac{K_1(\lambda) - K_1(\lambda_0)}{\lambda - \lambda_0}\right].$$

The function $(K_1(\lambda) - K_1(\lambda_0)) / (\lambda - \lambda_0)$ obviously belongs to $\mathfrak{L}_0$, while the function $(a_1(\lambda) - a_1(\lambda_0)) / (\lambda - \lambda_0)$ is a linear combination of functions of the form

$$\frac{e^{i\delta\lambda} - e^{i\delta\lambda_0}}{\lambda - \lambda_0} = \int_0^{\delta} ie^{i\lambda_0(\delta - t)}\, e^{i\lambda t}\, dt,$$

which obviously also belong to $\mathfrak{L}_0$. Thus $\mathscr{A}_2^{\pm}(\lambda) \in \mathfrak{G}$. Since $(\lambda - \lambda_0) / (\lambda \pm i) \in \mathfrak{G}^{\pm}$, equality (2.16) implies that the operator A_1 is representable in the form

$$A_1 = B_- A_2^-, \qquad A_1 = A_2^+ B_+, \tag{2.17}$$

where $A_2^{\pm}$ and $B_{\pm}$ are operators of $\mathfrak{G}$ with symbols equal respectively to $\mathscr{A}_2^{\pm}(\lambda)$ and $(\lambda - \lambda_0) / (\lambda \pm i)$

Let us assume that A is invertible on some side; then A_1 is invertible on the same side. By virtue of (2.17), this implies that at least one of the operators B_+ or B_- is invertible on some side, which contradicts Theorem 8.1 of Chapter I.

Analogously, if it is assumed that A is a Φ_+- or a Φ_--operator, we obtain the result that A_1 is the same. By virtue of proposition D), § 11, Chapter I, equality (2.17) implies that one of the operators B_+ or B_- is a Φ_+- or a Φ_--operator. The latter is impossible.

The theorem is proved.

§ 3. **Integral-difference operators with continuous symbols**

The results of § 2 are extended in this section to a wider (in some sense maximal) class of Wiener-Hopf integral-difference operators.

1. *Estimating the norm of integral-difference operators.* As already noted in § 2, for any operator $A \in \hat{\mathfrak{G}}$ operating in $L_p(0,\infty)$ $(1 \leqq p \leqq \infty)$ in accordance with the rule

$$(A\varphi)(t) = \sum_{j=-\infty}^{\infty} a_j\varphi(t - \delta_j) + \int_0^{\infty} k(t - s)\varphi(s)\, ds \tag{3.1}$$

the estimate

$$| A |_p \leqq \sum_{j=-\infty}^{\infty} | a_j | + \int_{-\infty}^{\infty} | k(t) | \, dt$$

holds, where $| A |_p$ denotes the norm of A in $L_p(0, \infty)$.

LEMMA 3.1. *Let an arbitrary operator A of $\hat{\mathfrak{G}}$ have the form* (3.1), *and let $\mathscr{A}(\lambda) \in \mathfrak{G}$ be its symbol. Then for any p, $1 \leqq p \leqq \infty$, the estimates*

$$\sup_{-\infty<\lambda<\infty} | \mathscr{A}(\lambda) | \leqq | A |_p \leqq \sum_{j=-\infty}^{\infty} | a_j | + \int_{-\infty}^{\infty} | k(t) | \, dt \tag{3.2}$$

are valid, where both of the following bounds are attained:

$$| A |_1 = \sum_{j=-\infty}^{\infty} | a_j | + \int_{-\infty}^{\infty} | k(t) | \, dt, \qquad | A |_2 = \sup_{-\infty<\lambda<\infty} | \mathscr{A}(\lambda) |. \tag{3.3}$$

The strict equality

$$r_A = \sup_{-\infty<\lambda<\infty} | \mathscr{A}(\lambda) | \tag{3.4}$$

holds for the spectral radius r_A of A in any space $L_p(0, \infty)$.

PROOF. Let ε be an arbitrary positive number. Let us choose a real number λ_0 so that

$$| \mathscr{A}(\lambda_0) | \geqq \sup_{-\infty<\lambda<\infty} | \mathscr{A}(\lambda) | - \varepsilon.$$

Since the symbol of the operator $A - \mathscr{A}(\lambda_0)I$ vanishes at the point λ_0, this operator is, by virtue of Theorem 2.4, noninvertible in $L_p(0, \infty)$ $(1 \leqq p \leqq \infty)$. Consequently $r_A \geqq | \mathscr{A}(\lambda_0) |$. Taking the arbitrariness of ε into account, we obtain

$$(| A |_p \geqq) \, r_A \geqq \sup_{-\infty<\lambda<\infty} | \mathscr{A}(\lambda) |. \tag{3.5}$$

Estimates (3.2) are thus established.

The operator $A - \mu I$ is invertible for all numbers μ satisfying the condition $| \mu | > \sup | \mathscr{A}(\lambda) |$. Indeed, we have

$$\inf_{-\infty<\lambda<\infty} | \mathscr{A}(\lambda) - \mu | > 0$$

in the case under consideration, and by virtue of (1.3)

$$\sup_{-\infty<\lambda<\infty} | a(\lambda) | \leqq \sup_{-\infty<\lambda<\infty} | \mathscr{A}(\lambda) | < | \mu |$$

(where $a(\lambda)$ is taken from the representation $\mathscr{A}(\lambda) = a(\lambda) + K(\lambda)$).

The last relation implies that

$$\nu(\mathscr{A}(\lambda) - \mu) = \nu(a(\lambda) - \mu) = 0.$$

Since the integer

$$n(\mathscr{A}(\lambda) - \mu) = (1/2\pi)[\arg(1 + (a(\lambda) - \mu)^{-1} K(\lambda))]_{\lambda=-\infty}^{\infty}$$

depends continuously on μ and equals zero for large values of $|\mu|$, it equals zero for all μ under consideration. By virtue of Theorem 2.3, the operator $A - \mu I$ is invertible. Thus $r_A \leqq \sup |\mathscr{A}(\lambda)|$. This, together with (3.5), yields equality (3.4).

Let us consider an operator $\tilde{A}$ defined in $L_2(-\infty, \infty)$ by the equality

$$(\tilde{A}\varphi)(t) = \sum_{j=-\infty}^{\infty} a_j\varphi(t - \delta_j) + \int_{-\infty}^{\infty} k(t-s)\varphi(s)\,ds$$
$$(-\infty < t < \infty).$$

By utilizing the Fourier transform it is easy to establish that

$$|\tilde{A}|_2 = \sup_{-\infty<\lambda<\infty} |\mathscr{A}(\lambda)|.$$

Since $|A|_2 = |P\tilde{A}P|_2$, where P is an orthoprojection operating in $L_2(-\infty, \infty)$ in accordance with the rule

$$(P\varphi)(t) = \begin{cases} \varphi(t), & t > 0, \\ 0, & t < 0, \end{cases}$$

it follows that $|A|_2 \leqq |\tilde{A}|_2$.

Thus the second equality of (3.3) is proved. The first equality of (3.3) is obtained from the easily proved relation

$$\lim_{n\to\infty} |A\varphi_n|_1 \geqq \sum_{j=-\infty}^{\infty} |a_j| + \int_{-\infty}^{\infty} |k(t)|\,dt,$$

where

$$\varphi_n(t) = \begin{cases} n, & 0 \leqq t \leqq 1/n, \\ 0, & 1/n < t < \infty. \end{cases}$$

The lemma is proved.

LEMMA 3.2. *Let $A_1, A_2 \in \hat{\mathfrak{G}}$ be operators with symbols $\mathscr{A}_1(\lambda)$, $\mathscr{A}_2(\lambda)$, and let $A \in \hat{\mathfrak{G}}$ be an operator with symbol $\mathscr{A}(\lambda) = \mathscr{A}_1(\lambda)\mathscr{A}_2(\lambda)$. Then*

$$|A|_p \leqq |A_1|_p \cdot |A_2|_p \qquad (1 \leqq p \leqq \infty). \tag{3.6}$$

PROOF. Obviously it can be supposed without loss of generality that A_1 and A_2 have the form

$$(A_m\varphi)(t) = \sum_{j=1}^{l} a_j^{(m)}\varphi(t - \delta_j^{(m)}) + \int_0^{\infty} k_m(t-s)\varphi(s)\,ds \quad (m = 1,2),$$

where the $k_m(t)$ are functions in $L_1(-\infty, \infty)$ with compact support. On this assumption it is easily proved that for sufficiently large δ the equality $A = U_{-\delta} A_1 A_2 U_\delta$ holds, which implies relation (3.6).

Let us recall that every operator A of $\widehat{\mathfrak{G}}$ can be written in the form

$$(A\varphi)(t) = \int_0^\infty \varphi(s)\,d\omega(t-s),$$

where $\omega(t)$ is a function of bounded variation without singular component.

LEMMA 3.3. *Let an operator $A \in \widehat{\mathfrak{G}}$ be written in the form*

$$(A\,\varphi)(t) = \int_0^\infty \varphi(s)\,d\omega(\overline{t-s}).$$

Then in any space $L_p(0, \infty)$, $1 \leqq p \leqq \infty$, the adjoint operator A^* has the form

$$(A^*\varphi)(t) = \int_0^\infty \varphi(s)\,d\omega(s-t),$$

and

$$|A|_p = |A^*|_p. \tag{3.7}$$

PROOF. The first assertion is verifiable without difficulty. To prove the second assertion, let us introduce the operators

$$(S_\tau\varphi)(t) = \begin{cases} \varphi(\tau - t), & 0 < t < \tau, \\ \varphi(t), & \tau < t < \infty, \end{cases}$$

$$(P_\tau\varphi)(t) = \begin{cases} \varphi(t), & 0 < t < \tau, \\ 0, & \tau < t < \infty \end{cases} \qquad (0 < \tau < \infty).$$

The operator S_τ is invertible and isometric, and $S_\tau^{-1} = S_\tau$, while P_τ is a projection with unit norm.

The following equality is immediately verifiable:

$$P_\tau A' P_\tau = S_\tau P_\tau A P_\tau S_\tau, \tag{3.8}$$

where A' is an operator transpose to A, i.e.

$$(A'\varphi)(t) = \int_0^\infty \varphi(s)\,d\omega(s-t).$$

Since $(\overline{A'\varphi})(t) = (A^*\bar{\varphi})(t)$,

$$|P_\tau A' P_\tau|_p = |P_\tau A^* P_\tau|_p. \tag{3.9}$$

Equalities (3.8) and (3.9) imply

$$|P_\tau A^* P_\tau|_p = |P_\tau A P_\tau|_p \qquad (0 < \tau < \infty),$$

which, together with the relation

$$|A|_p \leqq \liminf_{\tau\to\infty} |P_\tau A P_\tau|_p$$

implies (3.7)

2. *Operators with continuous symbols.* Let us denote by $\hat{\mathfrak{G}}_p$ the closure of the set $\mathfrak{G}$ in the operator norm $|\ |_p$, $1 \leqq p \leqq \infty$. Let A be an arbitrary operator of $\hat{\mathfrak{G}}_p$, and A_n, $n = 1,2,\cdots$, a sequence in $\mathfrak{G}$ which converges to A. By virtue of estimates (3.2) and (1.3), the sequence of functions $\mathscr{A}_n(\lambda) = a_n(\lambda) + K_n(\lambda)$ of $\mathfrak{G}$ is such that the sequence $a_n(\lambda)$ converges uniformly to some almost-periodic function $a(\lambda)$, and the sequence $K_n(\lambda)$, to some continuous function $K(\lambda)$ which tends to zero as $\lambda \to \pm\infty$.

It is easy to see that the function $\mathscr{A}(\lambda) = a(\lambda) + K(\lambda)$ does not depend on the choice of the sequence $A_n \in \mathfrak{G}$ converging to A. Thus with each operator $A \in \hat{\mathfrak{G}}_p$ is associated a continuous function $\mathscr{A}(\lambda)$, which we shall call the *symbol* of A.

It is easy to see that $\hat{\mathfrak{G}} = \hat{\mathfrak{G}}_1$.

Estimates (3.2) can now be extended to operators A of $\hat{\mathfrak{G}}_p$:

$$\sup_{-\infty<\lambda<\infty} |\mathscr{A}(\lambda)| \leqq |A|_p \leqq \sum_{j=-\infty}^{\infty} |a_j| + \int_{-\infty}^{\infty} |k(t)|\, dt \qquad (A \in \hat{\mathfrak{G}}_p),$$

while for $p = 2$ the second equality of (3.3) can be extended to operators of $\hat{\mathfrak{G}}_2$:

$$|A|_2 = \sup_{-\infty<\lambda<\infty} |\mathscr{A}(\lambda)| \qquad (A \in \hat{\mathfrak{G}}_2). \tag{3.10}$$

It is easy to see that $\hat{\mathfrak{G}}_{p_1} \subset \hat{\mathfrak{G}}_{p_2}$ for $1 \leqq p_1 \leqq p_2 \leqq 2$, and if $A \in \hat{\mathfrak{G}}_{p_1}$, then $|A|_{p_1} \geqq |A|_{p_2}$. Lemma 3.3 implies that $\hat{\mathfrak{G}}_p = \hat{\mathfrak{G}}_q$ $(p^{-1} + q^{-1} = 1)$ and $|A|_p = |A^*|_p$ for all A of $\hat{\mathfrak{G}}_p$.

Let us denote by $\mathfrak{G}_p$ the set of symbols of all of the operators of $\hat{\mathfrak{G}}_p$. The correspondence between the operators of $\hat{\mathfrak{G}}_p$ and their symbols of $\mathfrak{G}_p$ is one-to-one. Indeed, let us suppose that the symbol identically equal to zero corresponds to a nonzero operator $A \in \hat{\mathfrak{G}}_p$. Let the sequence $A_n \in \mathfrak{G}$ converge to A in the norm $|\ |_p$. By virtue of (3.2) and (3.3), the sequence A_n is fundamental in the operator norm $|\ |_2$ and consequently converges in that norm to some operator B. A and B coincide on the intersection $L_2 \cap L_p$; consequently $B \neq 0$. The symbol identically equal to zero corresponds to the operator B, which contradicts (3.10).

3. *Maximal ideal of the algebra* $\mathfrak{G}$.

THEOREM 3.1. *The set M_{λ_0} of all functions of $\mathfrak{G}$ which vanish at the point λ_0 $(-\infty < \lambda_0 < \infty)$ is a maximal ideal of the algebra $\mathfrak{G}$.*

The direct sum $M_{\mathfrak{P}} + \mathfrak{L}_0$, where $M_{\mathfrak{P}}$ is any maximal ideal of the algebra $\mathfrak{P}$, is also a maximal ideal of $\mathfrak{G}$.

All maximal ideals of $\mathfrak{G}$ are exhausted by the ideals of the two types mentioned.

PROOF. It is easily verifiable that M_{λ_0} and $M_{\mathfrak{P}} + \mathfrak{L}_0$ are maximal ideals of $\mathfrak{G}$.

Now let M be some maximal ideal of $\mathfrak{G}$, and let $M_1 = M \cap \mathfrak{P}$ and $M_2 = M \cap \mathfrak{L}$. It is easy to see that M_1 and M_2 are maximal ideals of $\mathfrak{P}$ and $\mathfrak{L}$, respectively, and that $M = M_1 + M_2$. If $M_2 = \mathfrak{L}_0$, the ideal M is of the second form mentioned.

Let $M_2 \neq \mathfrak{L}_0$. Then, as is well known, M_2 coincides with the set of all functions of $\mathfrak{L}$ which vanish at some real point λ_0(cf. Gel'fand, Raĭkov and Šilov [1], § 17). Let us show that M_1 coincides with the set of all functions of $\mathfrak{P}$ which vanish at this same point λ_0. Let us suppose that this is not so, and let $a(\lambda)$ and $K(\lambda)$ be functions of M_1 and $\mathfrak{L}_0$, respectively, such that $a(\lambda_0) \neq 0$ and $K(\lambda_0) \neq 0$. Since M and $\mathfrak{L}_0$ are ideals of $\mathfrak{G}$, the function $a(\lambda)K(\lambda) \in M \cap \mathfrak{L}_0 \subset M_2$. But this contradicts the fact that $a(\lambda_0)K(\lambda_0) \neq 0$.

The theorem is proved.

If $\mathfrak{R}$ is some Banach algebra of functions defined on the real axis, let us denote by $M_\lambda(\mathfrak{R})$, $-\infty < \lambda < \infty$, the set of all functions of $\mathfrak{R}$ which vanish at the point λ. Let us denote by $\mathfrak{M}(\mathfrak{R})$ the compact Hausdorff space of all maximal ideals of $\mathfrak{R}$.

The set $M_\lambda(\mathfrak{G})$, $-\infty < \lambda < \infty$, is a maximal ideal of the first type in $\mathfrak{G}$.

THEOREM 3.2. *The totality of all maximal ideals* $M_\lambda(\mathfrak{G})$, $-\infty < \lambda < \infty$, *is dense in* $\mathfrak{M}(\mathfrak{G})$.

PROOF. Let M_0 be an arbitrary ideal of the form $M_0 = M_{\mathfrak{P}} + \mathfrak{L}_0$, where $M_{\mathfrak{P}}$ is a maximal ideal of $\mathfrak{P}$. Let us consider the neighborhood of M_0 which consists of all ideals $M \in \mathfrak{M}(\mathfrak{G})$ for which

$$|\mathscr{A}_j(M) - \mathscr{A}_j(M_0)| < \varepsilon \qquad (j = 1,\cdots, n), \tag{3.11}$$

where $\varepsilon > 0$ and the $\mathscr{A}_j$ are some elements of $\mathfrak{G}$:

$$\mathscr{A}_j(\lambda) = a_j(\lambda) + K_j(\lambda) \qquad (a_j(\lambda) \in \mathfrak{P},\ K_j(\lambda) \in \mathfrak{L}_0).$$

Let us show that at least one ideal of the type $M_\lambda(\mathfrak{G})$ is contained in this neighborhood.

As is well known (cf. Gel'fand, Raĭkov and Šilov [1], § 29), the set of ideals $M_\lambda(\mathfrak{P})$ $(-\infty < \lambda < \infty)$ is dense in $\mathfrak{M}(\mathfrak{P})$; consequently a point λ_1 exists such that

$$|a_j(\lambda_1) - a_j(M_{\mathfrak{P}})| < \varepsilon/3 \qquad (j = 1,\cdots, n). \tag{3.12}$$

Let Δ be so large a positive number that $|K_j(\lambda)| < \varepsilon/3$ for $|\lambda| > \Delta$. Properties of almost-periodic functions imply the existence of a point λ_0 with $|\lambda_0| > \Delta$ such that

$$|a_j(\lambda_0) - a_j(\lambda_1)| < \varepsilon/3 \qquad (j = 1,\cdots, n). \tag{3.13}$$

The ideal $M_{\lambda_0}(\mathfrak{G})$ belongs to neighborhood (3.11). Indeed, $K_j(M_0) = 0$ because $K_j(\lambda) \in \mathfrak{L}_0 \subset M_0$, while $K_j(M_{\lambda_0}(\mathfrak{G})) = K_j(\lambda_0)$ and $a_j(M_{\lambda_0}(\mathfrak{G})) = a_j(\lambda_0)$ Consequently

$$\begin{aligned}|\mathscr{A}_j(M_{\lambda_0}(\mathfrak{G})) - \mathscr{A}_j(M_0)| &= |a_j(\lambda_0) + K_j(\lambda_0) - a_j(M_0)| \\ &\leqq |a_j(\lambda_0) - a_j(M_0)| + \varepsilon/3 \qquad (j = 1,\cdots, n).\end{aligned}$$

Taking into account relations (3.12) and (3.13) and the equality $a_j(M_0) = a_j(M_{\mathfrak{P}})$, we obtain

$$|\mathscr{A}_j(M_{\lambda_0}(\mathfrak{G})) - \mathscr{A}_j(M_0)| < \varepsilon \qquad (j = 1,\cdots, n).$$

The theorem is proved.

4. *The algebra $\mathfrak{G}_p$ and its maximal ideals.* Let us introduce a norm on the linear set of functions $\mathfrak{G}_p$ by setting

$$|\mathscr{A}(\lambda)|_p = |A|_p. \tag{3.14}$$

Lemma 3.2 implies that if the functions $\mathscr{A}_1(\lambda)$, $\mathscr{A}_2(\lambda) \in \mathfrak{G}_p$, then $\mathscr{A}(\lambda) = \mathscr{A}_1(\lambda)\mathscr{A}_2(\lambda)$ also belongs to $\mathfrak{G}_p$, and

$$|\mathscr{A}_1(\lambda)\mathscr{A}_2(\lambda)|_p \leqq |\mathscr{A}_1(\lambda)|_p |\mathscr{A}_2(\lambda)|_p.$$

Thus $\mathfrak{G}_p$, $1 \leqq p \leqq \infty$, turns out to be a commutative Banach algebra. By virtue of Lemma 3.3, $\mathfrak{G}_p$ is an algebra with a symmetric involution (cf. Gel'fand, Raĭkov and Šilov [1], § 8).

Relations (3.3) imply that $\mathfrak{G}_1 = \mathfrak{G}$, while $\mathfrak{G}_2$ represents the algebra of all functions of the form $\mathscr{A}(\lambda) = a(\lambda) + K(\lambda)$, where $a(\lambda)$ is any uniform almost-periodic function and $K(\lambda)$ $(-\infty < \lambda < \infty)$ is an arbitrary continuous function vanishing at infinity, with norm

$$|\mathscr{A}(\lambda)|_2 = \sup_{-\infty<\lambda<\infty} |\mathscr{A}(\lambda)|.$$

We need the following two simple propositions concerning Banach algebras.

1°. *Let $\mathfrak{R}_1$ and $\mathfrak{R}_2$ be commutative Banach algebras, and let $\mathfrak{R}_1$ be imbedded densely*[4] *into $\mathfrak{R}_2$. Then the space $\mathfrak{M}(\mathfrak{R}_2)$ is homeomorphic to a closed subset of $\mathfrak{M}(\mathfrak{R}_1)$.*

Indeed, a maximal ideal $M_1 \in \mathfrak{M}(\mathfrak{R}_1)$ defined by the equality $M_1 = M_2 \cap \mathfrak{R}_1$ can be placed into correspondence with each maximal ideal $M_2 \in \mathfrak{M}(\mathfrak{R}_2)$. Let us denote by $\mathfrak{N}$ the set of maximal ideals of $\mathfrak{M}(\mathfrak{R}_1)$ which is obtained under such a correspondence, i.e. $\mathfrak{N}$ is the set of maximal ideals of $\mathfrak{M}(\mathfrak{R}_1)$ which admits an extension (unique) to the maximal ideals of $\mathfrak{M}(\mathfrak{R}_2)$.

Let us show that under this correspondence the mapping of $\mathfrak{M}(\mathfrak{R}_2)$ onto $\mathfrak{N}$ is continuous. Let the ideal $M_0 \in \mathfrak{N}$, let

$$U = \{M \in \mathfrak{N}: |x_j(M) - x_j(M_0)| < \varepsilon;\ x_j \in \mathfrak{R}_1;\ j = 1,\cdots, n\}$$

be a neighborhood of it and $\tilde{M}_0 \in \mathfrak{M}(\mathfrak{R}_2)$ an extension of M_0. Then the neighborhood

[4] The algebra $\mathfrak{R}_1$ is said to be imbedded densely into the algebra $\mathfrak{R}_2$ if $\mathfrak{R}_1 \subset \mathfrak{R}_2$, $\bar{\mathfrak{R}}_1 = \mathfrak{R}_2$ and a constant $c > 0$ exists such that $|x|_2 \leqq c|x|_1$ for any $x \in \mathfrak{R}_1$.

$$V = \{\tilde{M} \in \mathfrak{M}(\mathfrak{R}_2) : |x_j(\tilde{M}) - x_j(\tilde{M}_0)| < \varepsilon\}$$

is mapped onto U. Thus the mapping is continuous, the set $\mathfrak{N}$ is closed, and consequently the sets $\mathfrak{M}(\mathfrak{R}_2)$ and $\mathfrak{N}$ are homeomorphic.

2°. *Let the commutative Banach algebras* $\mathfrak{R}_j$, $j = 1,2,3$, *be successively imbedded densely:* $\mathfrak{R}_1 \subset \mathfrak{R}_2 \subset \mathfrak{R}_3$. *If every maximal ideal of the algebra* $\mathfrak{R}_1$ *is extensible to a maximal ideal of the algebra* $\mathfrak{R}_3$, *then this is also true of the pair* $\mathfrak{R}_2$, $\mathfrak{R}_3$.

Indeed, let $M_2 \in \mathfrak{M}(\mathfrak{R}_2)$ and $M_1 = M_2 \cap \mathfrak{R}_1$. Then $M_1 \in M(\mathfrak{R}_1)$, and the ideal M_1 is consequently contained in some ideal $M_3 \in \mathfrak{M}(\mathfrak{R}_3)$. The closure of the ideal M_1 in the norm of $\mathfrak{R}_3$, and a fortiori in the norm of $\mathfrak{R}_2$, is contained in M_3. But the closure of M_1 in the norm of $\mathfrak{R}_2$ equals M_2.

THEOREM 3.3. *The spaces* $\mathfrak{M}(\mathfrak{G})$ *and* $\mathfrak{M}(\mathfrak{G}_2)$ *are homeomorphic. In other words, every maximal ideal of the algebra* $\mathfrak{G}$ *is extensible* (*uniquely*) *to a maximal ideal of the algebra* $\mathfrak{G}_2$.

Indeed, by virtue of proposition 1° the space $\mathfrak{M}(\mathfrak{G}_2)$ is homeomorphic to the closed set $\mathfrak{N} \subset \mathfrak{M}(\mathfrak{G})$ consisting of all maximal ideals of $\mathfrak{G}$ which admit extension to maximal ideals of $\mathfrak{G}_2$. All maximal ideals $M_\lambda(\mathfrak{G})$ $(-\infty < \lambda < \infty)$ obviously belong to $\mathfrak{N}$. Since the set of all ideals $M_\lambda(\mathfrak{G})$ is dense in $\mathfrak{M}(\mathfrak{G})$ by virtue of Theorem 3.2, it follows that $\mathfrak{N} = \mathfrak{M}(\mathfrak{G})$.

THEOREM 3.4. *Let* M *be a maximal ideal of the algebra* $\mathfrak{G}_2$. *Then* $\tilde{M} = M \cap \mathfrak{G}_p$ *is a maximal ideal of* $\mathfrak{G}_p$. *All of the maximal ideals of* $\mathfrak{G}_p$ *are exhausted by the ideals of this kind.*

This theorem is an immediate corollary of proposition 2°.

THEOREM 3.5. *For an element* $\mathscr{A}(\lambda) \in \mathfrak{G}_p$ *to be invertible it is necessary and sufficient that*

$$\inf_{-\infty<\lambda<\infty} |\mathscr{A}(\lambda)| > 0. \tag{3.15}$$

The necessity of condition (3.15) is obvious. Let us show its sufficiency. Let condition (3.15) be fulfilled for an element $\mathscr{A}(\lambda) = a(\lambda) + K(\lambda) \in \mathfrak{G}_p$ (here $a(\lambda)$ is an almost-periodic function and $K(\lambda)$ is a continuous function equal to zero at infinity). Then

$$\inf_{-\infty<\lambda<\infty} |a(\lambda)| \geqq \inf_{-\infty<\lambda<\infty} |\mathscr{A}(\lambda)| > 0;$$

consequently $\mathscr{A}(\lambda)$ is invertible in $\mathfrak{G}_2$. By virtue of Theorem 3.4, it is also invertible in $\mathfrak{G}_p$.

5. *Fundamental theorem.* The definitions of the indices $\nu(\mathscr{A})$ and $n(\mathscr{A})$ for nondegenerate functions $\mathscr{A}(\lambda) \in \mathfrak{G}$ obviously remain valid for an arbitrary

nondegenerate function $\mathscr{A}(\lambda) \in \mathfrak{G}_2$. Moreover, the indices $\nu(\mathscr{A})$ and $n(\mathscr{A})$ retain their stability property, i.e. under small (in the norm of the algebra $\mathfrak{G}_2$) perturbations of the nondegenerate function $\mathscr{A}(\lambda) \in \mathfrak{G}_2$ the indices do not vary.

Let us note in addition that, just as in the algebra $\mathfrak{G}$, in each of the algebras $\mathfrak{G}_p$, $1 < p \leqq \infty$, the correspondence between the operators of $\widehat{\mathfrak{G}}_p$ and their symbols is partially multiplicative in the following sense:

Let the operator A_1 belong to the closure of $\widehat{\mathfrak{G}}_+$ in the norm $|\ \ |_p$, let A_2 belong to the closure of $\widehat{\mathfrak{G}}_-$ in the same norm, and let A be an arbitrary operator of $\widehat{\mathfrak{G}}_p$.

Then the operator $B = A_1AA_2$ also belongs to $\widehat{\mathfrak{G}}_p$, and its symbol $\mathscr{B}(\lambda)$ equals the product $\mathscr{A}_1(\lambda)\mathscr{A}(\lambda)\mathscr{A}_2(\lambda)$.

THEOREM 3.6. *For an operator $A \in \widehat{\mathfrak{G}}_p$ to be invertible in the space $L_p(0, \infty)$, $1 \leqq p < \infty$, on at least one side, it is necessary and sufficient that condition* (3.15) *be fulfilled.*

If this condition is fulfilled, then for $\nu(\mathscr{A}) > 0$ A is invertible only on the left, for $\nu(\mathscr{A}) < 0$ it is invertible only on the right, and for $\nu(\mathscr{A}) = 0$ the invertibility of A is consistent with the index $n(\mathscr{A})$, i. e. A is invertible, only left invertible, or only right invertible depending on whether the index $n(\mathscr{A})$ is equal to zero, positive, or negative.

If condition (3.15) *is not fulfilled, A is neither a Φ_+- nor a Φ_--operator.*

PROOF. Let $A \in \widehat{\mathfrak{G}}_p$, and let its symbol satisfy condition (3.15). Then by virtue of Theorem 3.5, $\mathscr{A}^{-1}(\lambda) \in \mathfrak{G}_p$. By virtue of the density of $\mathfrak{G}$ in $\mathfrak{G}_p$, a nondegenerate function $\mathscr{A}_0(\lambda) \in \mathfrak{G}$ can be found such that the function $\mathscr{A}(\lambda)$ is representable in the form

$$\mathscr{A}(\lambda) = \mathscr{A}_0(\lambda)(1 + \mathscr{B}(\lambda)), \tag{3.16}$$

where $\mathscr{B}(\lambda) \in \mathfrak{G}_p$, and is of sufficiently small norm in $\mathfrak{G}_p$. It is easy to see that the equalities $\nu(\mathscr{A}_0) = \nu(\mathscr{A})$ and $n(\mathscr{A}_0) = n(\mathscr{A})$ hold. According to Theorem 1.1, the function $\mathscr{A}_0(\lambda)$ admits the factorization

$$\mathscr{A}_0(\lambda) = \mathscr{A}_-(\lambda)e^{i\nu\lambda}\left(\frac{\lambda-i}{\lambda+i}\right)^n\mathscr{A}_+(\lambda), \tag{3.17}$$

where $\mathscr{A}_+^{\pm1}(\lambda) \in \mathfrak{G}^+$, $\mathscr{A}_-^{\pm1}(\lambda) \in \mathfrak{G}^-$, $\nu = \nu(\mathscr{A})$ and $n = n(\mathscr{A})$.

Equalities (3.16) and (3.17) imply that A admits the following representation depending on the signs of the indices ν and n:

1) $A = A_-U_\nu V^{(n)}(I + B)A_+$ for $\nu \leqq 0$, $n \leqq 0$;
2) $A = A_-U_\nu(I + B)V^nA_+$ for $\nu < 0$, $n > 0$;
3) $A = A_-V^{(n)}(I + B)U_\nu A_+$ for $\nu > 0$, $n < 0$;
4) $A = A_-(I + B)U_\nu V^nA_+$ for $\nu \geqq 0$, $n \geqq 0$,

where $A_\pm$ and B are operators with the symbols $\mathscr{A}_\pm(\lambda)$ and $\mathscr{B}(\lambda)$, respectively.

It is possible to assume that $|B|_p < 1$ and therefore $I + B$ is invertible;

hence in the first case the operator

$$A^{(-1)} = A_{+}^{-1}(I + B)^{-1}V^{-n}U_{-\nu}A_{-}^{-1}$$

is a right inverse of A, while in the last case the operator

$$A^{(-1)} = A_{+}^{-1}V^{(-n)}U_{-\nu}(I + B)^{-1}A_{-}^{-1}$$

is a left inverse of A.

In the proof of Theorem 2.1 it was shown in particular that the operator $U_\nu V^n$ ($\nu < 0$, $n > 0$) is right invertible, while $V^{(n)}U_\nu$ ($\nu > 0$, $n < 0$) is left invertible. Since the operators $U_\nu BV^{(n)}$ and $V^{(n)}BU_\nu$ are of sufficiently small norm, the operator

$$U_\nu(I + B)V^{(n)} = U_\nu V^{(n)} + U_\nu BV^{(n)}$$

is right invertible for $\nu < 0$ and $n > 0$, while the operator

$$V^{(n)}(I + B)U_\nu = V^{(n)}U_\nu + V^{(n)}BU_\nu$$

is left invertible for $\nu > 0$ and $n < 0$. It follows immediately from this that A is right invertible in case 2) and left invertible in case 3).

The last assertion of the theorem is proved exactly as the corresponding proposition in Theorem 2.4.

The theorem is proved.

It is easy to derive from equalities (2.2) and (2.3) that $\dim \operatorname{Ker} A = \infty$ for $\nu < 0$ and $\dim \operatorname{Coker} A = \infty$ for $\nu > 0$, while for $\nu = 0$ we have $\dim \operatorname{Ker} A = -n$ if $n < 0$, and $\dim \operatorname{Coker} A = n$ if $n > 0$.

§ 4. Almost-periodic functions and semigroups

Let the semigroups of operators U_t and U_{-t}, $0 \leq t < \infty$, on the space $\mathfrak{B}$ satisfy the conditions of § 9, Chapter I. In the space $\mathfrak{B}$ let us consider the operator

$$A = \sum_{-\infty}^{\infty} a_j U_{\delta_j}, \tag{4.1}$$

where the δ_j are distinct real numbers and the a_j are complex numbers the series of which is absolutely convergent. Let us denote by $\widehat{\mathfrak{P}}$ the set of all operators of the form (4.1).

Let us put into correspondence with the operator A defined by (4.1) the almost-periodic function

$$\mathscr{A}(\lambda) = \sum_{j=-\infty}^{\infty} a_j e^{i\delta_j\lambda}$$

called, as previously, the *symbol* of A.

THEOREM 4.1. *For an operator $A \in \widehat{\mathfrak{P}}$ to be invertible, it is necessary and sufficient that*

$$\inf_{-\infty<\lambda<\infty} |\mathscr{A}(\lambda)| > 0. \tag{4.2}$$

If this condition is fulfilled, the invertibility of A is consistent with the index $\nu = \nu(\mathscr{A})$ of its symbol.

PROOF. Let us note first of all that the correspondence between the operators $A \in \widehat{\mathfrak{P}}$ and their symbols $\mathscr{A}(\lambda) \in \mathfrak{P}$ is partially multiplicative, as in § 2. The latter means that if the operators $A_1 A_2 \in \mathfrak{P}$ and their symbols are such that $\mathscr{A}_1(\lambda) \in \mathfrak{G}^- \cap \mathfrak{P}$ and $\mathscr{A}_2(\lambda) \in \mathfrak{G}^+ \cap \mathfrak{P}$ (cf. §1), then for any operator A of $\widehat{\mathfrak{P}}$ the operator $B = A_1 A A_2$ also belongs to $\widehat{\mathfrak{P}}$, and the equality $\mathscr{B}(\lambda) = \mathscr{A}_1(\lambda)\mathscr{A}(\lambda)\mathscr{A}_2(\lambda)$ is valid for its symbol.

Let condition (4.2) be fulfilled. Then, by virtue of factorization (1.5), the symbol $\mathscr{A}(\lambda)$ is representable in the form

$$\mathscr{A}(\lambda) = \mathscr{A}_-(\lambda)\, e^{i\nu\lambda}\, \mathscr{A}_+(\lambda),$$

where $\nu = \nu(\mathscr{A})$ and $\mathscr{A}_\pm(\lambda)$, $\mathscr{A}_\pm^{-1}(\lambda) \in \mathfrak{G}^\pm \cap \mathfrak{P}$.

By virtue of the partial multiplicativity of the correspondence between $\widehat{\mathfrak{P}}$ and $\mathfrak{P}$, A can be represented in the form $A = A_- U_\nu A_+$, where $A_\pm$ are operators of $\widehat{\mathfrak{P}}$ with symbols $\mathscr{A}_\pm(\lambda)$.

The operators $A_\pm$ are invertible, and their inverses are operators of $\widehat{\mathfrak{P}}$, with symbols $\mathscr{A}_\pm^{-1}(\lambda)$, respectively. Therefore A is invertible on the same side as U_ν.

Now let condition (4.2) not be fulfilled. Let us suppose nevertheless that A is invertible, perhaps only unilaterally. As in § 2, it can be assumed that $\mathscr{A}(\lambda)$ is an almost-periodic polynomial which vanishes at some real point λ_0. Following the plan of the proof of Theorem 2.4, let us consider the function

$$\mathscr{A}_\pm(\lambda) = \mathscr{A}(\lambda)\frac{\lambda \pm i}{\lambda - \lambda_0} = \mathscr{A}(\lambda) + (\lambda_0 \pm i)\frac{\mathscr{A}(\lambda)}{\lambda - \lambda_0}.$$

As we established in the proof of Theorem 2.4, the function $\mathscr{A}(\lambda)/(\lambda - \lambda_0)$ is the Fourier transform of some function $k(t) \in L_1(-\infty, \infty)$. It is not difficult now to verify that A admits the representations $A = A^+B_+$ and $A = B_-A^-$, where

$$A^\pm f = Af + (\lambda_0 \pm i)\int_{-\infty}^{\infty} k(t) U_t f\, dt,$$

and $B_\pm$ are operators of $\widehat{\mathfrak{G}}$ with symbols $(\lambda - \lambda_0)/(\lambda \pm i)$. We have arrived at a contradiction, since the operators $B_\pm$ do not have inverses on either side.

REMARK. If condition (4.2) is not fulfilled, then, by virtue of proposition D), § 11, Chapter I, A is neither a Φ_+- nor a Φ_--operator.

Let us note in addition that the symbol, being an almost-periodic function, can be associated in a natural way not only with operators of $\widehat{\mathfrak{P}}$ but also with operators of the closure of this set. Theorem 4.1 remains valid for the operators of this closure.

§ 5. Projection method for solving integral-difference equations

Let us denote by P_τ, $0 < \tau < \infty$, the projection defined in $L_p(0, \infty)$, $1 \leqq p < \infty$,[5] by the equality

$$(P_\tau \varphi)(t) = \begin{cases} \varphi(t), & 0 < t < \tau, \\ 0, & \tau < t < \infty. \end{cases}$$

If $A \in \widehat{\mathfrak{G}}$ and the operator $P_\tau A P_\tau$ is invertible on the subspace $P_\tau L_p$, let us denote by $(P_\tau A P_\tau)^{-1}$ the inverse operator of $P_\tau A P_\tau$ on the subspace $P_\tau L_p$. If the operator $P_\tau A P_\tau$ is noninvertible, let us set $|(P_\tau A P_\tau)^{-1}| = \infty$.

[5]The case $p = \infty$ is excluded since then P_τ does not converge (strongly) to the identity operator.

Here we shall take the operators A defined by the usual equality

$$(A\varphi)(t) = \sum_{j=-\infty}^{\infty} a_j\varphi(t-\delta_j) + \int_0^{\infty} k(t-s)\varphi(s)ds$$

only from the original set $\widehat{\mathfrak{G}}$.

THEOREM 5.1. *If*

$$\liminf_{\tau\to\infty} |(P_\tau A P_\tau)^{-1}|_p < \infty,$$

then the symbol $\mathscr{A}(\lambda)$ of A is nondegenerate, and $\nu(\mathscr{A}) = n(\mathscr{A}) = 0$.

Under fulfillment of these conditions the truncated equation

$$\sum_{t-\tau<\delta_j<t} a_j\varphi(t-\delta_j) + \int_0^{\tau} k(t-s)\varphi(s)ds = f(t) \qquad (0 < t < \tau), \tag{5.1}$$

beginning with some τ, has the unique solution $\varphi_\tau(t) \in P_\tau L_p$, and as $\tau \to \infty$ the functions

$$\tilde{\varphi}_\tau(t) = \begin{cases} \varphi_\tau(t), & t < \tau, \\ 0, & t > \tau \end{cases}$$

converge in the norm of $L_p(0, \infty)$ to a solution of the equation $A\varphi(t) = f(t)$, whatever the function $f(t) \in L_p(0, \infty)$ may be.

PROOF. The first assertion of the theorem is an immediate corollary of Lemma 1.1, Chapter III.

Let us use the plan of G. Baxter's proof of Theorem 2.1, Chapter III, to prove the second assertion. First let us carry out the proof for the case $L_1(0, \infty)$.

Let the function $\varphi(t)$ of $P_\tau L_1$ be a solution of equation (5.1), which can be written in the form $P_\tau A P_\tau \varphi = P_\tau f$. Let us set

$$g_1(t) = \begin{cases} -(AP_\tau\varphi)(t), & t > \tau, \\ 0, & t < \tau, \end{cases} \qquad g_2(t) = \begin{cases} -(AP_\tau\varphi)(t), & t < 0, \\ 0, & t > 0. \end{cases}$$

Then

$$\sum_{j=-\infty}^{\infty} a_j\varphi(t-\delta_j) + \int_{-\infty}^{\infty} k(t-s)\varphi(s)ds = f(t) + g_1(t) + g_2(t)$$

$(-\infty < t < \infty)$. Applying the Fourier transform to the latter equality, we obtain

$$\mathscr{A}(\lambda)\Phi(\lambda) = F(\lambda) + G_1(\lambda) + G_2(\lambda) \qquad (-\infty < \lambda < \infty).$$

Since the symbol $\mathscr{A}(\lambda)$ is nondegenerate and $\nu(\mathscr{A}) = n(\mathscr{A}) = 0$, by virtue of Theorem 1.1 the symbol admits the factorization

$$\mathscr{A}(\lambda) = \mathscr{A}_-(\lambda)\mathscr{A}_+(\lambda),$$

where $\mathscr{A}_\pm(\lambda)$, $\mathscr{A}_\pm^{-1}(\lambda) \in \mathfrak{G}_\pm$.

Let us introduce on the algebra $\mathfrak{G}$ (and in particular on the set $\mathfrak{L}_0$ of the Fourier transforms of all functions of $L_1(-\infty, \infty)$) the projection Q_γ, $-\infty < \gamma < \infty$, defined by the equality

$$Q_\gamma\left(\sum_{j=-\infty}^{\infty} a_j e^{i\delta_j\lambda} + \int_{-\infty}^{\infty} k(t)e^{i\lambda t}\,dt\right) = \sum_{\delta_j<\gamma} a_j e^{i\delta_j\lambda} + \int_{-\infty}^{\gamma} k(t)e^{i\lambda t}\,dt.$$

The equality

$$\mathscr{A}_+(\lambda)\Phi(\lambda) = \mathscr{A}_-^{-1}(\lambda)F(\lambda) + \mathscr{A}_-^{-1}(\lambda)G_1(\lambda) + \mathscr{A}_-^{-1}(\lambda)G_2(\lambda) \tag{5.2}$$

implies that $G_2\mathscr{A}_-^{-1} = -Q_0(F\mathscr{A}_-^{-1} + G_1\mathscr{A}_-^{-1})$.

It is easy to verify that $Q_0(G_1\mathscr{A}_-^{-1}) = Q_0(G_1Q_{-\tau}\mathscr{A}_-^{-1})$. Therefore

$$G_2\mathscr{A}_-^{-1} = -Q_0(F\mathscr{A}_-^{-1}) - Q_0(G_1Q_{-\tau}\mathscr{A}_-^{-1})$$

and

$$\|G_2\mathscr{A}_-^{-1}\| \leqq \|F\|\,\|\mathscr{A}_-^{-1}\| + \|G_1\mathscr{A}_+^{-1}\|\,\|\mathscr{A}_+Q_{-\tau}\mathscr{A}_-^{-1}\|.$$[6]

Whatever $\varepsilon > 0$ may be, the relation $\|\mathscr{A}_+Q_{-\tau}\mathscr{A}_-^{-1}\| < \varepsilon$ is fulfilled for sufficiently large τ; consequently

$$\|G_2\mathscr{A}_-^{-1}\| \leqq c\|F\| + \varepsilon\|G_1\mathscr{A}_+^{-1}\|, \tag{5.3}$$

where $c = \max(\|\mathscr{A}_-^{-1}\|, \|\mathscr{A}_+^{-1}\|)$. Analogously, proceeding from the equalities

$$\mathscr{A}_-(\lambda)\Phi(\lambda) = \mathscr{A}_+^{-1}(\lambda)F(\lambda) + \mathscr{A}_+^{-1}(\lambda)G_1(\lambda) + \mathscr{A}_+^{-1}(\lambda)G_2(\lambda)$$

and

$$(I - Q_\tau)(G_2\mathscr{A}_+^{-1}) = (I - Q_\tau)[G_2(I - Q_\tau)\mathscr{A}_+^{-1}],$$

we obtain the fact that for sufficiently large τ

$$\begin{aligned}\|G_1\mathscr{A}_+^{-1}\| &\leqq c\|F\| + \|(I + Q_\tau)G_2\mathscr{A}_-^{-1}\mathscr{A}_-(I - Q_\tau)\mathscr{A}_+^{-1}\| \\ &\leqq c\|F\| + \varepsilon\|G_2\mathscr{A}_-^{-1}\|.\end{aligned} \tag{5.4}$$

If τ is chosen so large that ε can be taken less than 1 in relations (5.3) and (5.4), we obtain

$$\|G_2\mathscr{A}_-^{-1}\| \leqq \frac{c}{1-\varepsilon}\|F\|, \qquad \|G_1\mathscr{A}_+^{-1}\| \leqq \frac{c}{1-\varepsilon}\|F\|.$$

The latter relations and equality (5.2) imply

$$\|\Phi\| \leqq \frac{c^2(3-\varepsilon)}{1-\varepsilon}\|F\|$$

[6] Here the symbol $\| \ \|$ denotes the norm of a *function* of $\mathfrak{G}$ introduced in § 1.

or

$$|P_\tau A P_\tau \varphi|_{L_1} \geqq \frac{1-\varepsilon}{c^2(3-\varepsilon)} |P_\tau \varphi|_{L_1} \qquad (\varphi(t) \in L_1(0, \infty)). \tag{5.5}$$

Thus condition (2.3) of Theorem 2.1, Chapter II, is fulfilled. It remains to ascertain that the operator $P_\tau AP_\tau$ maps the space $P_\tau L_1$ $(= L_1(0, \tau))$ onto the whole of $L_1(0, \tau)$. To this end let us observe that estimate (5.5) is valid for the adjoint operator A^* considered in $L_1(0, \infty)$. Therefore the equation $(P_\tau A P_\tau{}^*)\psi = 0$ has merely a trivial solution in $L_1(0, \tau)$, and a fortiori in the space $M(0, \tau)$.

The case $L_p(0, \infty)$ is considered analogously, with a few modifications.

The theorem is proved.

§ 6. Pair integral-difference equations

The following two types of equations are examined in this section:

$$\left.\begin{aligned} \sum_{j=-\infty}^{\infty} a_j^{(1)} \varphi(t-\alpha_j) + \int_{-\infty}^{\infty} k_1(t-s)\varphi(s)ds &= f(t) \qquad (0 < t < \infty), \\ \sum_{j=-\infty}^{\infty} a_j^{(2)} \varphi(t-\beta_j) + \int_{-\infty}^{\infty} k_2(t-s)\varphi(s)ds &= f(t) \qquad (-\infty < t < 0) \end{aligned}\right\} \tag{6.1}$$

and

$$\sum_{\alpha_j<t} a_j^{(1)} \varphi(t-\alpha_j) + \sum_{\beta_j>t} a_j^{(2)} \varphi(t-\beta_j) + \int_0^{\infty} k_1(t-s)\varphi(s)ds + \int_{-\infty}^{0} k_2(t-s)\varphi(s)ds = f(t) \quad (-\infty < t < \infty). \tag{6.2}$$

These equations will be considered in $L_p(-\infty, \infty)$ for the case when the respective symbols $\mathscr{A}_i(\lambda) = a^{(i)}(\lambda) + K_i(\lambda)$, $i = 1,2$, belong to the algebra $\mathfrak{G}$, and also, by analogy with § 3, in the more general case when they belong to the closure of this algebra in some weaker norm. If $\mathscr{A}_1(\lambda), \mathscr{A}_2(\lambda) \in \mathfrak{G}$, equations (6.1) and (6.2) can be rewritten in the form

$$\left.\begin{aligned} \int_{-\infty}^{\infty} \varphi(s)d\omega_1(t-s) &= f(t) \qquad (0 < t < \infty), \\ \int_{-\infty}^{\infty} \varphi(s)d\omega_2(t-s) &= f(t) \qquad (-\infty < t < 0) \end{aligned}\right\} \tag{6.3}$$

and

$$\int_0^{\infty} \varphi(s)d\omega_1(t-s) + \int_{-\infty}^{0} \varphi(s)d\omega_2(t-s) = f(t) \quad (-\infty < t < \infty), \tag{6.4}$$

where $\omega_1(t)$ and $\omega_2(t)$ are functions of bounded variation without singular component.

1. *The operator algebra* $\tilde{\mathfrak{G}}_p$. Let us denote by $\tilde{\mathfrak{G}}(=\tilde{\mathfrak{G}}_1)$ the set of all operators A operating in $L_p(-\infty,\infty)$, $1 \leqq p \leqq \infty$, in accordance with the formula

$$(A\varphi)(t) = \sum_{j=-\infty}^{\infty} a_j\varphi(t-\delta_j) + \int_{-\infty}^{\infty} k(t-s)\varphi(s)ds, \tag{6.5}$$

where $\sum_{-\infty}^{\infty}|a_j| < \infty$ and $K(t) \in L_1(-\infty,\infty)$.

Let $|A|_p$ be the norm of A in L_p; then, obviously,

$$|A|_p \leqq \sum_{-\infty}^{\infty} |a_j| + \int_{-\infty}^{\infty} |k(t)|dt. \tag{6.6}$$

That in the case $p = 1$ the equality sign is attained in (6.6) is proved exactly as for the operators of $\hat{\mathfrak{G}}$ (cf. § 3).

The set $\tilde{\mathfrak{G}}$(as distinguished from $\hat{\mathfrak{G}}$) is a commutative Banach algebra with the norm $|A|_1$.

Analogously to the preceding section, let us denote by $\tilde{\mathfrak{G}}_p$, $1 \leqq p \leqq \infty$, the closure (in the norm of $L_p(-\infty,\infty)$) of the algebra $\tilde{\mathfrak{G}}$.

There is obviously a one-to-one correspondence between the sets $\tilde{\mathfrak{G}}$ and $\hat{\mathfrak{G}}$ under which an operator of the form (2.1) is put into correspondence with an operator of the form (6.5).

LEMMA 6.1. *The equality* $|A|_p = |A_0|_p$ *holds for any operator* $A \in \tilde{\mathfrak{G}}$, *where the operator of* $\hat{\mathfrak{G}}$ *which corresponds to it is denoted by* A_0.

PROOF. Indeed, it is easy to see that $|A_0|_p \leqq |A|_p$. It is sufficient to prove the reverse inequality for the case when the sum in (6.5) is finite and the kernel $k(t)$ is finite (i.e. vanishes outside some finite interval), because the set of such operators is dense in both $\tilde{\mathfrak{G}}$ and $\hat{\mathfrak{G}}$.

Let us denote by $\varphi_n(t)$, $n = 1,2,\cdots$, the sequence of functions in $L_p(-\infty,\infty)$ having compact support, with $|\varphi_n|_p = 1$, for which

$$\lim_{n\to\infty} |A\varphi_n|_p = |A|_p.$$

Let us choose positive numbers ν_n, $n = 1,2,\cdots$, so large that the functions $\psi_n(t) = \varphi_n(t-\nu_n)$ and $\chi_n(t) = (A\psi_n)(t)$ vanish on the negative semiaxis. Then

$$(A\psi_n)(t) = \begin{cases} (A_0\psi)(t), & t > 0, \\ 0, & t < 0. \end{cases}$$

We have, moreover, $(A\varphi_n)(t) = \chi_n(t+\nu_n)$; consequently

$$|A\varphi_n|_p = |\chi_n(t)|_p = |A_0\psi_n|_p \leqq |A_0|_p,$$

because $|\psi|_p = 1$. Passing to the limit, we obtain $|A|_p \leqq |A_0|_p$.

The lemma is proved.

The lemma just proved implies that the Banach spaces $\hat{\mathfrak{G}}_p$ and $\tilde{\mathfrak{G}}_p$ are isomorphic and isometric. From this also follows the isomorphism and isometry of the commutative Banach operator algebra $\tilde{\mathfrak{G}}_p$ and the function algebra $\mathfrak{G}_p$. This isomorphism puts into correspondence with each operator $A \in \tilde{\mathfrak{G}}_p$ the function $\mathscr{A}(\lambda) \in \mathfrak{G}_p$, which is naturally called the *symbol of A*.

Let us denote by $\tilde{\mathfrak{G}}^+$ ($\tilde{\mathfrak{G}}^-$) the subalgebra of $\tilde{\mathfrak{G}}$ which consists of all operators A_+ (A_-) with symbols $\mathscr{A}_+(\lambda)$ of $\mathfrak{G}^+$ ($\mathscr{A}_-(\lambda)$ of $\mathfrak{G}^-$). Let us denote by $\tilde{\mathfrak{G}}_p^+$ ($\tilde{\mathfrak{G}}_p^-$) the subalgebra of $\tilde{\mathfrak{G}}_p$ which is the closure (in the norm of $\tilde{\mathfrak{G}}_p$) of $\tilde{\mathfrak{G}}^+$ ($\tilde{\mathfrak{G}}^-$).

2. *Fundamental theorem.* Let P be the projection defined in $L_p(-\infty, \infty)$ by the equality

$$(P\varphi)(t) = \begin{cases} \varphi(t), & t > 0. \\ 0, & t < 0, \end{cases}$$

and let Q be the complementary projection: $Q = I - P$.

It is easy to see that the equalities

$$A_+P = PA_+P, \qquad A_-Q = QA_-Q \tag{6.7}$$

hold for any operators $A_\pm \in \tilde{\mathfrak{G}}_p^\pm$.

The pair equation (6.1) can now be written in the form

$$PA_1\varphi + A_2Q\varphi = f,$$

and equation (6.2), in the form

$$A_1P\varphi + QA_2\varphi = f.$$

Here the A_j are operators of the algebra $\tilde{\mathfrak{G}}$ with symbols $\mathscr{A}_j(\lambda) = a^{(j)}(\lambda) + K_j(\lambda)$, $j = 1,2$.

In this section operators of this kind are considered in $L_p(-\infty, \infty)$ on the assumption that $A_1, A_2 \in \tilde{\mathfrak{G}}_p$, $1 \leqq p \leqq \infty$.

The following multiplication rules, which are analogous to rules (2.2) and (2.2′) of Chapter V, are valid for all operators $A_1, A_2 \in \tilde{\mathfrak{G}}_p$ and $A_\pm \in \tilde{\mathfrak{G}}_p^\pm$:

$$(A_1P + A_2Q)(A_+P + A_-Q) = A_1A_+P + A_2A_-Q \tag{6.8}$$

and

$$(PA_- + QA_+)(PA_1 + QA_2) = PA_-A_1 + QA_+A_2. \tag{6.8′}$$

These rules follow from (6.7).

THEOREM 6.1. *Let operators $A_1, A_2 \in \tilde{\mathfrak{G}}_p$. For the operator $A_1P + A_2Q$ ($PA_1 + QA_2$) to be invertible, even only unilaterally, it is necessary and sufficient that*

$$\inf_{-\infty<\lambda<\infty} |\mathscr{A}_j(\lambda)| > 0 \qquad (j = 1,2). \tag{6.9}$$

Let these conditions be fulfilled, and let the numbers ν and n be the indices of the function $\mathscr{A}_1(\lambda)/\mathscr{A}_2(\lambda)$. Then for $\nu > 0$ the operator $A_1P + A_2Q$ $(PA_1 + QA_2)$ is invertible only on the left; for $\nu < 0$ it is invertible only on the right; for $\nu = 0$ its invertibility is consistent with the index n, i. e. it is invertible, only left invertible, or only right invertible depending on whether the number n is equal to zero, positive, or negative.

PROOF. Let conditions (6.9) be fulfilled. Then the operator $A_1P + A_2Q$ can be represented in the form

$$A_1P + A_2Q = A_2(CP + Q),$$

where $C \in \tilde{\mathfrak{G}}_p$ is an operator with symbol $\mathscr{A}_1(\lambda)/\mathscr{A}_2(\lambda)$.

Since A_2 is invertible, it remains to analyze the operator $CP + Q$, which in turn can be represented as

$$CP + Q = (PCP + Q)(I + QCP).$$

The operator $I + QCP$ is invertible, and

$$(I + QCP)^{-1} = I - QCP.$$

If follows immediately from Theorem 3.6 that the operator $PCP + Q$ is invertible only on the left for $\nu > 0$, and only on the right for $\nu < 0$, while for $\nu = 0$ it is invertible only on the right, only on the left, or bilaterally, depending on the sign of the index n.

Let us pass to the proof of the necessity of the theorem's conditions.

Let us suppose that the operator $A_1P + A_2Q$ is invertible on some side. Just as in Lemma 6.1, it is sufficient to restrict ourselves to the case of finitary operators A_1, A_2, i. e. operators whose symbols $\mathscr{A}_1(\lambda)$ and $\mathscr{A}_2(\lambda)$ have the form

$$\mathscr{A}_j(\lambda) = \sum_{m=1}^{n} a_m^{(j)} e^{i\delta_m\lambda} + \int_a^b k_j(t)e^{i\lambda t}dt \qquad (k_j(t) \in L_1, j = 1,2).$$

Let r be so large a natural number that $U^rA_1 \in \tilde{\mathfrak{G}}^+$ and $U^{-r}A_2 \in \tilde{\mathfrak{G}}^-$, where $U \in \tilde{\mathfrak{G}}$ is a shift operator with symbol $e^{i\lambda}$. Then, by virtue of (6.8) and (6.8′), the factorizations

$$A_1P + A_2Q = U^{-r}(P + U^rA_2Q)(U^rA_1P + Q),$$
$$A_1P + A_2Q = U^r(U^{-r}A_1P + Q)(P + U^{-r}A_2Q) \qquad (6.10)$$

hold.

Let the operator $A_1P + A_2Q$ be, for example, right invertible. Then, by virtue of equalities (6.10), the operators $P + U^rA_2Q$ and $U^{-r}A_1P + Q$ are right invertible. The equalities

$$P + U^rA_2Q = (P + QU^rA_2Q)(I + PU^rA_2Q),$$
$$U^{-r}A_1P + Q = (PU^{-r}A_1P + Q)(I + QU^{-r}A_1P)$$

imply the right invertibility of the operators $P + QU^r A_2 Q$ and $PU^{-r}A_1P + Q$. Applying Theorem 3.6, we ascertain that conditions (6.9) are fulfilled.

The theorem is proved analogously for the operator $PA_1 + QA_2$.

REMARK. If even one of conditions (6.9) is not fulfilled, it is easy to show that the operator $A_1P + A_2Q$ $(PA_1 + QA_2)$ is neither a Φ_+- nor a Φ_--operator.

Let us note that, just as in § 2, the kernel of the operator $A_1P + A_2Q$ $(PA_1 + QA_2)$ and its image can be described under the condition $A_1, A_2 \in \bar{\mathfrak{G}}$. However, we shall not go into this.

3. *Relationship with a boundary value problem.* The results obtained above can be interpreted as results concerning a function-theoretic boundary value problem, with coefficients having a point of discontinuity of the second kind at infinity. Let us explain this in more detail. Let us denote by $\mathcal{F}_p^+$ $(\mathcal{F}_p^-)$ the totality of all Fourier transforms of the functions of $L_p(0, \infty)$. As is well known, functions of $\mathcal{F}_p^+$ $(\mathcal{F}_p^-)$ admit holomorphic continuation into the upper (lower) half-plane. The equations considered in §§ 3 and 6 are obviously equivalent to the following boundary value problem:

$$\mathcal{A}(\lambda)\Phi_+(\lambda) - \Phi_-(\lambda) = F_+(\lambda) \qquad (-\infty < \lambda < \infty),$$
$$\mathcal{A}_1(\lambda)\Phi_+(\lambda) + \mathcal{A}_2(\lambda)\Phi_-(\lambda) = F(\lambda) \qquad (-\infty < \lambda < \infty),$$
$$[\mathcal{A}_1(\lambda)\Phi(\lambda)]_+ + [\mathcal{A}_2(\lambda)\Phi(\lambda)]_- = F(\lambda) \qquad (-\infty < \lambda < \infty),$$

where $\mathcal{A}(\lambda), \mathcal{A}_1(\lambda), \mathcal{A}_2(\lambda) \in \mathfrak{G}_p$; $F_+(\lambda), \Phi_+(\lambda) \in \mathcal{F}_p^+$; $\Phi_-(\lambda) \in \mathcal{F}_p^-$; $\Phi = \Phi_+ + \Phi_-$; $F(\lambda)$ is the Fourier transform of the function $f(t) \in L_p(-\infty, \infty)$; and $[F(\lambda)]_\pm$ are defined by the equalities

$$[F(\lambda)]_+ = \int_0^\infty f(t)e^{i\lambda t}\,dt, \qquad [F(\lambda)]_- = \int_{-\infty}^0 f(t)e^{i\lambda t}\,dt.$$

The coefficients $\mathcal{A}(\lambda)$, $\mathcal{A}_1(\lambda)$ and $\mathcal{A}_2(\lambda)$ have a point of discontinuity of the second kind at infinity, but of a special form.

CHAPTER VIII
SYSTEMS OF EQUATIONS

A matrix Wiener-Hopf equation is considered in this chapter.

Let us note that many questions connected with projection methods of solving system of convolution equations remain unsolved. This can be explained by the fundamental difficulties which usually arise when passing from a single convolution equation to systems of such equations.

Theorems concerning factorization of matrix functions are presented in the first section. These results play an important rôle in the subsequent sections.

§ 1. General theorems concerning factorization of matrix functions

Let us introduce the following notation. M is the set of all bounded measurable functions on the unit circle Γ; H_p^+ (H_p^-), $p \geqq 1$, is the set of all functions of $L_p(\Gamma)$ whose Fourier coefficients with negative (positive) indices equal zero.

If $\mathfrak{E}$ is some linear set, the set of all n-dimensional vectors with coordinates in $\mathfrak{E}$ is denoted everywhere in the sequel by $\mathfrak{E}_n$, and the set of all n-order matrices with elements in $\mathfrak{E}$, by $\mathfrak{E}_{n\times n}$.

If $\mathfrak{E}$ is a normed space, the vectors of $\mathfrak{E}_n$ are provided with a norm which, for example, equals the sum of the norms of the components. In $\mathfrak{E}_{n\times n}$ the norm of the element $A = \| a_{jk} \|$ can be defined, for example, by the equality

$$|A| = \max_k \sum_{j=1}^n |a_{jk}|.$$

The representation of a matrix function $A(\zeta) \in M_{n\times n}$ in the form

$$A(\zeta) = A_-(\zeta) D(\zeta) A_+(\zeta) \tag{1.1}$$

is called a *right factorization* of it, where $D(\zeta)$ is a diagonal matrix function of the form $D(\zeta) = \| \zeta^{\kappa_j} \delta_{jk} \|_1^n$, $\kappa_1 \geqq \kappa_2 \geqq \cdots \geqq \kappa_n$ are some integers, $A_+^{\pm 1}(\zeta) \in (H_p^+)_{n\times n}$, $A_-^{\pm 1}(\zeta) \in (H_p)^-_{n\times n}$, and $p \geqq 2$.[1)]

If the factors $A_\pm(\zeta)$ exchange places in (1.1), the factorization of the matrix function $A(\zeta)$ obtained in the process is called a *left factorization*. The factor $D(\zeta)$

[1)]Let us note that the definition of factorization permits the factors $A_+(\zeta)$ and $A_-(\zeta)$ to originate from $M_{n\times n}$.

is called *diagonal.* It is easy to see that every *right* factorization of the matrix function $A(\zeta)$ generates a *left* factorization of the transpose matrix function $A'(\zeta)$.

THEOREM 1.1. *Let the matrix function* $A(\zeta) \in M_{n\times n}$ *admit right factorization. Then all right factorizations of the matrix function* $A(\zeta)$ *have one and the same diagonal factor.*

The same theorem is valid for left factorizations.

PROOF. Let two right factorizations of $A(\zeta)$ be given: the factorization (1.1) and

$$A(\zeta) = \tilde{A}_-(\zeta)\,\tilde{D}(\zeta)\,\tilde{A}_+(\zeta), \tag{1.2}$$

where $\tilde{D}(\zeta) = \| \zeta^{\tilde{\kappa}_j}\delta_{jk} \|_1^n$. Equalities (1.1) and (1.2) imply that

$$B_-(\zeta)D(\zeta) = \tilde{D}(\zeta)\,B_+(\zeta), \tag{1.3}$$

where

$$B_-(\zeta) = \tilde{A}_-^{-1}(\zeta)A_-(\zeta), \qquad B_+(\zeta) = \tilde{A}_+(\zeta)A_+^{-1}(\zeta), \tag{1.4}$$

with $B_\pm(\zeta) \in (H_1^\mp)_{n\times n}$. Equality (1.3) signifies that

$$b_{jk}^-(\zeta)\,\zeta^{\kappa_k} = \zeta^{\tilde{\kappa}_j} b_{jk}^+(\zeta) \qquad (j, k = 1,\cdots, n),$$

where $b_{jk}^\mp(\zeta)$ are elements of the matrices $B_\pm(\zeta)$.

The equality $b_{jk}^-(\zeta) = b_{jk}^+(\zeta) = 0$ holds in all cases when $\kappa_k < \tilde{\kappa}_j$. Indeed, since the sets H_1^+ and H_1^- intersect only at constants, the equality

$$b_{kj}^-(\zeta) = \zeta^{\tilde{\kappa}_j-\kappa_k} b_{jk}^+(\zeta) \qquad (\kappa_k < \tilde{\kappa}_j)$$

implies that both of its sides equal zero.

Now let us suppose that $D(\zeta) \neq \tilde{D}(\zeta)$. Then for some natural number r, $1 \leqq r \leqq n$, we have $\kappa_r \neq \tilde{\kappa}_r$. Without loss of generality it can be assumed that $\kappa_r < \tilde{\kappa}_r$. It is easy to see that the inequality $\kappa_k < \tilde{\kappa}_j$ holds for all subscripts j and k taking the values $j = 1,\cdots, r$; $k = r,\cdots,n$, and therefore $b_{jk}^+(\zeta) = 0$ for those values of j and k.

The latter equalities imply that every minor of order r which is composed of the first r rows of the matrix $B_+(\zeta)$ is identically equal to zero. Hence, according to Laplace's theorem, $\det B_+(\zeta) \equiv 0$, which is impossible.

The theorem is proved.

The theorem just proved implies that if the matrix function $A(\zeta) \in M_{n\times n}$ admits right (left) factorization, the numbers κ_j, $j = 1,\cdots,n$, are *uniquely* determined by the matrix function $A(\zeta)$. These numbers are called, in correspondence with the type of factorization, the *right* (*left*) *indices* of the matrix function $A(\zeta)$, or *partial indices.*

Let us note that in general the right and left indices of a matrix function $A(\zeta) \in M_{n\times n}$ do not coincide.

THEOREM 1.2. *If a matrix function* $A(\zeta) \in M_{n\times n}$ *admits some right factorization* (1.1), *the general form of the factors* $\tilde{A}_{\pm}(\zeta)$ *in every such factorization is given by the equalities*

$$\begin{aligned}\tilde{A}_+(\zeta) &= B_+(\zeta)\,A_+(\zeta),\\ \tilde{A}_-(\zeta) &= A_-(\zeta)\,D(\zeta)\,B_+^{-1}(\zeta)\,D^{-1}(\zeta),\end{aligned} \tag{1.5}$$

where $B_+(\zeta)$ *is an arbitrary nonsignular matrix function whose elements satisfy the following conditions*:

1) $b_{jk}^+(\zeta) = 0$ *if* $\kappa_k < \kappa_j$.
2) $b_{jk}^+(\zeta)$ *is a constant if* $\kappa_k = \kappa_j$.
3) $b_{jk}^+(\zeta)$ *is a polynomial in* ζ *of degree* $\leqq \kappa_k - \kappa_j$ *if* $\kappa_k > \kappa_j$.

PROOF. Let us begin the proof with a simple heuristic argument which explains equalities (1.5).

Let us take the factorization

$$A(\zeta) = A_-(\zeta)\,D(\zeta)\,A_+(\zeta)$$

and replace the factor $A_+(\zeta)$ in it by $B_+(\zeta)A_+(\zeta)$. To preserve the factorization, a compensating factor $B_+^{-1}(\zeta)$ must be "carried" through $D(\zeta)$ and, together with its inverse, fall into $(H_p^+)_{n\times n}$, i.e.

$$\begin{aligned}A(\zeta) &= A_-(\zeta)\,D(\zeta)\,A_+(\zeta) = A_-(\zeta)\,D(\zeta)\,B_+^{-1}(\zeta)\,B_+(\zeta)\,A_+(\zeta)\\ &= A_-(\zeta)\,B_-(\zeta)\,D(\zeta)\,B_+(\zeta)\,A_+(\zeta),\end{aligned}$$

where $B_-(\zeta)D(\zeta) = D(\zeta)B_+^{-1}(\zeta)$.

This explains the form of the factor in (1.5).

Let equalities (1.1) and (1.2) give two factorizations of the matrix function $A(\zeta)$. As was established in the preceding theorem, $D(\zeta) = \tilde{D}(\zeta)$. Let us define matrix functions $B_{\pm}(\zeta)$ by equalities (1.4) and denote their elements by $b_{jk}^{\mp}(\zeta)$. The equality

$$B_-(\zeta)D(\zeta) = D(\zeta)B_+(\zeta) \tag{1.6}$$

implies that

$$b_{jk}^-(\zeta) = \zeta^{\kappa_j-\kappa_k}b_{jk}^+(\zeta) \qquad (j,k = 1,\cdots,n). \tag{1.7}$$

Since the sets H_1^+ and H_1^- intersect only at constants, equalities (1.7) imply that the functions $b_{jk}^+(\zeta)$ satisfy conditions 1) — 3). Equalities (1.5) follow from (1.4) and (1.6)

Let us observe that the matrix function $B_+(\zeta)$ has the following form:

$$B_+(\zeta) = \left\|\begin{matrix} Q_1 & 0 & \cdots & 0\\ * & Q_2 & \cdots & 0\\ \cdots & \cdots & \cdots & \cdots\\ * & * & \cdots & Q_m\end{matrix}\right\|,$$

where the Q_j, $j = 1, \cdots, m$, are nonsingular constant matrices and the places at which there are matrices with elements which are polynomials of appropriate degrees in ζ are marked by asterisks; consequently the determinant of $B_+(\zeta)$ is a constant:

$$\det B_+(\zeta) = \det Q_1 \det Q_2 \cdots \det Q_m.$$

It is not difficult to verify that, for some choice of a nonsingular matrix function $B_+(\zeta)$ with regard for conditions 1) — 3), formulas (1.5) transform a given right factorization (1.1) of the matrix function $A(\zeta)$ into some new right factorization of it.

The theorem is proved.

An analogous theorem is valid for left factorizations.

§ 2. Canonical factorization of matrix functions in *R*-algebras

We shall use the definitions of § 5, Chapter I, in the present section.

Let $\mathscr{C}$ be an arbitrary *R*-algebra. The right factorization (1.1) of a nonsingular matrix function $A(\zeta) \in \mathscr{C}_{n\times n}$ is called *canonical* if $A_+^{\pm 1}(\zeta) \in \mathscr{C}^+_{n\times n}$ and $A_-^{\pm 1}(\zeta) \in \mathscr{C}^-_{n\times n}$. A left *canonical* factorization of a matrix function $A(\zeta) \in \mathscr{C}_{n\times n}$ is defined analogously.

Canonical factorizations are distinguished from the usual ones by the fact that the factors $A_\pm^{\pm 1}(\zeta)$ remain within the limits of the same algebra.

Let the matrix function $A(\zeta) \in \mathscr{C}_{n\times n}$ admit right canonical factorization of the form (1.1). Then

$$\det A(\zeta) = \det A_-(\zeta) \det D(\zeta) \det A_+(\zeta),$$

whence

$$\sum_{j=1}^{n} \kappa_j = \operatorname{ind} \det A(\zeta),$$

where the κ_j are right indices of $A(\zeta)$.

THEOREM 2.1. *Let $\mathscr{C}$ be an arbitrary R-algebra. For every nonsingular matrix function $A(\zeta) \in \mathscr{C}_{n\times n}$ to admit right (left) canonical factorization, it is necessary and sufficient that the algebra $\mathscr{C}$ be decomposing.*

We need the following lemma to prove this theorem.

LEMMA 2.1. *Let $\mathscr{C}$ be an arbitrary R-algebra. If the function $a(\zeta)$ belongs to $\mathscr{C}^+$ ($\mathscr{C}^-$) and vanishes at some point ζ_0, $|\zeta_0| < 1$ ($|\zeta_0| > 1$), then the function $a(\zeta)/(\zeta - \zeta_0)$ also belongs to $\mathscr{C}^+$ ($\mathscr{C}_-$).*

PROOF. Let us denote by $p_n(\zeta)$, $n = 1, 2, \cdots$, a sequence of polynomials in nonnegative powers of ζ which tends to $a(\zeta)$ in the norm of the algebra $\mathscr{C}$. Then the sequence $P_n(\zeta)$ converges uniformly to $a(\zeta)$ on the unit disk, and in particular $P_n(\zeta_0) \to 0$. Since $(\zeta - \zeta_0)^{-1} \in \mathscr{C}$, the sequence of polynomials

$$\tilde{P}_n(\zeta) = \frac{P_n(\zeta) - P_n(\zeta_0)}{\zeta - \zeta_0}$$

tends to the function $a(\zeta)/(\zeta - \zeta_0)$ in the norm of $\mathscr{C}$.

The lemma is proved.

PROOF OF THEOREM 2.1. Let us consider first the case when the nonsingular matrix function $A(\zeta)$ is contained in $\mathscr{C}^+_{n\times n}$. Let us denote by $\zeta_1,\cdots,\zeta_q$ all zeros of the function det $A(\zeta)$ in the disk $|\zeta| < 1$, and by $m_1,\cdots,m_q$ their multiplicities. For convenience in the sequel let us suppose that $\zeta_q = 0$. If this point is not a zero of the function det $A(\zeta)$, let us set $m_q = 0$.

Let $f_j(\zeta)$ be the jth row of the matrix $A(\zeta) = \| a_{jk}(\zeta) \|^n_{j,k=1}$, i.e. $f_j(\zeta) = \{a_{j1}(\zeta),\cdots, a_{jn}(\zeta)\}$, $j = 1,\cdots,n$. Let us denote by p_j the multiplicity of the zero $\zeta = \zeta_1$ of the vector function $f_j(\zeta)$. Obviously $\sum_1^n p_j \leqq m_1$.

It can be assumed without loss of generality that $p_1 \geqq p_2 \geqq \cdots \geqq p_n$. Let $\sum_1^n p_j < m_1$. Then complex numbers $c_1,\cdots, c_l$ $(l \leqq n;\ c_l = 1)$ can be found such that

$$\sum_{j=1}^{l} \frac{c_j f_j(\zeta)}{(\zeta - \zeta_1)^{p_j}}\Bigg|_{\zeta=\zeta_1} = 0.$$

Consequently the vector function

$$f(\zeta) = \sum_{j=1}^{l} \frac{c_j f_j(\zeta)}{(\zeta - \zeta_1)^{-p_j}}$$

vanishes at the point $\zeta = \zeta_1$. The components of this vector function belong to the algebra $\mathscr{C}^+$, by virtue of Lemma 2.1. The vector function

$$\hat{f}(\zeta) = f(\zeta)(\zeta - \zeta_1)^{p_l} = \sum_{j=1}^{l} \frac{c_j f_j(\zeta)}{(\zeta - \zeta_1)^{p_j - p_l}}$$

obviously has a zero of multiplicity $\hat{p}_l > p_l$ at the point $\zeta = \zeta_1$.

Let us form a matrix function $\hat{B}_1(\zeta)$:

$$\hat{B}_1(\zeta) = \left\| \begin{matrix} 1 & 0 & \cdots 0 \cdots & 0 \\ 0 & 1 & \cdots 0 & 0 \\ \cdots & \cdots & \cdots & \cdots \\ \dfrac{c_1}{(\zeta - \zeta_1)^{p_1 - p_l}} & \dfrac{c_2}{(\zeta - \zeta_1)^{p_2 - p_l}} & \cdots 1 \cdots & 0 \\ \cdots & \cdots & \cdots & \cdots \\ 0 & 0 & \cdots 0 \cdots & 1 \end{matrix} \right\| \quad l\text{th row}$$

for which det $\hat{B}_1(\zeta) = 1$ and $\hat{B}_1^{\pm 1}(\zeta) \in \mathscr{C}^-_{n\times n}$.

It is easy to see that the matrix $\hat{B}_1(\zeta)A(\zeta)$ is obtained from the matrix $A(\zeta)$ by replacing the latter's lth row with the vector $\hat{f}(\zeta)$. Consequently $\hat{B}_1(\zeta)A(\zeta) \in \mathscr{C}^+_{n\times n}$. If, moreover, $\sum_{j\neq l} p_j + \hat{p}_l < m_1$, then by successively repeating the described operation we arrive, after a finite number of steps, at the matrix function $B_1(\zeta)\,A(\zeta) \in \mathscr{C}^+_{n\times n}$ whose rows have zeros with multiplicities $\hat{p}_j$, $j = 1,\cdots,n$, where $\hat{p}_1 \geqq \hat{p}_2 \geqq \cdots \geqq \hat{p}_n$ and $\sum_1^n \hat{p}_j = m_1$, at the point $\zeta = \zeta_1$. In this connection $\det B_1(\zeta) = 1$ and $B_1^{\pm 1}(\zeta) \in \mathscr{C}^-_{n\times n}$.

Let us form the diagonal matrix function

$$D_1(\zeta) = \left\| \left(\frac{\zeta}{\zeta - \zeta_1} \right)^{\hat{p}_j} \delta_{jk} \right\|_{j,k=1}^{n} \qquad (D_1^{\pm 1}(\zeta) \in \mathscr{C}^-_{n\times n}).$$

It is easy to see that the determinant of the matrix $A_1(\zeta) = D_1(\zeta)B_1(\zeta)A(\zeta)$ does not vanish at the point $\zeta = \zeta_1$. Thus in the disk $|\zeta| < 1$ the points $\zeta_2, \zeta_3, \cdots, \zeta_q$ exhaust all of the zeros of the function $\det A_1(\zeta)$, their multiplicities being equal respectively to the numbers $m_2, m_3, \cdots, m_q + m_1$.

Let us carry out an analogous operation on the matrix $A_1(\zeta)$ just obtained and its zero $\zeta = \zeta_2$. We obtain as the result the matrix function $A_2(\zeta) = D_2(\zeta)B_2(\zeta)A_1(\zeta)$ whose determinant vanishes only at the points $\zeta_3, \zeta_4, \cdots, \zeta_q$, the multiplicities of these zeros being equal respectively to $m_3, m_4, \cdots, m_q + m_1 + m_2$. Continuing these constructions in the same way, we arrive at the matrix function

$$A_{q-1}(\zeta) = D_{q-1}(\zeta)B_{q-1}(\zeta)A_{q-2}(\zeta) \in \mathscr{C}^+_{n\times n}$$
$$(B_{q-1}^{\pm 1}(\zeta),\ D_{q-1}^{\pm 1}(\zeta) \in \mathscr{C}^-_{n\times n}),$$

whose determinant has the unique zero $\zeta = 0$, of multiplicity $\sum_1^q m_j$. Finally, let us construct the matrix function $B(\zeta)$ $(B^{\pm 1}(\zeta) \in \mathscr{C}^-_{n\times n})$ by means of the above-described operation so that the matrix function $B(\zeta)\,A_{q-1}(\zeta) \in \mathscr{C}^+_{n\times n}$ and the function $\det\,(B(\zeta)\,A_{q-1}(\zeta))$ have the unique zero $\zeta = 0$ of multiplicity $\sum m_j$, where the sum of the multiplicities $\kappa_1 \geqq \kappa_2 \geqq \cdots \geqq \kappa_n$ of the zero $\zeta = 0$ of the corresponding vector rows of the matrix $B(\zeta)A_{q-1}(\zeta)$ equals $\sum m_j$.

Let us form the diagonal matrix $D(\zeta) = \| \zeta^{\kappa_j} \delta_{jk} \|_1^n$. It is easy to see that

$$D^{-1}(\zeta)\,B(\zeta)A_{q-1}(\zeta) = A_+(\zeta) \in \mathscr{C}^+_{n\times n}, \tag{2.1}$$

where $\det A_+(\zeta) \neq 0$ $(|\zeta| \leqq 1)$. Let us rewrite equality (2.1) in the form

$$B(\zeta)D_{q-1}(\zeta)B_{q-1}(\zeta) \cdots D_1(\zeta)B_1(\zeta)A(\zeta) = D(\zeta)A_+(\zeta).$$

This implies

$$A(\zeta) = A_-(\zeta)D(\zeta)A_+(\zeta) \qquad (|\zeta| = 1),$$

where

$$A_-(\zeta) = [B(\zeta)D_{q-1}(\zeta)B_{q-1}(\zeta)\cdots D_1(\zeta)B_1(\zeta)]^{-1},$$

with $A_-^{\pm 1}(\zeta) \in \mathscr{C}^-_{n\times n}$ and $A_+^{\pm 1}(\zeta) \in \mathscr{C}^+_{n\times n}$.

Let us now prove sufficiency in the general case. Let $A(\zeta)$ be an arbitrary nonsingular matrix function of $\mathscr{C}_{n\times n}$. Obviously a matrix $\Phi(\zeta)$ can be chosen whose elements are polynomials in integral powers of ζ such that the matrix $B(\zeta) = I - A(\zeta)\Phi^{-1}(\zeta)$ is of sufficiently small norm. Then, by virtue of Lemma 5.1, Chapter *I*, the matrix function $A(\zeta)\Phi^{-1}(\zeta)$ admits the right factorization

$$A(\zeta)\Phi^{-1}(\zeta) = B_-(\zeta)B_+(\zeta) \qquad (B_-^{\pm 1}(\zeta) \in \mathscr{C}_{n\times n}^-, \; B_+^{\pm 1}(\zeta) \in \mathscr{C}_{n\times n}^+).$$

Consequently

$$B_+(\zeta)\Phi(\zeta) = B_-^{-1}(\zeta)A(\zeta). \tag{2.2}$$

Let us denote by r a natural number so large that the matrix $\zeta^r\Phi(\zeta)$ falls into $\mathscr{C}_{n\times n}^+$. Then the matrix function $\tilde{A}(\zeta) = \zeta^r B_+(\zeta)\Phi(\zeta)$ will belong to $\mathscr{C}_{n\times n}^+$, and $\det \tilde{A}(\zeta) \neq 0$ ($|\zeta| = 1$). According to what was proved in the first part, the matrix $\tilde{A}(\zeta)$ admits the right factorization

$$\tilde{A}(\zeta) = \tilde{A}_-(\zeta)\tilde{D}(\zeta)\tilde{A}_+(\zeta) \qquad (|\zeta| = 1).$$

This and equality (2.2) imply that the matrix $A(\zeta)$ admits the right *canonical* factorization

$$A(\zeta) = A_-(\zeta)D(\zeta)A_+(\zeta), \tag{2.3}$$

where $A_-(\zeta) = B_-(\zeta)\tilde{A}_-(\zeta)$, $A_+(\zeta) = \tilde{A}_+(\zeta)$ and $D(\zeta) = \zeta^{-r}\tilde{D}(\zeta)$.

Passing to the proof of the necessity of the theorem's conditions, let us observe that it is sufficient to confine ourselves to the case $n = 1$. Indeed, factorization (2.3) implies the equality

$$\det A(\zeta) = \det A_-(\zeta) \det D(\zeta) \det A_+(\zeta).$$

It can be concluded from this that if for some n every nonsingular matrix function $A(\zeta) \in \mathscr{C}_{n\times n}$ admits factorization (2.3), then every function $a(\zeta) \in \mathscr{C}$ which does not vanish on the unit circle admits the factorization

$$a(\zeta) = a_-(\zeta)\zeta^\kappa a_+(\zeta) \qquad (a_+^{\pm 1}(\zeta) \in \mathscr{C}^+, \; a_-^{\pm 1}(\zeta) \in \mathscr{C}^-).$$

Applying Theorem 5.1 of Chapter I, we obtain the fact that the algebra $\mathscr{C}$ is decomposing.

A theorem follows immediately from the theorem just proved.

THEOREM 2.2. *Every nonsingular matrix function $A(\zeta) \in W_{n\times n}$*[2] *admits right (left) canonical factorization.*

Now let $\tilde{\mathscr{C}}$ be an arbitrary R-algebra on the *real axis* (cf. § 8.3, Chapter I). The concept of canonical factorization on the real axis is obtained by means of the usual bilinear transformation of the corresponding concept on the circle. Namely, the representation of a matrix function $A(\lambda) \in \tilde{\mathscr{C}}_{n\times n}$ in the form

[2] As usual, W denotes the algebra of all functions expanding into absolutely convergent Fourier series. Obviously W is a decomposing R-algebra.

$$A(\lambda) = A_-(\lambda)D(\lambda)A_+(\lambda)$$

is called a *canonical factorization* of it, where $D(\lambda)$ is a diagonal matrix function of the form

$$D(\lambda) = \left\| \left(\frac{\lambda - i}{\lambda + i}\right)^{\kappa_j} \delta_{jk} \right\|_{j,k=1}^{n},$$

$\kappa_1 \geqq \kappa_2 \geqq \cdots \geqq \kappa_n$ are integers, and $A_+^{\pm 1}(\lambda) \in \tilde{\mathscr{C}}_{n\times n}^{+}$ and $A_-^{\pm 1}(\lambda) \in \tilde{\mathscr{C}}_{n\times n}^{-}$.

THEOREM 2.3. *Let $\tilde{\mathscr{C}}$ be an arbitrary R-algebra on the real axis. For every matrix function $A(\lambda) \in \tilde{\mathscr{C}}_{n\times n}$ saftisying the condition*

$$\det A(\lambda) \neq 0 \qquad (-\infty < \lambda < \infty) \tag{2.4}$$

to admit right (left) canonical factorization, it is necessary and sufficient that the algebra $\tilde{\mathscr{C}}$ be decomposing.

Theorem 2.3 is obtained from Theorem 2.1 by means of the substitution $\zeta = (\lambda - i)/(\lambda + i)$, which transforms the real axis into the unit circle. In turn, a theorem follows from Theorem 2.3:

THEOREM 2.4. *If the matrix function $A(\lambda) \in \mathfrak{L}_{n\times n}$*[3] *satisfies condition* (2.4), *it admits right (left) canonical factorization.*

§ 3. Factorization of continuous matrix functions

Since the algebra of continuous functions on the unit circle is not decomposing, canonical factorization is not in general possible here; however, the following theorem does hold.

THEOREM 3.1. *Every nonsingular continuous nth order matrix function $A(\zeta)$ ($|\zeta| = 1$) admits the right factorization* (1.1). *In this connection the following assertions are true.*

a) *The factors $A_+^{\pm 1}(\zeta)$ and $A_-^{\pm 1}(\zeta)$ belong respectively to the spaces $(H_p^+)_{n\times n}$ and $(H_p^-)_{n\times n}$, whatever the number p $(1 < p < \infty)$ may be.*

b) *The operator $(G\varphi)(\zeta) = A_-(\zeta)\mathscr{P}A_-^{-1}(\zeta)\varphi(\zeta)$, where $\mathscr{P} = \|\delta_{jk}P\|_1^n$ and P is the operator of the natural projection of the space L_p onto H_p^+, is a bounded linear operator in the space $(L_p)_n$ for any p $(1 < p < \infty)$.*

c) *The equality $\sum_1^n \kappa_j = \operatorname{ind} \det A(\zeta)$ holds, where the κ_j, $j = 1,\cdots,n$, are right indices of the matrix function $A(\zeta)$.*

PROOF. Let $M(\zeta)$ be some nth order continuous matrix function. Let us form the operators $\mathscr{P}_M$ and $\mathscr{Q}_M$ by setting

[3] For the definition of the algebra $\mathfrak{L}$ cf. § 8 of Chapter I.

$$(\mathscr{P}_M\Phi)(\zeta) = \mathscr{P}(M(\zeta)\Phi(\zeta)), \qquad (\mathscr{Q}_M\Phi)(\zeta) = \mathscr{Q}(\Phi(\zeta)M(\zeta)),$$

where $\mathscr{Q} = I - \mathscr{P}$ and $\Phi \in (L_p)_{n\times n}$ $(1 < p < \infty)$. They are bounded linear operators for any p, $1 < p < \infty$.

Let us denote by $|\mathscr{P}_M|_p$ and $|\mathscr{Q}_M|_p$ the norms of the respective operators in the space $(L_p)_{n\times n}$.

Let us show first that if the norms of the elements of the matrix function $M(\zeta)$ in the space C are so small that

$$\max_{p_1 \leqq r \leqq p_2} \{|\mathscr{P}_M|_r, |\mathscr{Q}_M|_r\} < 1, \tag{3.1}$$

where p_1, p_2 are a pair of numbers satisfying the condition $1 < p_1 \leqq 2 \leqq p_2 < \infty$, then the matrix function $I + M(\zeta)$ admits factorization (1.1), with the factors possessing properties a) and b) only for $p_1 \leqq p \leqq p_2$. Indeed, under condition (3.1) the operators $I + \mathscr{P}_M$ and $I + \mathscr{Q}_M$ are invertible in the space $(L_r)_{n\times 1}$, $p_1 \leqq r \leqq p_2$. Consequently solutions X and Y of the equations

$$X + \mathscr{P}(MX) = I \qquad \text{and} \qquad Y + \mathscr{Q}(YM) = I$$

exist in $(L_r)_{n\times n}$. These equalities imply that $X \in (H_r^+)_{n\times n}$, $Y \in (H_r^-)_{n\times n}$, $(I + M)X = Z_-$ and $Y(I + M) = Z_+$, where $Z_\pm \in (H_r^\pm)_{n\times n}$. The last two equalities imply the equality $Z_+X = YZ_-$. It follows from this that each of these products is a constant matrix. Taking into account, moreover, that $Z_-(\infty) = Y(\infty) = I$, we obtain $Z_+X + YZ_- = I$.

Thus the equality

$$I + M = Z_-Z_+ \tag{3.2}$$

holds, in which $Z_-^{\pm 1} \in (H_r^-)_{n\times n}$, $Z_+^{\pm 1} \in (H_r^+)_{n\times n}$ for $p_1 \leqq r \leqq p_2$.

Now let us show that the operator $K = Z_+^{-1}\mathscr{P}Z_-^{-1}$ is bounded in all $(L_r)_n$ spaces. Since the equality

$$(KF)(\zeta) = Z_+^{-1}(\zeta)Z_-^{-1}(\infty)F(\infty)$$

is valid for any vector $F(\zeta) \in (H_r^-)_n$, it is sufficient to show that the operator K, as an operator operating in $(H_r^+)_n$, is bounded.

By virtue of condition (3.1), for any $F \in (H_r^+)_n$ the equation $\Phi + \mathscr{P}(M\Phi) = F$ has the unique solution $\Phi \in (H_r^+)_n$ $(p_1 \leqq r \leqq p_2)$ for which the equality

$$(I + M)\Phi = F + G \qquad (G(\zeta) \in (H_r)_n)$$

is valid. Taking (3.2) into account, we obtain

$$\Phi = Z_+^{-1}\mathscr{P}Z_-^{-1}F.$$

Thus

$$(I + \mathscr{P}_M)^{-1}|(H_r^+)_n = Z_+^{-1}\mathscr{P}Z_-^{-1}|(H_r^+)_n,$$

which implies the boundedness of the restriction $Z_+^{-1}\mathscr{P}Z_-^{-1}|(H_r^+)_n$; and the operator $Z_+^{-1} \mathscr{P} Z_-^{-1}$ is consequently bounded. Since

$$Z_- \mathscr{P} Z_-^{-1} = (I + M) Z_+^{-1} \mathscr{P} Z_-^{-1},$$

the operator $Z_- \mathscr{P} Z_-^{-1}$ is bounded in $(L_r)_n$ for $p_1 \leqq r \leqq p_2$.

Now let $A(\zeta)$ be an arbitrary nonsingular continuous matrix function, and let p_1, $1 < p_1 \leqq 2$, be an arbitrary fixed number, while $1/p_2 = 1 - 1/p_1$. It is possible to choose a matrix function $R(\zeta)$ whose elements are polynomials in integral powers of ζ and such that the matrix function $M(\zeta) = A(\zeta)R^{-1}(\zeta) - I$ satisfies conditions (3.1). By virtue of what has been proved, the matrix function $I + M(\zeta)$ admits factorization (3.2); therefore

$$A(\zeta) = (I + M(\zeta))R(\zeta) = Z_-(\zeta)Z_+(\zeta)R(\zeta).$$

Let us note that the construction utilized in the proof of Theorem 2.1 for the nonsingular matrix function $A(\zeta) \in \mathscr{C}^+_{n\times n}$ can be carried out for any matrix function $A(\zeta) \in (H_r^+)_{n\times n}$ such that $A^{-1}(\zeta) \in (L_r)_{n\times n}$. By virtue of this, the matrix function $Z_+(\zeta)R(\zeta)$ admits the factorization

$$Z_+(\zeta)R(\zeta) = B_-(\zeta)D(\zeta)A_+(\zeta),$$

where $D(\zeta)$ is the diagonal factor, $B_-^{\pm 1}(\zeta)$ $(\in (H_r^-)_{n\times n})$ are matrix functions with rational elements, and $A_+^{\pm 1}(\zeta) \in (H_r^+)_{n\times n}$. Thus the matrix function $A(\zeta)$ admits the factorization

$$A(\zeta) = A_-(\zeta)D(\zeta)A_+(\zeta), \tag{3.3}$$

where $A_-(\zeta) = Z_-(\zeta)B_-(\zeta)$, and consequently $A_-^{\pm 1}(\zeta) \in (H_r^-)_{n\times n}$ $(p_1 \leqq r \leqq p_2)$.

Let us now show that the operator $A_- \mathscr{P} A_-^{-1} = Z_- B_- \mathscr{P} B_-^{-1} Z_-^{-1}$ is bounded in $(L_r)_n$. Since $\mathscr{P} = (I + \mathscr{S})/2$, where $\mathscr{S}$ is a singular integration operator in $(L_r)_n$, the problem reduces to proving the boundedness of the operator

$$Z_-B_-\mathscr{S}B_-^{-1}Z_-^{-1} = Z_-\mathscr{S}Z_-^{-1} + Z_-B_-(\mathscr{S}B_-^{-1} - B_-^{-1}\mathscr{S})Z_-^{-1}.$$

The boundedness in $(L_r)_n$ of the first addend has already been established. It is not difficult to see that the operator

$$((\mathscr{S}B_-^{-1} - B_-^{-1}\mathscr{S}) Z_-^{-1}\varphi)(\zeta) = \frac{1}{\pi i} \int_{|\tau|=1} \frac{B_-^{-1}(\tau) - B_-^{-1}(\zeta)}{\tau - \zeta} Z_-^{-1}(\tau)\varphi(\tau)\, d\tau$$

is a bounded operator from $(L_r)_n$ into $(L_\infty)_n$, which implies the boundedness of the second addend.

Let us show that the factors $A_\pm(\zeta)$ in factorization (3.3) possess properties a) and b) for all p, $1 < p < \infty$. Let p' be an arbitrary number less than p_1, and let $1/p_2' = 1 - 1/p_1'$. It follows from what has been proved that generally speaking another factorization of the matrix function $A(\zeta)$ exists:

$$A(\zeta) = \tilde{A}_-(\zeta)\tilde{D}(\zeta)\tilde{A}_+(\zeta),$$

whose factors possess properties a) and b) for all p of the segment $p_1' \leqq p \leqq p_2'$.

By virtue of Theorems 1.1 and 1.2, $\tilde{D}(\zeta) = D(\zeta)$, and the factors $\tilde{A}_{\pm}(\zeta)$ equal respectively the factors $\tilde{A}_{\pm}(\zeta)$ multiplied by polynomial matrix functions. It is easy to derive from this that the factors of factorization (3.3) possess properties a) and b) also in the segment $p_1' \leqq p \leqq p_2'$.

Property c) is established by means of property a) and the equality $\det A(\zeta) = \det A_-(\zeta)\zeta^{\Sigma\varkappa_j} \det A_+(\zeta)$.

The theorem is proved.

COROLLARY 3.1. *Let* $A(\zeta)$ $(|\zeta| = 1)$ *be an arbitrary nonsingular continuous nth order matrix function, and let equality* (3.3) *be a right factorization of it. Then the operator* $B = A_+^{-1}(\zeta)\mathscr{P}D^{-1}(\zeta)\mathscr{P}A_-^{-1}(\zeta)$ *is bounded in* $(L_p)_n$, $1 < p < \infty$.

Indeed, equality (3.3) implies that, together with the operator $A\mathscr{P}A_-^{-1}$, the operator

$$A_+^{-1}D^{-1}\mathscr{P}A_-^{-1} = A^{-1}A_-\mathscr{P}A_-^{-1}$$

is also bounded.

It is not difficult to verify that the operator $A_+^{-1}D^{-1}\mathscr{P}A_-^{-1} - B$ is finite-dimensional, and therefore B is bounded.

Let us note that a theorem analogous to Theorem 3.1 holds for the left factorization of a continuous matrix function $A(\zeta)$. In this connection property b) must be replaced by the property that the operator $A_+\mathscr{Q}A_+^{-1}$ is bounded in $(L_p)_n$, $1 < p < \infty$.

§ 4. General theorems

1. Let an operator $V \in \mathfrak{A}(\mathfrak{B})$ which is invertible only on the left satisfy conditions (Σ I) and (Σ II) of §§1.3 and 3.1, Chapter I. Let us assume in addition that

(Σ III) dim Coker $V < \infty$.[4]

In accordance with the stipulation which has been made, we denote by $\mathfrak{B}_n$ the Banach space of all n-dimensional vectors with components of $\mathfrak{B}$, and by $\mathfrak{R}_{n\times n}(V)$ the set of all nth order matrices with elements of $\mathfrak{R}(V)$. The elements of $\mathfrak{R}_{n\times n}(V)$ can be interpreted as bounded linear operators operating in $\mathfrak{B}_n$. Let us put into correspondence with each operator $A = \|A_{jk}\|_1^n \in \mathfrak{R}_{n\times n}(V)$ a matrix function $A(\zeta) = \|A_{jk}(\zeta)\|_1^n$ $(|\zeta| = 1)$, its symbol, which is continuous on the unit circle. Here the $A_{jk}(\zeta)$ are the symbols of the A_{jk}.

LEMMA 4.1. *Let* $A = \|A_{jk}\|_1^n \in \mathfrak{A}(\mathfrak{B}_n)$, *and let the operators* $A_{jk} \in \mathfrak{A}(\mathfrak{B})$, $j, k = 1, \cdots, n$, *commute with each other to within completely continuous addends. For A to be a* Φ*-operator, it is necessary and sufficient that the operator* det A[5] *be a* Φ*-operator.*

[4] This condition has already been encountered, but now it plays an important role.

[5] The determinant det A is formed in the usual way. The order of the factors in each term plays no part in this connection, since the various possible results differ from each other by completely continuous addends.

PROOF. Let us denote by $\hat{\mathfrak{A}}$ and $\hat{\mathfrak{A}}_{n\times n}$, respectively, the factor algebras of the algebras $\mathfrak{A} = \mathfrak{A}(\mathfrak{B})$ and $\mathfrak{A}_{n\times n} = \mathfrak{A}(\mathfrak{B}_n)$ modulo the ideal of all completely continuous operators operating in $\mathfrak{B}$ and $\mathfrak{B}_n$, and let us denote by $\hat{A}_{jk}$ $(\in \hat{\mathfrak{A}})$ and $\hat{A}$ $(\in \hat{\mathfrak{A}}_{n\times n})$ the residue classes containing the operators A_{jk} and A. The elements $\hat{A}_{jk}$, $j, k = 1,\cdots,n$, obviously commute with each other.

Let us now show that the element $\hat{A} = \| \hat{A}_{jk} \|_1^n$ is invertible in $\hat{\mathfrak{A}}_{n\times n}$ if and only if the element $\det \hat{A}$ is invertible in $\hat{\mathfrak{A}}$. Indeed, if $\det \hat{A}$ is invertible, the element $\| (\det \hat{A})^{-1}\hat{B}_{jk} \|_1^n$ where the $\hat{B}_{jk}$ are cofactors of the elements $\hat{A}_{jk}$, is $\hat{A}$'s inverse.

Now let $\hat{A}$ be invertible and $\hat{A}^{-1} = \|\hat{C}_{jk}\|_1^n$. The elements $\hat{C}_{jk}$ commute with each other, as well as with the $\hat{A}_{jk}$.

Indeed, the equalities

$$\hat{A}_{jk}\hat{A}^{-1} = \hat{A}^{-1}\hat{A}\hat{A}_{jk}\hat{A}^{-1} = \hat{A}^{-1}\hat{A}_{jk}$$

imply that the $\hat{A}_{jk}$ commute with the $\hat{C}_{ml}$, while the equalities

$$\hat{C}_{jk}\hat{A}^{-1} = \hat{A}^{-1}\hat{A}\hat{C}_{jk}\hat{A}^{-1} = \hat{A}^{-1}\hat{C}_{jk}$$

imply that the $\hat{C}_{jk}$ commute with each other. Therefore the equality $\det \hat{A} \times \det \hat{A}^{-1} = \det \hat{A}\hat{A}^{-1}$ is valid, which implies the invertibility of $\det \hat{A}$.

In order to complete the lemma's proof, it remains to make use of proposition E), § 11, Chapter I, concerning the fact that A (respectively, $\det A$) is a Φ-operator if and only if the element $\hat{A}$ ($\det \hat{A}$) is invertible.

THEOREM 4.1. *Let an operator V which is invertible only on the left satisfy conditions* $(\Sigma\mathrm{I})-(\Sigma\mathrm{III})$. *For an operator $A \in \mathfrak{R}_{n\times n}(V)$ to be a Φ-operator, it is necessary and sufficient that*

$$\det A(\zeta) \neq 0 \qquad (|\zeta| = 1). \tag{4.1}$$

PROOF. Let us show that $\det A$ can be represented in the form

$$\det A = B + T, \tag{4.2}$$

where $B \in \mathfrak{R}(V)$, $B(\zeta) = \det A(\zeta)$, and T is completely continuous. It is sufficient for this to show that if $A \in \mathfrak{R}(V)$, then $A_1A_2 = A_1 \circ A_2 + T_1$, where $A_1 \circ A_2$ is a product of the elements A_1 and A_2 in the sense of the algebra $\mathfrak{R}_0(V)$ (cf. §3, Chapter I) and the operator T_1 is completely continuous. Let R_{jn}, $n = 1, 2,\cdots$, be a sequence of operators of $\mathfrak{K}(V)$ which converges to the operator A_j $(j = 1, 2)$. Then, obviously,

$$R_{1n}R_{2n} = R_{1n} \circ R_{2n} + K_n,$$

where the operator K_n is finite-dimensional. Passing to the limit in the latter equality and utilizing the continuity of the new multiplication, we obtain the desired relation.

Equality (4.2) and Lemma 4.1 imply that $A \in \mathfrak{R}_{n\times n}(V)$ is a Φ-operator if and only if $B \in \mathfrak{R}(V)$ is one. It is easy to obtain from Theorem 3.1, Chapter I, the fact that B is a Φ-operator if and only if $B(\zeta) = \det A(\zeta) \neq 0$.

The theorem is proved.

Theorem 4.1 can be sharpened as follows with respect to its necessary part.[6)]

1°. *Let an operator V which is invertible only on the left satisfy conditions* (ΣI) *and* (Σ II), *and let the operator $A \in \mathfrak{R}_{n\times n}(V)$ be a Φ_+- (Φ_--) operator. Then relation* (4.1) *is fulfilled.*

As a matter of fact, by means of a device already used repeatedly the proof can be reduced to the case when $A \in \mathfrak{R}_{n\times n}(V)$. Let us suppose that $\det A(\zeta_0) = 0$, $|\zeta_0| = 1$. Then there is an nth order constant nonsingular matrix C such that all of the elements of some column (for example, the *first*) of the matrix function $A(\zeta)C$ vanish at ζ_0. The operator $AC \in \mathfrak{R}_{n\times n}(V)$ can be represented in the form

$$AC = B \begin{Vmatrix} V - \zeta_0 I & & & 0 \\ & I & & \\ & & \ddots & \\ 0 & & & I \end{Vmatrix} \tag{4.3}$$

where $B \in \mathfrak{R}_{n\times n}(V)$. By virtue of proposition D), §11, Chapter I, equality (4.3) implies that AC, and consequently also A, is not a Φ_+-operator. It can be shown analogously that under violation of condition (4.1) A is not a Φ_--operator.

2. In what follows it is assumed that operators V and $V_{(-1)}$ satisfy, in addition, the following condition.

(Σ IV) *The radical of the algebra $\mathfrak{R}_0(V)$ consists only of zero, i.e. to each function $a(\zeta) \in R(\zeta)$ there corresponds a unique operator $A \in \mathfrak{R}(V)$ whose symbol $\mathscr{A}(\zeta)$ coincides with $a(\zeta)$.*

By $\mathfrak{R}_{n\times n}(\zeta)$ we denote the algebra of matrix symbols $A(\zeta)$ corresponding to the various operators $A \in R_{n\times n}(V)$.

THEOREM 4.2. *Let the operator V satisfy conditions* (Σ I), (Σ II) *and* (Σ IV), *and let the symbol $A(\zeta) \in \mathfrak{R}_{n\times n}(\zeta)$ of an operator $A \in \mathfrak{R}_{n\times n}(V)$ admit the right canonical factorization*

$$A(\zeta) = A_-(\zeta) D(\zeta) A_+(\zeta) \qquad (D(\zeta) = \| \zeta^{\kappa_j} \delta_{jk} \|_1^n). \tag{4.4}$$

Then the following assertions are true.

1) *A is normally solvable.*

2)
$$\dim \operatorname{Ker} A = - \sum_{\kappa_j < 0} \kappa_j \dim \operatorname{Coker} V, \tag{4.5}$$

[6)]The sufficient part of Theorem 4.1 can be established by another method, without applying Lemma 4.1.

$$\dim \operatorname{Coker} A = \sum_{\kappa_j > 0} \kappa_j \dim \operatorname{Coker} V. \tag{4.5'}$$

3)
$$\kappa(A) = -\operatorname{ind} \det A(\zeta) \dim \operatorname{Coker} V. \tag{4.6}$$

If the equation

$$Ag = f \qquad (f \in \mathfrak{B}_n) \tag{4.7}$$

is solvable, one of its solutions is given by the formula

$$g = A_+^{-1} D^{[-1]} A_-^{-1} f,$$

where $A_\pm$ *are invertible operators with symbols* $A_\pm(\zeta) \in \mathfrak{R}^\pm_{n\times n}(\zeta)$, *and* $D^{[-1]} = \| V^{(-k_j)} \delta_{jk} \|_1^n$.

If κ_r *is the largest negative index, subspaces* $\mathfrak{L}_j \subset \mathfrak{B}_n$, $j = r, r+1, \cdots, n$, *exist such that*

$$\begin{aligned} \operatorname{Ker} A = {} & \mathfrak{L}_r \dotplus \tilde{V}\mathfrak{L}_r \dotplus \cdots \dotplus \tilde{V}^{|\kappa_r|-1}\mathfrak{L}_r \\ & \dotplus \mathfrak{L}_{r+1} \dotplus \tilde{V}\mathfrak{L}_{r+1} \dotplus \cdots \dotplus \tilde{V}^{|\kappa_{r+1}|-1}\mathfrak{L}_{r+1} \dotplus \\ & \cdots \dotplus \mathfrak{L}_n \dotplus \tilde{V}\mathfrak{L}_n \dotplus \cdots \dotplus \tilde{V}^{|\kappa_n|-1}\mathfrak{L}_n, \end{aligned} \tag{4.8}$$

where $\tilde{V} = \| \delta_{jk} V \|_1^n$.

PROOF. Equality (4.4) and proposition 1° of § 3, Chapter I, imply that A can be represented in the form $A = A_- D A_+$.

Since the operators $A_\pm$ are invertible, A is normally solvable, and

$$\dim \operatorname{Ker} A = \dim \operatorname{Ker} D, \qquad \dim \operatorname{Coker} A = \dim \operatorname{Coker} D.$$

This implies equalities (4.5) and (4.6). It is not difficult to verify the following equality:

$$AA^{[-1]}A = A,$$

where

$$A^{[-1]} = A_+^{-1} D^{[-1]} A_-^{-1}, \tag{4.9}$$

which implies the assertion concerning the solvability of equation (4.7).

In order to prove the last assertion of the theorem, let us observe that the equation $A\varphi = 0$ is equivalent to the equation $DA_+\varphi = 0$. Let us denote by $\mathfrak{L} \in \mathfrak{B}$ the subspace $\operatorname{Ker} V^{(-1)}$. Theorem 6.1 of Chapter I implies that the subspace $\operatorname{Ker} D$ is the direct sum of all the subspaces

$$\mathfrak{L}_{(j)} \dotplus \tilde{V}\mathfrak{L}_{(j)} \dotplus \cdots \dotplus \tilde{V}^{|\kappa_j|-1}\mathfrak{L}_{(j)} \qquad (j = r, r+1, \cdots, n),$$

where $\mathfrak{L}_{(j)} \subset \mathfrak{B}_n$ is the subspace of all vectors of the form

$$f = \{\underbrace{0, \cdots, 0}_{j-1}, \varphi, 0, \cdots, 0\} \qquad (\varphi \in \mathfrak{L}).$$

Taking into consideration the equality $A_+^{-1}\tilde{V}^p = \tilde{V}^p A_+^{-1}$ $(p \geqq 0)$, we obta the fact that the subspace Ker $A = A_+^{-1}$ Ker D can be represented in the form (4.8), where $\mathfrak{L}_j = A_+^{-1}\mathfrak{L}^{(j)}$.

COROLLARY 4.1. *Let the conditions of Theorem 4.2 be fulfilled. For the operator A to be invertible (left invertible, right invertible), it is nececessary and sufficient that all right indices of the matrix function $A(\zeta)$ be equal to zero (nonnegative, nonpositive). If A is invertible on some side, an operator $A^{[-1]}$ inverse to A on the corresponding side is defined by* (4.9).

REMARK. All results cited above (as well as Theorem 4.3 formulated below) remain valid if, instead of condition (Σ II), it is required that A belong to $\hat{\mathfrak{R}}_{n\times n}(U)$ (cf. the second scheme of § 3, Chapter I).

3. In the case of Hilbert space the propositions of this section can be supplemented by the following:

2°. *Let V be an isometric operator operating in Hilbert space $\mathfrak{H}$, and let $A \in \mathfrak{R}_{n\times n}(V)$. If*

$$\det A(\zeta) \neq 0 \qquad (|\zeta| = 1), \tag{4.10}$$

then A is normally solvable, and the equalities

$$\dim \operatorname{Ker} A = -\mathfrak{m} \sum_{\kappa_j<0} \kappa_j \quad \text{and} \quad \dim \operatorname{Coker} A = \mathfrak{m} \sum_{\kappa_j>0} \kappa_j$$

hold, where $\kappa_j, j = 1,\cdots, n$, are the right indices of the matrix function $A(\zeta)$ and $\mathfrak{m} = \dim \operatorname{Coker} V$.

Let condition (4.10) *be fulfilled, and let the equality*

$$A(\zeta) = A_-(\zeta)D(\zeta)A_+(\zeta) \qquad (|\zeta| = 1)$$

give a right factorization of $A(\zeta)$.

The operator $A^{[-1]}$ defined by the equality[7]

$$A^{[-1]} = A_+^{-1}(V)D^{-1}(V)A_-^{-1}(V) \tag{4.11}$$

is bounded and possesses the property $AA^{[-1]}A = A$.

In particular, if the equation $A\varphi = f$ is solvable, one of the solutions is obtained by the formula $\varphi = A^{[-1]}f$.

Let us prove this proposition first for the case when

$$\dim \operatorname{Coker} V = 1 \quad \text{and} \quad \bigcap_1^\infty V_j\mathfrak{H} = 0. \tag{4.12}$$

As is easy to see, V is unitarily equivalent in this case to an operator of multiplication by the independent variable ζ in the Hardy space H_2^+, and all of the assertions are easily verified. The operators $A_\pm^{\pm 1}(V)$ are defined by the equalities

[7]The definition of the operators appearing in this equality will be given below.

$$(A_+^{\pm 1}(V)f)(\zeta) = A_+^{\pm 1}(\zeta)f(\zeta)$$

and

$$(A_-^{\pm 1}(V)f)(\zeta) = \mathscr{P}A_-^{\pm 1}(\zeta)f(\zeta),$$

where $\mathscr{P}$ is an orthoprojection projecting $(L_2(\Gamma))_n$ onto $(H_2^+)_n$. Each of these operators, generally speaking, is unbounded. However, the operator $A^{[-1]}$ defined by equality (4.11), or, what is the same, by the equality

$$A^{[-1]} = A_+^{-1}(\zeta)\mathscr{P}D^{-1}(\zeta)\mathscr{P}A_-^{-1}(\zeta),$$

is bounded in $(H_2^+)_n$ by virtue of Corollary 3.1.

In the general case, as is well known (cf., for example, K. Hoffman's book [1]), the space $\mathfrak{H}$ can be decomposed into an orthogonal sum of a finite or infinite number of subspaces $\mathfrak{H}_j$, $j = 0, 1, \cdots$, such that the operators $V|\mathfrak{H}_j$, $j = 1,2,\cdots$, satisfy conditions (4.12), and the operator $V|\mathfrak{H}_0$ is unitary.

The operator $A^{[-1]}$ is defined on $\mathfrak{H}_j$, $j = 1,2,\cdots$, by the method just described, while it can be defined on $\mathfrak{H}_0$ by the equality

$$A^{[-1]}|\mathfrak{H}_0 = A^{-1}(V|\mathfrak{H}_0).$$

Let us note, moreover, that if condition (4.10) is fulfilled, Corollary 4.1 remains valid.

4. Let the projections P_τ $(\in \mathfrak{A};\ \tau \in \Omega)$ converge strongly to the identity operator as $\tau \to \infty$ and satisfy, as in §1 of Chapter III, the following conditions:

$$P_\tau V P_\tau = P_\tau V, \qquad P_\tau V^{(-1)} P_\tau = V^{(-1)} P_\tau \qquad (\tau \in \Omega). \tag{4.13}$$

Let us denote by $\mathscr{P}_\tau$ the projection defined in the space $\mathfrak{B}_n$ by the matrix $\|\delta_{jk}P_\tau\|_1^n$. Let us recall that the operator V satisfies conditions (Σ I)—(Σ IV). The following theorem is valid under these assumptions.

THEOREM 4.3. *Let the symbol $A(\zeta) \in \mathfrak{R}_{n\times n}(\zeta)$ of an operator $A \in \mathfrak{R}_{n\times n}(V)$ admit left canonical factorization. If A is invertible and all left indices of the symbol $A(\zeta)$ equal zero, then $A \in \Pi\{\mathscr{P}_\tau, \mathscr{P}_\tau\}$.*

PROOF. Let $A_+(\zeta) = \|A_{jk}^+(\zeta)\|_1^n$ and $A_-(\zeta) = \|A_{jk}^-(\zeta)\|_1^n$ be the factors of the left factorization of the matrix function $A(\zeta) = A_+(\zeta)A_-(\zeta)$. Then

$$A(\zeta) = \left\| \sum_{k=1}^n A_{jk}^+(\zeta)A_{kl}^-(\zeta) \right\|_{j,l=1}^n.$$

The latter equality and proposition 1° of § 3, Chapter I, imply that A admits the representation

$$A = \left\| \sum_{k=1}^n A_{kl}^- A_{jk}^+ \right\|_{j,l=1}^n,$$

where A^+_{jk} $(\in \mathfrak{R}(V))$ and $A^-_{kl}(\in \mathfrak{R}(V))$ are operators with symbols $A^+_{jk}(\zeta)$, $A^-_{kl}(\zeta)$.

As was established in the proof of Theorem 1.1, Chapter III, the operators of $\mathfrak{R}(V)$ commute to within a completely continuous addend. Therefore, transposing the factors A^-_{kl} and A^+_{jk}, let us represent A in the form

$$A = \left\| \sum_{k=1}^{n} A^+_{jk} A^-_{lk} \right\|_{j,l=1}^{n} + T,$$

where T is a completely continuous operator in $\mathfrak{B}_n$, and, consequently,

$$A = \| A^+_{jk} \|_1^n \cdot \| A^-_{jk} \|_1^n + T.$$

By virtue of Theorem 2.2, Chapter II, the operator $B = \| A^+_{jk} \|_1^n \cdot \| A^-_{jk} \|_1^n$ belongs to the class $\Pi\{\mathscr{P}_\tau, \mathscr{P}_\tau\}$. Since A is invertible, it also belongs to $\Pi\{\mathscr{P}_\tau, \mathscr{P}_\tau\}$ by virtue of Theorem 3.1, Chapter II.

Let us note in addition that the conditions of Theorem 4.3 are fulfilled if the matrix function $A(\zeta)$ admits left and right canonical factorizations and its left and right indices equal zero.

§ 5. Systems of discrete Wiener-Hopf equations

1. As in §7 of Chapter I, let E be one of the Banach spaces l_p, $p \geqq 1$, and let operators V and $V^{(-1)}$ be defined in E by the equalities

$$V\{\xi_j\}_1^\infty = \{0, \xi_1, \xi_2, \cdots\}, \qquad V^{(-1)}\{\xi_j\}_1^\infty = \{\xi_2, \xi_3, \cdots\}.$$

The operators V and $V^{(-1)}$ satisfy conditions (Σ I)—(Σ IV) of § 4. If the matrix function $A(\zeta)$ belongs to $\mathfrak{R}_{n\times n}(\zeta)$, the operator $A \in \mathfrak{R}_{n\times n}(V)$ corresponding to it is defined in the space E by the block matrix

$$A = \| a_{j-k} \|_{j,k=1}^\infty, \tag{5.1}$$

where the blocks $a_j, j = 0, \pm 1, \cdots$, are the matrix Fourier coefficients of the matrix function $A(\zeta)$.

All results of § 4[8] are applicable to the operator A.

Let us cite one of these applications as an example.

THEOREM 5.1. *Let* $A(\zeta)$ $(|\zeta| = 1)$ *be an arbitrary continuous nth order matrix function, and let* $a_j, j = 0, \pm 1, \cdots$, *be its matrix Fourier coefficients. For the operator* A *defined in* $(l_2)_n$ *by equality* (5.1) *to be a* Φ_+- (Φ_--) *operator, it is necessary and sufficient that* $\det A(\zeta) \neq 0$ *for* $|\zeta| = 1$.

Let this condition be fulfilled, and let the equality

$$A(\zeta) = A_-(\zeta) D(\zeta) A_+(\zeta) \qquad (D(\zeta) = \| \zeta^{\kappa_j} \delta_{jk} \|_1^n)$$

[8] All results of § 4, except Theorem 4.3, are also applicable in the case when A operates in the space m or c.

give a right factorization of $A(\zeta)$. *Then*

$$\dim \operatorname{Ker} A = -\sum_{\kappa_j<0} \kappa_j, \qquad \dim \operatorname{Coker} A = \sum_{\kappa_j>0} \kappa_j.$$

If the equation $Ag = f$ *is solvable, one of its solutions is given by the formula*

$$g = A_+^{-1} D^{[-1]} A_-^{-1} f,$$

where A_+^{-1}, $D^{[-1]}$ *and* A_-^{-1} *are matrix Toeplitz operators composed of the matrix Fourier coefficients of the matrix functions* $A_+^{-1}(\zeta)$, $D^{-1}(\zeta)$ *and* $A_-^{-1}(\zeta)$.

This theorem is a corollary of propositions 1° and 2° of § 4.

2. Let us denote by P_m, $m = 1, 2, \cdots$, a family of projections defined in E by the equalities

$$P_m\{\xi_j\}_1^\infty = \{\xi_1, \cdots, \xi_m, 0, 0, \cdots\}.$$

Let us note that V, $V^{(-1)}$ and P satisfy conditions (4.13)

LEMMA 5.1. *Let an operator* $A \in \mathfrak{R}(V)$ *be defined in the space E by the matrix* $A = \|a_{j-k}\|_{j,k=1}^\infty$. *Then the operator* $A' = \|a_{k-j}\|_1^\infty$ *also belongs to* $\mathfrak{R}(V)$. *In other words, if* $A(\zeta) \in \mathfrak{R}(\zeta)$, *then* $A(\zeta^{-1}) \in \mathfrak{R}(\zeta)$.

PROOF. Let us first establish the boundedness of A'. It is sufficient for this to show that the norms of the operators $P_m A' P_m$, $m = 1, 2, \cdots$, are bounded in the aggregate (cf. Kantorovič and Akilov [1], Chapter X, § 1.2). Let us denote by S_m the operator defined in E by the equality

$$S_m\{\xi_j\}_1^\infty = \{\xi_m, \xi_{m-1}, \cdots, \xi_1, \xi_{m+1}, \cdots\}.$$

S_m is invertible and isometric, with $S_m^{-1} = S_m$. It is easy to see that

$$P_m A' P_m = S_m P_m A P_m S_m,$$

which implies that

$$|P_m A' P_m| = |P_m A P_m|. \tag{5.2}$$

Thus the boundedness of A' is established. It is not difficult to derive also the equality

$$|A'| = |A| \tag{5.3}$$

from (5.2) and the strong convergence of the projections P_m.

Let the sequence of operators R_m, $m = 1, 2, \cdots$, of $\mathfrak{R}(V)$ converge to A. Then, by virtue of (5.3), the sequence of operators $R_m' \in \mathfrak{R}(V)$ corresponding to the functions $R_m(\zeta^{-1})$ converges to A'.

The lemma is proved.

Let us return to considering the matrix case.

Along with the operator A with the block matrix (5.1), let us consider an operator A^τ defined in E_n by the block transpose matrix

$$A^\tau = \left\| a_{k-j} \right\|_{j,k=1}^{\infty}. \tag{5.4}$$

By virtue of Lemma 5.1 we have $A^\tau \in \mathfrak{R}_{n\times n}(V)$, and $A_\tau(\zeta) = A(\zeta^{-1})$.

Let us denote by $\mathscr{P}_m$ the projection defined in E_n by the matrix $\mathscr{P}_m = \left\| \delta_{jk} P_m \right\|_1^n$.

THEOREM 5.2. *Let the matrix function a* (ζ) *belong to* $\mathfrak{R}_{n\times n}(\zeta)$, *and let the* a_j, $j = 0, \pm 1, \cdots$, *be its matrix Fourier coefficients. If*

$$\liminf_{m\to\infty} \left| (\mathscr{P}_m A \mathscr{P}_m)^{-1} \right| < \infty,$$

the operators A and A^τ *defined by* (5.1) *and* (5.4) *are invertible in* E_n. *Under fulfillment of these conditions,* $A \in \Pi\{\mathscr{P}_m, \mathscr{P}_m\}$.

PROOF. The invertibility of A follows from Lemma 1.1, Chapter III. Let us denote by the previous symbol S_m the operator now defined in the space E_n by the equality

$$S_m\{\xi_j\}_1^\infty = \{\xi_m, \xi_{m-1}, \cdots, \xi_1, \xi_{m+1}, \cdots\} \qquad (m = 1, 2, \cdots)$$

(ξ_j are n-dimensional vectors). It is easy to see that S_m is invertible and isometric, $S_m^{-1} = S_m$, and

$$\mathscr{P}_m A^\tau \mathscr{P}_m = S_m \mathscr{P}_m A \mathscr{P}_m S_m \qquad (m = 1, 2, \cdots).$$

This implies that

$$\liminf_{m\to\infty} \left| (\mathscr{P}_m A^\tau \mathscr{P}_m)^{-1} \right| < \infty$$

and, by virtue of Lemma 1.1 of Chapter III, A^τ is invertible.

Now let A and A^τ be invertible in E_n. Let us denote by $\tilde{A}$ the operator defined in the space $\tilde{E}_n$ of bilateral sequences of n-dimensional vectors $\xi = \{\xi_j\}_{-\infty}^{\infty}$ by the block matrix $\left\| a_{j-k} \right\|_{j,k=-\infty}^{\infty}$, and let us denote by $\mathscr{Q}_m$ the projection operating in $\tilde{E}_n$ in accordance with the rule[9]

$$\mathscr{Q}_m\{\xi_j\}_{-\infty}^{\infty} = \{\cdots, \xi_1, \xi_0, \xi_1, \cdots, \xi_m, 0, \cdots\} \qquad (m = 0, 1, \cdots).$$

It is easy to see that the operator $\mathscr{Q}_m \tilde{A} \mathscr{Q}_m$ $(m = 0, 1, \cdots)$ considered in the space $\mathscr{Q}_m E_n$ is similar to the operator A^τ, the similarity being realized by an isometric operator. This implies that $\mathscr{Q}_m \tilde{A} \mathscr{Q}_m$ is invertible in $\mathscr{Q}_m E_n$, and

$$\left| (\mathscr{Q}_m \tilde{A} \mathscr{Q}_m)^{-1} \right| = \left| (A^\tau)^{-1} \right|.$$

By virtue of Theorem 2.1 of Chapter II, $\tilde{A} \in \Pi\{\mathscr{Q}_m, \mathscr{Q}_m\}$. By virtue of Lemma 1.1 of Chapter VI, the operator

$$\tilde{T} = \mathscr{Q}_0 \tilde{A} (I - \mathscr{Q}_0) + (I - \mathscr{Q}_0) \tilde{A} \mathscr{Q}_0$$

[9] Unlike the notation adopted earlier, here $\mathscr{Q}_m \neq I - \mathscr{P}_m$.

is completely continuous (here I is the identity operator in $\tilde{E}_n$). The restriction of the operator $(I - \mathcal{Q}_0)\tilde{A}(I - \mathcal{Q}_0)$ to the subspace $(I - \mathcal{Q}_0)\tilde{E}_n$ $(= E_n)$ coincides with A; therefore this restriction is an invertible operator in $(I - \mathcal{Q}_0)\tilde{E}_n$. This implies that the operator

$$\tilde{C} = \mathcal{Q}_0\tilde{A}\mathcal{Q}_0 + (I - \mathcal{Q}_0)\tilde{A}(I - \mathcal{Q}_0) \tag{5.5}$$

is invertible in $\tilde{E}_n$.

Since the equality $\tilde{C} = \tilde{A} - \tilde{T}$ is valid, $\tilde{C} \in \Pi\{\mathcal{Q}_m, \mathcal{Q}_m\}$ by virtue of Theorem 3.1 of Chapter II. The operator $\mathcal{Q}_m\tilde{C}\mathcal{Q}_m$ is, by virtue of (5.5), representable in the form

$$\mathcal{Q}_m\tilde{C}\mathcal{Q}_m = \mathcal{Q}_0\tilde{A}\mathcal{Q}_0 + \mathcal{Q}_m(I - \mathcal{Q}_0)\tilde{A}(I - \mathcal{Q}_0)\mathcal{Q}_m.$$

Since the restriction of the operator $\mathcal{Q}_m(I - \mathcal{Q}_0)A(I - \mathcal{Q}_0)\mathcal{Q}_m$ to $\mathcal{Q}_m(I - \mathcal{Q}_0)\tilde{E}_n$ coincides with the operator $\mathcal{P}_mA\mathcal{P}_m$, the latter is invertible in the subspace $\mathcal{P}_mE_n$, and so

$$\sup_m |(\mathcal{P}_mA\mathcal{P}_m)^{-1}| < \infty.$$

The theorem is proved.

3. THEOREM 5.3. *Let $A(\zeta)$ be an arbitrary matrix function of the algebra $W_{n\times n}$, let a_j, $j = 0, \pm 1, \cdots$, be its matrix Fourier coefficients, and let*

$$A_m = \|a_{j-k}\|_{j,k=1}^m \qquad (m = 1, 2, \cdots) \tag{5.6}$$

be a truncated block matrix. If

$$\liminf_{m\to\infty} |A_m^{-1}| < \infty, \tag{5.7}$$

then the following conditions are fulfilled.

a) $\det A(\zeta) \neq 0$ $(|\zeta| = 1)$.

b) *The left and right indices of $A(\zeta)$ equal zero.*

Under fulfillment of conditions a) *and* b) *the system*

$$\sum_{k=1}^m a_{j-k}\xi_k = \eta_j \qquad (j = 1, \cdots, m), \tag{5.8}$$

beginning with some m, has the unique solution $\{\xi_j^{(m)}\}_{j=1}^m$, and as $m \to \infty$ the sequence

$$\xi^{(m)} = \{\xi_1^{(m)}, \xi_2^{(m)}, \cdots, \xi_m^{(m)}, 0, \cdots\} \qquad (m = 1, 2, \cdots) \tag{5.9}$$

converges in the norm of E_n to a solution of the full system

$$\sum_{k=1}^\infty a_{j-k}\xi_k = \eta_j \qquad (j = 1, 2, \cdots), \tag{5.10}$$

whatever the element $\{\eta_j\}_1^\infty \in E_n$ may be.

PROOF. Since the matrix function $A(\xi) \in W_{n\times n}$, it belongs to the algebra $\mathfrak{R}_{n\times n}(\zeta)$ for each of the spaces E_n. If condition (5.7) is fulfilled, then, by virtue of Theorem 5.2, the operators A and A^τ defined by equalities (5.1) and (5.4) are invertible, and, by virtue of Theorem 4.1, $\det A(\zeta) \neq 0$ for $|\zeta| = 1$.

Theorem 2.2 implies that the matrix functions $A(\zeta)$ and $A^\tau(\zeta) = A(\zeta^{-1})$ admit right and left canonical factorizations in the algebra $W_{n\times n}$. The factors in these factorizations obviously belong to the corresponding algebras $\mathfrak{R}^{\pm}_{n\times n}(\zeta)$. Let us denote by $\kappa_j^{(r)}(A(\zeta))$ and $\kappa_j^{(l)}(A(\zeta))$, $j = 1,\cdots,n$, the corresponding right and left indices of the matrix function $A(\zeta)$. Now applying Corollary 4.1, we obtain

$$\kappa_j^{(r)}(A(\zeta)) = \kappa_j^{(r)}(A(\zeta^{-1})) = 0 \qquad (j = 1,\cdots,n).$$

Since $\kappa_j^{(r)}(A(\zeta^{-1})) = -\kappa_{n-j+1}^{(l)}(A(\zeta))$, condition b) is fulfilled.

The second assertion of the theorem is a corollary of both Theorem 4.3 and Theorem 5.2.

If we confine ourselves to considering the space $(l_2)_n$, the algebra of all continuous matrix functions can be taken instead of the algebra $W_{n\times n}$. That is, the following theorem is valid.

THEOREM 5.4. *Let $A(\zeta)$ ($|\zeta| = 1$) be an arbitrary continuous nth order matrix function, and let a_j, $j = 0, \pm 1,\cdots$, be its matrix Fourier coefficients. If relation* (5.7) *holds, all of the conclusions of Theorem* (5.3) *are valid.*

Clearly, here all norms must be understood in the sense of $(l_2)_n$, and the element η must also be taken from $(l_2)_n$.

This theorem is an immediate corollary of Theorems 5.1 and 5.2.

§ 6. Systems of Wiener-Hopf integral equations

1. The basic results of § 8, Chapter I, and § 3, Chapter III, admit generalizations to systems of the form

$$\varphi(t) - \int_0^\infty k(t-s)\varphi(s)\,ds = f(t) \qquad (0 \leqq t < \infty). \tag{6.1}$$

Here $k(t)$ is an nth order matrix function with elements of $\tilde{L}_1$; $\varphi(t)$ and $f(t)$ are n-dimensional vector functions. For the sake of brevity, we shall everywhere write $\tilde{L}$ instead of $\tilde{L}_1$ and $\tilde{L}_{n\times n}$ instead of $(\tilde{L}_1)_{n\times n}$.

As previously (cf. § 8 of Chapter I), let operators V and $V(-1)$ be defined by the equalities

$$\begin{aligned}(Vx)(t) &= x(t) - \int_0^t e^{s-t}x(s)\,ds,\\ (V^{(-1)}x)(t) &= x(t) - \int_t^\infty e^{t-s}\,x(s)\,ds\end{aligned} \qquad (0 < t < \infty).$$

In the case of the spaces $L_p(0, \infty)$, $1 \leqq p < \infty$, these operators satisfy conditions $(\Sigma\,\mathrm{I})$—$(\Sigma\,\mathrm{IV})$ of § 4. The operator A defined in $(L_p)_n$ by equation (6.1) belongs to the set $\mathfrak{R}_{n\times n}(V)$, and its symbol $A(\zeta)$ $(|\zeta| = 1)$ is defined by the equality

$$A(\zeta) = I - K\left(i\frac{1+\zeta}{1-\zeta}\right),$$

where $K(\lambda)$ is the Fourier transform of the matrix function $k(t)$. If the matrix function $I - K(\lambda)$ of $\mathfrak{L}_{n\times n}$ satisfies the condition

$$\det\,(I - K(\lambda)) \neq 0 \qquad (-\infty < \lambda < \infty), \tag{6.2}$$

it admits, by virtue of Theorem 2.4, right (left) canonical factorization.

THEOREM 6.1. *Let* $k(t) \in \tilde{L}_{n\times n}$. *For the operator* A *defined in* $(L_p)_n$, $1 \leqq p < \infty$, by *equation* (6.1) *to be a* Φ_+- (Φ_--) *operator, it is necessary and sufficient that condition* (6.2) *be fulfilled. If this condition is fulfilled, then*

$$\dim\operatorname{Ker} A = -\sum_{\kappa_j<0}\kappa_j, \qquad \dim\operatorname{Coker} A = \sum_{\kappa_j>0}\kappa_j,$$

where the κ_j, $j = 1,\cdots,n$, *are the right indices of the matrix function* $I - K(\lambda)$.

This theorem is easily derived from proposition 1° of § 4 and Theorem 4.2.

2. Let us consider in addition the projections

$$(P_\tau\varphi)(t) = \begin{cases}\varphi(t), & 0 < t < \tau,\\ 0, & \tau < t < \infty\end{cases} \qquad (\varphi(t) \in L_p),$$

$$\mathscr{P}_\tau = \|\delta_{jk}P_\tau\|_1^n \qquad (0 < \tau < \infty).$$

Let us note that the operator $\mathscr{P}_\tau A\mathscr{P}_\tau$ is defined in the space $\mathscr{P}_\tau(L_p)_n$ by the equality

$$(P_\tau AP_\tau\varphi)(t) = \varphi(t) - \int_0^\tau k(t-s)\varphi(s)\,ds \qquad (0 < t < \tau).$$

THEOREM 6.2. *Let* $k(t) \in \tilde{L}_{n\times n}$. *If*

$$\liminf_{\tau\to\infty}\left|(\mathscr{P}_\tau A\mathscr{P}_\tau)^{-1}\right| < \infty, \tag{6.3}$$

then the following assertions are true.

a) $\det\,(I - K(\lambda)) \neq 0$ $(-\infty < \lambda < \infty)$, *where* $K(\lambda)$ *is the Fourier transform of* $k(t)$.

b) *The left and right indices of the matrix function* $I - K(\lambda)$ *equal zero.*

Under fulfillment of conditions a) *and* b) *the truncated equation*

$$\varphi(t) - \int_0^\tau k(t-s)\varphi(s)\,ds = f(t) \qquad (0 < t < \tau),$$

beginning with some τ, has the unique solution $\varphi_\tau(t)$, and as $\tau \to \infty$ the vector functions

$$\tilde{\varphi}_\tau(t) = \begin{cases} \varphi_\tau(t), & 0 < t < \tau, \\ 0, & \tau < t < \infty \end{cases}$$

converge in the norm of $(L_p)_n$ to a solution of equation (6.1), *whatever the vector function $f(t) \in (L_p)_n$ may be.*

PROOF. Lemma 1.1 of Chapter III implies that under condition (6.3) A is invertible. Let us now show that the auxiliary operator

$$(B\varphi)(t) = \varphi(t) - \int_0^\infty k(s - t)\varphi(s)\, ds$$

is invertible in $(L_p)_n$. Let us denote by S_τ the operator defined in $(L_p)_n$ by the equality

$$(S_\tau\varphi)(t) = \begin{cases} \varphi(\tau - t), & 0 < t < \tau, \\ \varphi(t), & \tau < t < \infty, \end{cases} \qquad (0 < \tau < \infty).$$

S_τ is invertible and isometric, and $S_\tau^{-1} = S_\tau$. It is easy to see that

$$\mathscr{P}_\tau B \mathscr{P}_\tau = S_\tau \mathscr{P}_\tau A \mathscr{P}_\tau S_\tau,$$

which implies that

$$\liminf_{\tau \to \infty} |(\mathscr{P}_\tau B \mathscr{P}_\tau)^{-1}| < \infty.$$

By virtue of Lemma 1.1, Chapter III, B is invertible. Let us note that the symbol of B equals $I - K(-\lambda)$.

The invertibility of A and B implies, by virtue of Theorem 6.1, that condition a) is fulfilled and all right indices of the matrix functions $I - K(\lambda)$ and $I - K(-\lambda)$ equal zero. Condition b) follows from the obvious relationship between the right indices of $I - K(-\lambda)$ and the left indices of $I - K(\lambda)$. The second assertion of the theorem is proved just as in Theorem 5.2. It can also be derived from Theorem 4.3. The continuous analogues of the remaining theorems of § 5 are established analogously.

3. Let us now establish the applicability of the Galerkin method to the solution of equation (6.1) for the case of Hilbert space.

As previously, let the functions $\psi_j(t)$, $j = 0, 1, \cdots$, be defined by the equalities

$$\psi_0(t) = \sqrt{2}e^{-t}, \qquad \psi_{j+1}(t) = (V\psi_j)(t) \qquad (j = 0, 1, \cdots)$$

(cf. § 3.2 of Chapter III).

As in § 3, Chapter III, it is easy to obtain the fact that equation (6.1) is equivalent to the infinite system

$$\sum_{k=0}^{\infty} a_{j-k}\xi_k = \eta_j \qquad (j = 0, 1, \cdots), \tag{6.4}$$

where the a_j, $j = 0, \pm 1, \cdots$, are the matrix Fourier coefficients of the matrix function $I - K(i(1+\zeta)/(1-\zeta))$; the η_j, $j \geqq 0$, are the vectorial Fourier coefficients of the vector function $F(i(1+\zeta)/(1-\zeta))$ (here $F(\lambda)$ is the Fourier transform of the vector function $f(t)$); and the ξ_j are n-dimensional vectors. If in this connection $\{\xi_j\}_0^\infty \in (l_2)_n$ is a solution of system (6.4), the vector function $\varphi(t) = \sum_0^\infty \zeta_j \psi_j(t)$ gives a solution of equation (6.1).

An approximate solution of equation (6.1) is sought in the form

$$\varphi_m(t) = \sum_{j=0}^{m-1} \xi_j \psi_j(t),$$

where the ξ_j are n-dimensional vectors satisfying the block system of equations

$$\sum_{k=0}^{m-1} (A\psi_k, \psi_j)\xi_k = (f, \psi_j) \qquad (j = 0, 1, \cdots, m-1). \tag{6.5}$$

The following notation is adopted here:

$$(A\psi_k, \psi_j) = \|(A_{lr}\psi_k, \psi_j)\|_{l,r=1}^n,$$

where $A = \|A_{lr}\|_1^n$ is the matrix representation of the operator A defined by equation (6.1), and for the vector $f = \{f_k\}_1^n \in (L_2)_n$ it is assumed that $(f, \psi_j) = \{(f_k, \psi_j)\}_{k=1}^n$.

It is easy to see that system (6.5) coincides with the block system

$$\sum_{k=0}^{m-1} a_{j-k}\xi_k = \eta_j \qquad (j = 0, 1, \cdots, m-1).$$

System (6.5) can be regarded as a system of mn equations in mn unknowns. Let us denote the matrix of this system by A_m.

THEOREM 6.3. *Let* $k(t) \in \tilde{L}_{n\times n}$. *If*

$$\lim_{m\to\infty} |A_m^{-1}| < \infty,$$

conditions a) *and* b) *of Theorem* 6.2 *are fulfilled. Under fulfillment of these conditions system* (6.5), *beginning with some* m, *has the unique solution* $\{\xi_j^{(m)}\}_{j=0}^{m-1}$, *and as* $m \to \infty$ *the vector functions*

$$\varphi_m(t) = \sum_{j=0}^{m-1} \xi_j^{(m)} \psi_j(t)$$

converge in the norm of $(L_2)_n$ *to a solution of equation* (6.1), *whatever the vector function* $f(t) \in (L_2)_n$ *may be.*

This theorem follows from the above-mentioned relationship between the solutions of equation (6.1) and those of system (6.4), and from Theorem 5.4.

§ 7. Systems of finite difference equations

1. Certain results of § 10, Chapter I, and § 6, Chapter IV, admit generalization to systems of finite difference equations of the form

$$\sum_{j=-\infty}^{\infty} a_j \varphi(t-j) = f(t) \qquad (0 < t < \infty), \tag{7.1}$$

where the a_j, $j = 0, \pm 1, \cdots$, are constant nth order matrices, while $\varphi(t)$ and $f(t)$ are n-dimensional vector functions.

As in the above-named sections, let an isometric operator V be defined in Hilbert space $L_2(0, \infty)$ by the equality

$$(V\varphi)(t) = \begin{cases} \varphi(t-1), & t > 1, \\ 0, & t < 1. \end{cases}$$

V satisfies conditions (Σ I), (Σ II) and (Σ IV), but does not satisfy condition (Σ III). The algebra $\mathfrak{R}_{n\times n}(\zeta)$ coincides in this case with the set of all continuous nth order matrix functions.

If $A(\zeta)$ ($|\zeta| = 1$) is a continuous nth order matrix function and the a_j, $j = 0, \pm 1, \cdots$, are its matrix Fourier coefficients, the corresponding operator $A \in \mathfrak{R}_{n\times n}(V)$ operating in $(L_2)_n$ is defined by (7.1).[10] The results of § 4 are applicable to A. Let us, for example, cite the following proposition.

THEOREM 7.1. *Let $A(\zeta)$ ($|\zeta| = 1$) be an arbitrary continuous nth order matrix function, and let the a_j ($j = 0, \pm 1, \cdots$) be its matrix Fourier coefficients. For the operator A defined in $(L_2)_n$ by equation* (7.1) *to be a Φ_+- (Φ_--) operator, it is necessary and sufficient that*

$$\det A(\zeta) \neq 0 \qquad (|\zeta| = 1) \tag{7.2}$$

and that all right indices of the matrix function $A(\zeta)$ be nonnegative (nonpositive).

This theorem is a corollary of propositions 1° and 2° of § 4. A number of corollaries can be obtained in addition from proposition 2°:

1) *Under fulfillment of condition* (7.2) *the operator A is normally solvable.*

2) *If among the right indices of $A(\zeta)$ there are positive (negative) ones, then* dim Coker $A = \infty$ (dim Ker $A = \infty$).

3) *If the equation $A\varphi = f$ is solvable, one of its solutions is obtained from the formula $\varphi = A^{[-1]}f$, where the operator $A^{[-1]}$ is defined in the following way.*

Let the equality $A(\zeta) = A_-(\zeta)D(\zeta)A_+(\zeta)$ give a right factorization of $A(\zeta)$. Let us form the operators

$$A_+^{-1}\varphi = \sum_{j=0}^{\infty} a_j^+ \varphi(t-j), \qquad A_-^{-1}\varphi = \sum_{j=-\infty}^{0} a_j^- \varphi(t-j),$$

---

[10] Cf. § 6 of Chapter IV concerning the convergence of the series in the left side of (7.1).

$$D^{[-1]}\varphi = \sum_{j=-\infty}^{\infty} d_j \varphi(t - j),$$

where the $a_j^{\mp}$ $(j = 0, 1, \cdots)$ and d_j $(j = 0, \pm 1, \cdots)$ are the matrix Fourier coefficients of the matrix functions $A_{\pm}^{-1}(\zeta)$ and $D^{-1}(\zeta)$, and let us set

$$A^{[-1]} = A_{+}^{-1} D^{[-1]} A_{-}^{-1}.$$

Let us note, moreover, that $A^{[-1]}$ is bounded and possesses the property $AA^{[-1]}A = A$.

If all right indices of $A(\zeta)$ are nonnegative (nonpositive), $A^{[-1]}$ is a left (right) inverse of A.

2. We need the projections

$$(P_m\varphi)(t) = \begin{cases} \varphi(t), & 0 < t < m, \\ 0, & m < t < \infty \end{cases} \qquad (\varphi(t) \in L_2),$$

$$\mathcal{P}_m = \left\| \delta_{jk} P_m \right\|_{j,k=1}^{n} \qquad (m = 1, 2, \cdots).$$

THEOREM 7.2. *Let $A(\zeta)$ $(|\zeta| = 1)$ be an arbitrary continuous nth order matrix function, let a_j, $j = 0, \pm 1, \cdots$, be its matrix Fourier coefficients and let A be defined by the left side of* (7.1). *If*

$$\liminf_{m \to \infty} \left| (\mathcal{P}_m A \mathcal{P}_m)^{-1} \right| < \infty,$$

then the following conditions are fulfilled.

a) det $A(\zeta) \neq 0$ *for* $|\zeta| = 1$.

b) *The left and right indices of* $A(\zeta)$ *equal zero.*

Under fulfillment of conditions a) *and* b) *the truncated equation*

$$\sum_{t-m<j<t} a_j \varphi(t - j) = f(t) \qquad (0 < t < m) \tag{7.3}$$

has, beginning with some m, the unique solution $\varphi_m(t) \in (L_2(0, m))_n$, and as $m \to \infty$ the vector functions

$$\tilde{\varphi}_m(t) = \begin{cases} \varphi_m(t), & 0 < t < m, \\ 0, & m < t < \infty \end{cases}$$

converge in the norm of the space $(L_2)_n$ to a solution of equation (7.1), *whatever the vector function $f(t) \in (L_2)_n$ may be.*

Proof is carried out under the same scheme as the proof of Theorem 6.2, Chapter IV. Equation (7.3), or, what is the same, the equation $\mathcal{P}_m A \mathcal{P}_m \varphi = \mathcal{P}_m f$ is equivalent to the system

$$\sum_{k=1}^{m} a_{j-k} \varphi_k(t) = f_j(t) \qquad (j = 1, \cdots, m),$$

where the vector functions $\varphi_j(t) = \varphi(t + j - 1)$ and $f_j(t) = f(t + j - 1)$ $(0 \leqq t \leqq 1; j = 1,\cdots,m)$ belong to $(L_2(0, 1))_n$. By virtue of this equivalence and Lemma 6.1 of Chapter IV,

$$|(\mathscr{P}_m A \mathscr{P}_m)^{-1}| = |A_m^{-1}|$$

where the matrix A_m is defined by (5.6).

In order to complete the proof, it remains to apply Theorem 5.4 of this chapter and Theorem 2.1 of Chapter II.

The results of the present section remain valid in $(L_p)_n$, $1 \leqq p < \infty$, if the symbol $A(\zeta)$ belongs to the corresponding algebra $\mathfrak{R}_{n\times n}(\zeta)$ and admits canonical factorization.

§ 8. Systems of pair equations

As in § 2 of Chapter V, let an invertible operator $U \in \mathfrak{A}(\mathfrak{B})$ and a projection $P \in \mathfrak{A}(\mathfrak{B})$ satisfy the following conditions:

1) *The spectral radii of U and U^{-1} equal unity.*

2) $UP = PUP$, $UP \neq PU$ *and* $PU^{-1} = PU^{-1}P$.

Let us assume in addition that

3) $\dim \operatorname{Coker} (U|P\mathfrak{B}) < \infty$.

Let us denote by $\mathfrak{B}_n$, as previously, the space of all n-dimensional vectors with elements of $\mathfrak{B}$, by $\mathfrak{R}_{n\times n}(U)$ the set of all nth order matrices with elements of $\mathfrak{R}(U)$, and by $\mathscr{P}$ and $\mathscr{Q}$ the projections defined in $\mathfrak{B}_n$ by the equalities $\mathscr{P} = \|\delta_{jk}P\|_1^n$ and $\mathscr{Q} = I - \mathscr{P}$.

THEOREM 8.1. *Let conditions* 1), 2) *and* 3) *be fulfilled, and let the operators $A, B \in \mathfrak{R}_{n\times n}(U)$. For the operator $A\mathscr{P} + B\mathscr{Q}$ $(\mathscr{P}A + \mathscr{Q}B)$ to be a Φ-operator it is necessary and sufficient that*

$$\det A(\zeta) \neq 0 \quad \textit{and} \quad \det B(\zeta) \neq 0 \quad (|\zeta| = 1). \tag{8.1}$$

This theorem is a corollary of Theorem 1.2, Chapter V, and Theorem 4.1 (cf. the remark in § 4). The necessary part of Theorem 8.1 can be sharpened as follows.

1°. *Let conditions* 1) *and* 2) *be fulfilled, and let the operators $A, B \in \mathfrak{R}_{n\times n}(U)$. If the operator $A\mathscr{P} + B\mathscr{Q}$ $(\mathscr{P}A + \mathscr{Q}B)$ is a Φ_+- or a Φ_--operator, relations* (8.1) *are fulfilled.*

This proposition follows from Theorem 1.2 of Chapter V and proposition 1° of § 4.

Let us suppose in addition the following condition is fulfilled:

4) *The algebra $\mathfrak{R}(U)$ does not have a radical.*

If conditions (8.1) are fulfilled and the matrix function $C(\zeta) = B^{-1}(\zeta)A(\zeta)$ of $\mathfrak{R}_{n\times n}(\zeta)$ admits the right canonical factorization

$$C(\zeta) = C_-(\zeta)D(\zeta)C_+(\zeta) \qquad (D(\zeta) = \|\zeta^{\kappa_j}\delta_{jk}\|_1^n),$$

it is not difficult to verify that

$$\dim \operatorname{Ker}(A\mathscr{P} + B\mathscr{Q}) = -\mathfrak{m} \sum_{\kappa_j<0} \kappa_j,$$
$$\dim \operatorname{Coker}(A\mathscr{P} + B\mathscr{Q}) = \mathfrak{m} \sum_{\kappa_j>0} \kappa_j,$$

where $\mathfrak{m} = \dim \operatorname{Coker}(U|P\mathfrak{B})$ and the operator $A\mathscr{P} + B\mathscr{Q}$ can be represented in the form

$$A\mathscr{P} + B\mathscr{Q} = BC_-(D\mathscr{P} + \mathscr{Q})(C_+\mathscr{P} + C_-^{-1}\mathscr{Q}),$$

where $C_\pm, D \in \mathfrak{R}_{n\times n}(U)$ are operators with symbols $C_\pm(\zeta)$ and $D(\zeta)$. Let us form the operator

$$(A\mathscr{P} + B\mathscr{Q})^{[-1]} = (C_+^{-1}\mathscr{P} + C_-\mathscr{Q})(D^{[-1]}\mathscr{P} + \mathscr{Q})\, C_-^{-1}B^{-1}, \tag{8.2}$$

where $D^{[-1]} = \|U^{-\kappa_j}\delta_{jk}\|_1^n$.

If all right indices κ_j are equal to zero (nonnegative, nonpositive), the operator $(A\mathscr{P} + B\mathscr{Q})^{[-1]}$, as is easy to see, is the inverse (a left inverse, a right inverse) of the operator $A\mathscr{P} + B\mathscr{Q}$. In any case, the equality

$$(A\mathscr{P} + B\mathscr{Q})(A\mathscr{P} + B\mathscr{Q})^{[-1]}(A\mathscr{P} + B\mathscr{Q}) = A\mathscr{P} + B\mathscr{Q} \tag{8.3}$$

is valid, which implies that if the equation $(A\mathscr{P} + B\mathscr{Q})\varphi = f$ is solvable, one of its solutions is given by the formula

$$\varphi = (A\mathscr{P} + B\mathscr{Q})^{[-1]}f.$$

In the case when U is a unitary operator operating in Hilbert space $\mathfrak{H}$ and P is an othogonal projection in $\mathfrak{H}$ which satisfies condition 2), these results can be supplemented with the following proposition.

2°. *Let $A(\zeta)$ and $B(\zeta)$, $|\zeta| = 1$, be arbitrary continuous nth order matrix functions satisfying conditions* (8.1). *Then the operator $A\mathscr{P} + B\mathscr{Q}$ is normally solvable, and*

$$\dim \operatorname{Ker}(A\mathscr{P} + B\mathscr{Q}) = -\mathfrak{m} \sum_{\kappa_j<0} \kappa_j,$$
$$\dim \operatorname{Coker}(A\mathscr{P} + B\mathscr{Q}) = \mathfrak{m} \sum_{\kappa_j>0} \kappa_j;$$

here the κ_j, $j = 1,\cdots,n$, are the right indices of the matrix function $C(\zeta) = B^{-1}(\zeta)A(\zeta)$, and $\mathfrak{m} = \dim \operatorname{Coker}(U|P\mathfrak{H})$.

This proposition is a simple corollary of Theorem 1.2 of Chapter V and proposition 2° of § 4.

If conditions (8.1) are fulfilled, an operator $(A\mathscr{P} + B\mathscr{Q})^{[-1]} \in \mathfrak{A}(\mathfrak{H}_n)$ exists such that equality (8.3) is valid. This operator is defined by means of factoring the matrix function $C(\zeta)$, analogously to the way this was done above. The validity of the corresponding constructions follows from proposition 2° of §4.

Let us note, moreover, that results analogous to those just cited hold for the operator $\mathscr{P}A + \mathscr{Q}B$.

It is possible to obtain as corollaries of the results of this section theorems concerning systems of pair integral equations, transpose systems, their discrete and

difference analogues, and systems of singular integral equations on the circle.
Let us cite, for example, the following proposition.

THEOREM 8.2 *Let the matrix functions* $A(\zeta)$ *and* $B(\zeta)$ *belong to* $W_{n\times n}$, *and let* a_j *and* b_j, $j = 0, \pm 1, \cdots$, *be their matrix Fourier coefficients. For an operator* Z *defined in the space* $(\tilde{L}_p)_n$, $1 \leqq p < \infty$, *by the equality*

$$(Z\varphi)(t) = \sum_{j<t} a_j\varphi(t-j) + \sum_{j>t} b_j\varphi(t-j) \qquad (-\infty < t < \infty)$$

to be a Φ_+- (Φ_--) *operator it is necessary and sufficient that conditions* (8.1) *be fulfilled and all right indices of the matrix function* $C(\zeta) = B^{-1}(\zeta)A(\zeta)$ *be nonnegative* (*nonpositive*).

Let us note that if conditions (8.1) are fulfilled and all right indices of $C(\zeta)$ are nonnegative (nonpositive), then Z is left (right) invertible, and an inverse operator $Z^{(-1)}$ on the corresponding side is defined by (8.2).

§ 9. **Projection methods for the solution of pair equations**

Projection methods are presented in this section for the solution of discrete, continual, and difference pair systems, their transposes, and systems of singular integral equations on the circle.

1. *General theorem.* Let us suppose that an invertible operator $U \in \mathfrak{A}(\mathfrak{B})$ and a projection $P \in \mathfrak{A}(\mathfrak{B})$ satisfy conditions 1) – 4) of § 8 (the notation of the latter section is utilized in what follows).

Let the projections $P_\tau \in \mathfrak{A}(\mathfrak{B})$ $(\tau \in \Omega)$ converge strongly to the identity operator as $\tau \to \infty$. As in § 1 of Chapter VI, it will be assumed in the sequel that the following conditions are satisfied.

5) *The projections* P_τ, $\tau \in \Omega$, *and* P *are commutative.*

6) *The subspaces* $P_\tau P\mathfrak{B}$, $(I - P_\tau)P\mathfrak{B}$, $P_\tau Q\mathfrak{B}$ *and* $(I - P_\tau)Q\mathfrak{B}$, $\tau \in \Omega$, *are invariant relative to the operators* PU^{-1}, UP, QU *and* $U^{-1}Q$, *respectively* (*here* $Q = I - P$).

Let us denote by $\mathscr{P}_\tau$ the projection defined in $\mathfrak{B}_n$ by the equality $\mathscr{P}_\tau = \|\delta_{jk}P_\tau\|_1^n$.

THEOREM 9.1. *Let the following conditions be fulfilled for operators* $A,B \in \mathfrak{R}_{n\times n}(U)$:

a) $\det A(\zeta) \neq 0 \neq \det B(\zeta)$ *for* $|\zeta| = 1$.

b) *The matrix function* $A(\zeta)$ *admits left canonical factorization and its left indices equal zero, while the matrix function* $B(\zeta)$ *admits right canonical factorization and its right indices equal zero.*

c) *The operators* $\mathscr{P}A\mathscr{P}$ *and* $\mathscr{Q}B\mathscr{Q}$ *are respectively invertible in* $\mathscr{P}\mathfrak{B}_n$ *and* $\mathscr{Q}\mathfrak{B}_n$.

If the operator $A\mathscr{P} + B\mathscr{Q}$ ($\mathscr{P}A + \mathscr{Q}B$) *is invertible, it belongs to the class* $\Pi\{\mathscr{P}_\tau, \mathscr{P}_\tau\}$.

PROOF. $A\mathscr{P} + B\mathscr{Q}$ can be represented in the form

$$A\mathscr{P} + B\mathscr{Q} = \mathscr{P}A\mathscr{P} + \mathscr{Q}B\mathscr{Q} + T,$$

where the operator T is completely continuous (cf. Lemma 1.1, Chapter VI). Theorem 4.3 (cf. the remark in § 4) and proposition 2° of § 2, Chapter II, imply that $\mathscr{P}A\mathscr{P} + \mathscr{Q}B\mathscr{Q} \in \Pi\{\mathscr{P}_\tau, \mathscr{P}_\tau\}$. Since $A\mathscr{P} + B\mathscr{Q}$ is invertible, it belongs, by virtue of Theorem 3.1, Chapter II, to the class $\Pi\{\mathscr{P}_\tau, \mathscr{P}_\tau\}$.

Let us note that all conditions of Theorem 9.1 are fulfilled if: a) the matrix functions $A(\zeta)$ and $B(\zeta)$ admit left and right canonical factorizations, their left and right indices being equal to zero; b) the matrix function $B^{-1}(\zeta)A(\zeta)$ admits right canonical factorization, and its right indices equal zero.

As corollaries of Theorem 4.2 various propositions can be obtained concerning projection methods for the solution of systems of pair integral equations, transpose systems, their discrete and difference analogues, and systems of singular equations on the unit circle. Part of these propositions and some refinements of them are cited in the following subsections.

2. *Systems of discrete equations.*

THEOREM 9.2. *Let $A(\zeta)$ and $B(\zeta)$, $|\zeta| = 1$, be arbitrary continuous nth order matrix functions, and let the a_j and b_j, $j = 0, \pm 1, \cdots$, be their matrix Fourier coefficients. If* $\det A(\zeta) \neq 0 \neq \det B(\zeta)$ *for* $|\zeta| = 1$ *and if the left and right indices of the matrix functions $A(\zeta)$ and $B(\zeta)$ equal zero, while the right indices of the matrix function $B_{-1}(\zeta)\, A(\zeta)$ also equals zero, the system*

$$\sum_{k=0}^{m} a_{j-k}\xi_k + \sum_{k=-m}^{-1} b_{j-k}\xi_k = \eta_j \qquad (j = 0, \pm 1, \cdots, \pm m),$$

beginning with some m, has the unique solution $\{\xi_j^{(m)}\}_{j=-m}^{m}$, and as $m \to \infty$ the sequence

$$\xi^{(m)} = \{\cdots, 0, \xi_{-m}^{(m)}, \cdots, \xi_m^{(m)}, 0, \cdots\} \tag{9.1}$$

converges in the norm of $(\bar{l}_2)_n$ to a solution of the system

$$\sum_{k=0}^{\infty} a_{j-k}\xi_k + \sum_{k=-\infty}^{-1} b_{j-k}\xi_k = \eta_j \qquad (j = 0, \pm 1, \cdots), \tag{9.2}$$

whatever the element $\{\eta_j\}_{-\infty}^{\infty} \in (\bar{l}_2)_n$ may be.

PROOF. Let us adopt the notation of § 2, Chapter VI, and let the projections $\mathscr{P}$ and $\mathscr{P}_m$ be defined in $(\bar{l}_2)_n$ by the equalities

$$\mathscr{P} = \|\delta_{jk}P\|_1^n, \qquad \mathscr{P}_m = \|\delta_{jk}P_m\|_1^n.$$

System (9.2) can be written in the form

$$(A\mathscr{P} + B\mathscr{Q})\xi = \eta \qquad (\xi = \{\xi_j\}_{-\infty}^{\infty},\ \eta = \{\eta_j\}_{-\infty}^{\infty},\ \mathscr{Q} = I - \mathscr{P}).$$

The operators $A\mathscr{P} + B\mathscr{Q}$, $\mathscr{P}A\mathscr{P}$ and $\mathscr{Q}B\mathscr{Q}$ are respectively invertible in the spaces $(\bar{l}_2)_n$, $\mathscr{P}(\bar{l}_2)_n$ and $\mathscr{Q}(\bar{l}_2)_n$ by virtue of proposition 2°, § 4, and proposition 2°, § 8.

It is easy to see that operators $\mathscr{P}A\mathscr{P}$ and $\mathscr{Q}B\mathscr{Q}$ satisfy the conditions of Theorem 5.4. Therefore $\mathscr{P}A\mathscr{P} + \mathscr{Q}B\mathscr{Q} \in \Pi\{\mathscr{P}_m, \mathscr{P}_m\}$. Since the operator $A\mathscr{P} + B\mathscr{Q}$ is invertible and differs from the operator $\mathscr{P}A\mathscr{P} + \mathscr{Q}B\mathscr{Q}$ by completely continuous addends, by virtue of Theorem 3.1 of Chapter II we have $A\mathscr{P} + B\mathscr{Q} \in \Pi\{\mathscr{P}_m, \mathscr{P}_m\}$.

An analogous theorem holds for the pair system

$$\sum_{k=-\infty}^{\infty} a_{j-k}\xi_k = \eta_j \qquad (j = 0, 1, \cdots),$$

$$\sum_{k=-\infty}^{\infty} b_{j-k}\xi_k = \eta_j \qquad (j = -1, -2, \cdots).$$

In the case when $A(\zeta) = B(\zeta)$ Theorem 9.2 admits the following refinement.

THEOREM 9.3. *Let $A(\zeta)$, $|\zeta| = 1$, be an arbitrary continuous nth order matrix function, and let a_j, $j = 0, \pm 1, \cdots$, be its matrix Fourier coefficients. If* $\liminf_{m\to\infty}|A_m^{-1}| < \infty$, *where $A_m = \|a_{j-k}\|_{j,k=-m}^{m}$, then the following conditions are fullfilled*:

a) $\det A(\zeta) \neq 0$ *for* $|\zeta| = 1$.

b) *The left and right indices of $A(\zeta)$ equal zero.*

Under fulfillment of conditions a) *and* b) *the system*

$$\sum_{k=-m}^{m} a_{j-k}\xi_k = \eta_j \qquad (j = 0, \pm 1, \cdots, \pm m),$$

beginning with some m, has the unique solution $\{\xi_j^{(m)}\}_{j=-m}^{m}$, and as $m \to \infty$ the sequence (9.1) *converges in the norm of $(\tilde{l}_2)_n$ to a solution of the system*

$$\sum_{k=-\infty}^{\infty} a_{j-k}\xi_k = \eta_j \qquad (j = 0, \pm 1, \cdots), \tag{9.3}$$

whatever the element $\{\eta_j\}_{-\infty}^{\infty} \in (\tilde{l}_2)_n$ may be.

The first assertion of the theorem follows from Theorem 5.4, since the mth order truncation of system (9.3) coincides with the $(2m + 1)$th order truncation of system (5.10). The second assertion is a corollary of Theorem 9.2.

Let us note that if the respective matrix functions expand into absolutely convergent Fourier series, the results of the present subsection remain valid for the spaces $(\tilde{l}_p)_n$, $p \geqq 1$, and $(\tilde{c}_0)$.

3. *Systems of integral equations.* All results of the preceding subsection carry over to the continuous case. We shall not make the corresponding formulations. Let us go into the question of the applicability of the Galerkin method to the solution of the equation

$$\varphi(t) - \int_0^{\infty} k_1(t-s)\varphi(s)\,ds - \int_{-\infty}^{0} k_2(t-s)\varphi(s)\,ds = f(t) \tag{9.4}$$

$(-\infty < t < \infty)$ in $(\tilde{L}_2)_n$, where $k_1(t)$ and $k_2(t)$ are nth order matrix functions with elements of $\tilde{L}_1$.

Equation (9.4) can be written in the form $A\mathscr{P}\varphi + B\mathscr{Q}\varphi = f$, where $A,B \in \mathfrak{R}_{n\times n}(U)$, $\mathscr{P} = \|\delta_{jk}P\|_1^n$ and the operators U and P are the same as in § 4, Chapter V.

Let $\psi_j(\ )$, $j = 0, \pm 1, \cdots$, be the system of functions which is defined by equalities (3.6) of Chapter VI.

We seek an approximate solution of equation (9.4) in the form

$$\varphi_m(t) = \sum_{j=-m}^{m} \xi_j \psi_j(t),$$

where the ξ_j are n-dimensional vectors satisfying the block system of equations

$$\sum_{k=0}^{m} (A\psi_k, \psi_j)\xi_k + \sum_{k=-m}^{-1} (B\psi_k, \psi_j)\xi_k = (f, \psi_j)$$
$$(j = 0, \pm 1, \cdots, \pm m) \tag{9.5}$$

(the same notation as in § 6.3 is adopted here).

THEOREM 9.4. *Let $k_j(t)$, $j = 1, 2$, be nth order matrix functions with elements of $\tilde{L}_1$ which satisfy the following conditions:*

a) $\det(I - K_j(\lambda)) \neq 0$ $(j = 1, 2; -\infty < \lambda < \infty)$, *where $K_j(\lambda)$ is the Fourier transform of $k(t)$.*

b) *The left and right indices of the matrix functions $I - K_j(\lambda)$, $j = 1,2$, equal zero.*

c) *The right indices of the matrix function $(I - K_2(\lambda))^{-1}(I - K_1(\lambda))$ equal zero.*

Then system (9.5), *beginning with some m, has the unique solution $\{\xi_j^{(m)}\}_{j=-m}^{m}$, and as $m \to \infty$ the vector functions*

$$\varphi_m(t) = \sum_{j=-m}^{m} \xi_j^{(m)} \psi_j(t)$$

converge in the norm of $(\tilde{L}_2)_n$ to a solution of equation (9.4), *whatever the vector function $f(t) \in (\tilde{L}_2)_n$ may be.*

This theorem, as in the case $n = 1$ (cf. § 3 of Chapter VI), is a corollary of Theorem 9.2.

An analogous theorem holds for the pair system

$$\varphi(t) - \int_{-\infty}^{\infty} k_1(t - s)\varphi(s)\, ds = f(t) \qquad (0 < t < \infty),$$

$$\varphi(t) - \int_{-\infty}^{\infty} k_2(t - s)\varphi(s)\, ds = f(t) \qquad (-\infty < t < 0).$$

Let us note that in the case $k_1(t) = k_2(t)$ Theorem 9.4 can be supplemented in the following way: the left and right indices of the matrix function $I - K_1(\lambda)$ being equal to zero is also a necessary condition for the applicability of the Galerkin method.

4. *Systems of singular integral equations.* Let us adopt the notation of § 4, Chapter VI. In the space $(L_p(\Gamma))_n$ let us consider a system of singular integral equations which can be written in vector-matrix notation as the single equation

$$A(\zeta)\varphi(\zeta) + \frac{B(\zeta)}{\pi i}\int_\Gamma \frac{\varphi(z)}{z-\zeta}\,dz = f(\zeta) \qquad (|\zeta| = 1), \tag{9.6}$$

where $A(\zeta)$ and $B(\zeta)$ are arbitrary nth order matrix functions continuous on the contour Γ.

THEOREM 9.5. *Let $A(\zeta)$ and $B(\zeta)$ be arbitrary nth order matrix functions continuous on Γ, let $C(\zeta) = A(\zeta) + B(\zeta)$ and $D(\zeta) = A(\zeta) - B(\zeta)$, and let c_j and d_j be their matrix Fourier coefficients. If* $\det C(\zeta) \neq 0 \neq \det D(\zeta)$ *for $|\zeta| = 1$, if the left and right indices of $C(\zeta)$ and $D(\zeta)$ equal zero, and if the right indices of the matrix function $D^{-1}(\zeta)C(\zeta)$ equal zero, then, beginning with some m, the system of equations*

$$\sum_{k=0}^{m} c_{j-k}\xi_k + \sum_{k=-m}^{-1} d_{j-k}\xi_k = f_j \qquad (j = 0, \pm 1, \cdots, \pm m),$$

where the f_j are the Fourier coefficients of the vector function $f(\zeta)$, has the unique solution $\{\xi_j^{(m)}\}_{j=-m}^{m}$, and as $m \to \infty$ the vector functions

$$\varphi_m(\zeta) = \sum_{-m}^{m} \xi_j^{(m)}\zeta^j$$

converge in the norm of $(L_p(\Gamma))_n$ to a solution of equation (9.6), *whatever the vector function $f(\zeta) \in (L_p(\Gamma))_n$ may be.*

This theorem is derived from Theorem 9.2, just as in the case $n = 1$ (cf. § 4 of Chapter VI).

5. *Systems of difference equations.*

THEOREM 9.6. *Let $A(\zeta)$ and $B(\zeta)$, $|\zeta| = 1$, be arbitrary continuous nth order matrix functions, and let the a_j and b_j be their matrix Fourier coefficients. If* $\det A(\zeta) \neq 0 \neq \det B(\zeta)$ *for $|\zeta| = 1$, if all left and right indices of $A(\zeta)$ and $B(\zeta)$ equal zero, and if the right indices of $B^{-1}(\zeta)A(\zeta)$ equal zero, then the equation*

$$\sum_{t-m<j<t} a_j f(t-j) + \sum_{t<j<t+m} b_j f(t-j) = g(t) \qquad (-m < t < m),$$

beginning with some m, has the unique solution $f_m(t) \in (L_2(-m, m))_n$, and as $m \to \infty$ the vector functions

$$\tilde{f}_m(t) = \begin{cases} f_m(t), & |t| < m, \\ 0, & |t| > m, \end{cases}$$

converge in the norm of $(\tilde{L}_2)_n$ to a solution of the equation

$$\sum_{j<t} a_j f(t-j) + \sum_{j>t} b_j f(t-j) = g(t) \qquad (-\infty < t < \infty),$$

whatever the vector function $g(t) \in (\tilde{L}_2)_n$ may be.

Theorem 9.6 is derived from Theorem 9.2 analogously to the way Theorem 7.2 is derived from Theorem 5.4.

§ 10. Generalized projection method for composite operators

A certain generalization of the projection method is analyzed in this section. It arises in connection with operators of the form

$$A = \sum_{j=1}^{l} A_{j1} \cdots A_{jm}, \tag{10.1}$$

where the $A_{jk} \in \mathfrak{A}(\mathfrak{B})$ are operators in some sense more elementary than A.

Let P_τ be projections of $\mathfrak{A}(\mathfrak{B})$. Here, as previously, $\tau \in \Omega$, where Ω is some unbounded set of positive numbers. We shall say that a *generalized projection method* is applicable to an operator A of the form (10.1), or that A belongs to the set $\tilde{\Pi}\{P_\tau\}$, if, beginning with some τ, the operators

$$A_\tau = \sum_{j=1}^{l} P_\tau A_{j1} P_\tau A_{j2} P_\tau \cdots P_\tau A_{jm} P_\tau$$

are invertible in the subspaces $P_\tau\mathfrak{B}$, and for some $x \in \mathfrak{B}$ the relation[11]

$$\lim_{\tau\to\infty} \left| A_\tau^{-1} P_\tau x - A^{-1}x \right| = 0$$

holds.

1. *Auxiliary propositions.* As usual, we denote by $\mathfrak{A}_{r\times r}(\mathfrak{B})$ the algebra of all rth order matrices with elements of $\mathfrak{A}(\mathfrak{B})$, and by $\mathfrak{B}_r$ the Banach space of all r-dimensional vectors with components of $\mathfrak{B}$. The algebra $\mathfrak{A}_{r\times r}(\mathfrak{B})$ is identified in a natural way with the algebra $\mathfrak{A}(\mathfrak{B}_r)$.

The analysis made below of the generalized projection method for an operator A of the form (10.1) is based on reducing this method to the usual method for some special operator acting in the space $\mathfrak{B}_r$. If A is prescribed by equality (10.1), the corresponding special operator is defined in the space $\mathfrak{B}_r$, with $r = l(m + 1) + 1$, by the matrix

$$\Xi\{A_{jk}\} = \begin{Vmatrix} M & F \\ G & 0 \end{Vmatrix}.$$

The principal block M here is the matrix of order $l(m + 1)$ which is defined by the equality

$$M = \begin{Vmatrix} I_l & B_1 & 0 & \cdots & 0 \\ 0 & I_l & B_2 & \cdots & 0 \\ \cdot & \cdot & \cdot & \cdot & \cdot \\ 0 & 0 & \cdots & I_l & B_m \\ 0 & 0 & \cdots & 0 & I_l \end{Vmatrix},$$

[11] Let us recall that A_τ^{-1} denotes the operator inverse to A_τ on the subspace $P_\tau\mathfrak{B}$.

in which I_l in turn is the unit matrix of $\mathfrak{A}_{l\times l}(\mathfrak{B})$, and

$$B_j = \begin{Vmatrix} A_{1j} & & & 0 \\ & A_{2j} & & \\ & & \ddots & \\ 0 & & & A_{lj} \end{Vmatrix} \qquad (j = 1, 2, \cdots, m);$$

F is a column of length $l(m + 1)$ whose upper lm elements equal zero and lower l elements equal the identity operator; finally,

$$G = \|\underbrace{I, \cdots, I}_{l}, \underbrace{0, \cdots, 0}_{lm}\|.$$

In explaining the structure of the matrix $\Xi\{A_{jk}\}$, let us note that it is the matrix of the following system of linear equations:

$$\begin{aligned} x_1 - A_{11}x_{l+1} &= 0, & x_{r-2l} - A_{1m}x_{r-l} &= 0, \\ x_2 - A_{21}x_{l+2} &= 0, & \cdots\cdots\cdots\cdots & \\ \cdots\cdots\cdots\cdots & & x_{r-l-1} - A_{lm}x_{r-1} &= 0, \\ x_l - A_{l1}x_{2l} &= 0, & x_{r-l} - x &= 0, \\ x_{l+1} - A_{12}x_{2l+1} &= 0, & \cdots\cdots\cdots\cdots & \\ \cdots\cdots\cdots\cdots & & x_{r-1} - x &= 0, \\ x_{2l} - A_{l2}x_{3l} &= 0, & x_1 + x_2 + \cdots + x_l &= y, \end{aligned}$$

which emerges in a natural way if we are freed of the operator products in the equation

$$\sum_{j=1}^{l} A_{j1}A_{j2} \cdots A_{jm}x = y.$$

Clearly this is not the only system solving this problem.

The following proposition establishing the relationship between the operator A and its matrix $\Xi\{A_{jk}\}$ plays an important role.

LEMMA 10.1. *Let an operator $A \in \mathfrak{A}(\mathfrak{B})$ have the form* (10.1). *Then the expansion*[12)]

$$\Xi\{A_{jk}\} = \Xi_1\{A_{jk}\}\Xi_0\{A_{jk}\}\Xi_2\{A_{jk}\} \tag{10.2}$$

holds for the matrix $\Xi\{A_{jk}\} \in \mathfrak{A}_{r\times r}(\mathfrak{B})$, where the $\Xi_s\{A_{jk}\}$, $s = 0, 1, 2$, are matrices of $\mathfrak{A}_{r\times r}(\mathfrak{B})$ with the following properties:

a) *$\Xi_1\{A_{jk}\}$ is a subtriangular matrix with diagonal elements equal to I.*

b) *$\Xi_2\{A_{jk}\}$ is a supertriangular matrix with diagonal elements also equal to I.*

c) *$\Xi_0\{A_{jk}\}$ is a diagonal matrix defined by the equality*

---

12) Equality (10.2) is a factorization of the matrix $\Xi\{A_{jk}\}$. For more detail concerning the factorization of operator matrices, cf. M. A. Barkar′ and I. C. Gohberg [1]. Unlike the case considered in that article, here we factor, in general, noninvertible matrices, but of a special kind.

$$\Xi_0\{A_{jk}\} = \begin{Vmatrix} I & 0 & \cdots & 0 \\ 0 & I & \cdots & 0 \\ \cdot & \cdot & \cdots & \cdot \\ 0 & 0 & \cdots & A \end{Vmatrix}.$$

PROOF. Let us set

$$\Xi_1\{A_{jk}\} = \begin{Vmatrix} I_{l(m+1)} & 0 \\ H & I \end{Vmatrix}, \qquad \Xi_2\{A_{jk}\} = \begin{Vmatrix} M & F \\ 0 & I \end{Vmatrix},$$

where M and F are defined above and H is a row of length $l(m+1)$. In order to describe this row, let us divide it into $m+1$ parts, each of whose lengths equals l. These parts, in succession, have the form

$$\begin{gathered} I, I, \cdots\cdots I, \\ A_{11}, A_{21}, \cdots\cdot, A_{l1}, \\ A_{11}A_{12}, A_{21}A_{22}, \cdots\cdot, A_{l1}A_{l2}, \\ \cdots\cdots\cdots\cdots\cdots\cdots \\ A_{11}A_{12}\cdots A_{1m}\, A_{21}A_{22}\cdots A_{2m}\cdots A_{l1}A_{l2}\cdots A_{lm}. \end{gathered}$$

That identity (10.2) holds for the factors defined in this way is established by immediate verification.

The lemma is proved.

Since the end factors $\Xi_1\{A_{jk}\}$ and $\Xi_2\{A_{jk}\}$ in (10.2) are invertible operators in the space $\mathfrak{B}_r$, (10.2) implies that A is invertible if and only if the matrix $\Xi\{A_{jk}\}$ is invertible.

Let us note, moreover, that the equality

$$\Xi'\{A_{jk}\} = \begin{Vmatrix} M' & 0 \\ F' & I \end{Vmatrix} \begin{Vmatrix} I_{l(m+1)} & 0 \\ 0 & A' \end{Vmatrix} \begin{Vmatrix} I_{l(m+1)} & H' \\ 0 & I \end{Vmatrix} \tag{10.3}$$

holds for the matrix $\Xi'\{A_{jk}\}$, the transpose of $\Xi\{A_{jk}\}$, where M' and F' are the transpose matrices of M and F, H' is the matrix (column) obtained from H by transposing it and replacing each component of the form $A_{\alpha 1}, A_{\alpha 2}, \cdots, A_{\alpha\beta}$ with the product $A_{\alpha\beta}\, A_{\alpha\beta-1} \cdots A_{\alpha 1}$, while the operator A' is defined by the equality

$$A' = \sum_{j=1}^{l} A_{jm}A_{jm-1} \cdots A_{j1}.$$

Identity (10.3) can be obtained by immediate verification, or can be derived from (10.2) by passing to the transpose matrices.

2. *Conditions for applicability of the generalized projection method.* Let the family of projections $P_\tau \in \mathfrak{A}(\mathfrak{B})$, $t \in \Omega$, converge strongly to the identity operator as $\tau \to \infty$. Let us define the projection $\mathscr{P}_\tau$ in the space $\mathfrak{B}_r$ by the equality $\mathscr{P}_\tau = \|\delta_{jk}P_\tau\|_{j,k=1}^{r}$.

THEOREM 10.1. *Let an invertible operator A of $\mathfrak{A}(\mathfrak{B})$ have the form* (10.1). *For it to belong to the class $\tilde{\Pi}\{P_\tau\}$, it is necessary and sufficient that the operator B defined in the space $\mathfrak{B}_r$ by the matrix $\Xi\{A_{jk}\}$ belong to the class $\Pi\{\mathscr{P}_\tau, \mathscr{P}_\tau\}$.*

PROOF. The theorem is proved with the help of the following two equalities:

$$B = \Xi_1\Xi_0\Xi_2 \tag{10.4}$$

and

$$\mathscr{P}_\tau B\mathscr{P}_\tau = \Xi_{1,\tau}\Xi_{0,\tau}\Xi_{2,\tau}, \tag{10.5}$$

where $\Xi_s = \Xi_s\{A_{jk}\}$ and $\Xi_{s,\tau} = \Xi_s\{\mathscr{P}_\tau A_{jk}P_\tau\}$ $(s = 0, 1, 2)$.

Equality (10.4) coincides with (10.2), while (10.5) follows from the same equality (10.2) if the operators A_{jk} in it are replaced by the operators $P_\tau A_{jk}P_\tau$ and it is taken into account that $\mathscr{P}_\tau B\mathscr{P}_\tau = \Xi\{P_\tau A_{jk}P_\tau\}$.

Obviously the operators $\Xi_{s,\tau}$ converge strongly to the operators Ξ_s, $s = 0, 1, 2$, as $\tau \to \infty$.

Since $(\Xi_{s,\tau} - \mathscr{P}_\tau)^r = 0$ for $s = 1, 2$, the restrictions of the operators $\Xi_{s,\tau}$ $(s = 1, 2)$ to the subspaces $\mathscr{P}_\tau\mathfrak{B}_r$ are invertible, their inverses having the form

$$\Xi_{s,\tau}^{-1} = \mathscr{P}_s + \sum_{k=1}^{r-1} (\mathscr{P}_\tau - \Xi_{s,\tau})^k \qquad (s = 1, 2).$$

This implies that the operators $\Xi_{s,\tau}^{-1}\mathscr{P}_\tau$, $s = 1,2$, converge strongly to Ξ_s^{-1} as $\tau \to \infty$.

From what the been said and from equalities (10.4) and (10.5) it follows immediately that the operator B belongs to the class $\Pi\{\mathscr{P}_\tau, \mathscr{P}_\tau\}$ if and only if the operator $\Xi_0 = \Xi_0\{A_{jk}\}$ belongs to the same class. Taking into account the form of the matrices Ξ_0 and $\Xi_{0,\tau}$, we conclude that Ξ_0 will belong to the class $\Pi\{\mathscr{P}_\tau, \mathscr{P}_\tau\}$ if and only if A belongs to the class $\tilde{\Pi}\{P_\tau\}$.

The theorem is proved.

3. *Applications.* By means of Theorem 10.1 and previously established propositions it is possible to find the conditions for applicability of generalized projection methods to operators of the form (10.1), where the A_{jk} are operators in convolutions of one of the types considered above. Let us go into only one example.

Let $t_{jk}(\zeta)$ $(|\zeta| = 1; j = 1,\cdots, l; k = 1,\cdots,n)$ be arbitrary continuous functions. Let us denote by T_{jk} the operator defined in l_2 by the Toeplitz matrix composed of the Fourier coefficients of the functions $t_{jk}(\zeta)$, and by P_n the usual truncation projection in l_2:

$$P_n\{\xi_j\}_1^\infty = \{\xi_1, \xi_2, \cdots, \xi_n, 0, \cdots\}.$$

THEOREM 10.2. *For the operator*

$$T = \sum_{j=1}^{l} T_{j1}T_{j2}\cdots T_{jm}$$

to belong to the class $\tilde{\Pi}\{P_n\}$, *it is necessary and sufficient that* T *and*

$$T' = \sum_{j=1}^{l} T_{jm}T_{jm-1}\cdots T_{j1}$$

be invertible.

PROOF. By virtue of Theorem 10.1, the condition $T \in \tilde{\Pi}\{P_n\}$ is equivalent to the condition $\Xi\{T_{jk}\} \in \Pi\{\mathscr{P}_n, \mathscr{P}_n\}$. The latter condition, according to Theorem 5.2 of Chapter VIII, is equivalent to the invertibility of each of the operators $\Xi\{T_{jk}\}$ and $\Xi'\{T_{jk}\}$. Finally, by virtue of equalities (10.2) and (10.3), this is equivalent to the invertibility of T and T'.

The theorem is proved.

Let us note that the invertibility of T and T' follows immediately from the condition

$$\lim_{n\to\infty} |T_n^{-1}| < \infty,$$

where

$$T_n = \sum_{j=1}^{l} P_n T_{j1} P_n T_{j2} P_n \cdots P_n T_{jm} P_n.$$

Let us further note that the invertibility of T and T' is equivalent to the condition

$$\det \Xi\{t_{jk}(\zeta)\} \neq 0 \qquad (|\zeta| = 1)$$

and to all right and left indices of the matrix function $\Xi\{a_{jk}(\zeta)\}$ being equal to zero.

APPENDIX

ASYMPTOTICS OF SOLUTIONS OF HOMOGENEOUS CONVOLUTION EQUATIONS

Here we take up the asymptotics of solutions of a system of homogeneous integral equations which is written in vector-matrix notation as

$$\varphi(t) - \int_0^\infty k_{11}(t-s)\varphi(s)\,ds - \int_{-\infty}^0 k_{12}(t-s)\varphi(s)\,ds = 0 \qquad (0 < t < \infty),$$
$$\varphi(t) - \int_0^\infty k_{21}(t-s)\varphi(s)\,ds - \int_{-\infty}^0 k_{22}(t-s)\varphi(s)\,ds = 0 \qquad (-\infty < t < 0), \quad (1)$$

where $\varphi(t)$ is the n-dimensional vector-function sought, and the $k_{jr}(t), j, r = 1, 2$, are nth order matrix functions. System (1) is a generalization of both pair equation systems (the case when $k_{11} = k_{12}$ and $k_{21} = k_{22}$) and their transpose systems (the case when $k_{11} = k_{21}$ and $k_{12} = k_{22}$).

The first section is auxiliary; certain properties of the operator defined by the left side of equation (1) are established in it. By means of these properties we prove that equation (1) has the same solutions in each space (of some class).

Asympotic expansions of solutions of equation (1) in exponential functions with polynomial coefficients are found in § 2. Several general theorems concerning the operators and results of § 1 play a vital role in this connection. The asymptotics of solutions of a system of Wiener-Hopf equations, a system of pair integral equations, and its transpose system are obtained as a corollary.

More precise results are obtained in §§ 3 and 4 for the transpose equation of a pair and for a system of Wiener-Hopf equations. Here we manage, in addition, to determine the polynomial coefficients in the principal term of the asymptotics for functions forming a basis of the kernel of the corresponding operator.

All results of this appendix can be carried over to the discrete case.

§ 1. Some auxiliary propositions

1. In the sequel E and $\tilde{E}$ denote the Banach spaces introduced in § 8 of Chapter I. The notation $L = L_1(0, \infty)$ and $\tilde{L} = L_1(-\infty, \infty)$ is also utilized.

Let $\mathfrak{B}$ be some Banach space of functions defined on the real axis (semiaxis), and let $\sigma(t)$ be a continuous function which does not vanish on that axis (semiaxis).

Let us agree to write $f(t) \in \sigma(t)\mathfrak{B}$ if $\sigma^{-1}(t)f(t) \in \mathfrak{B}$. The set $\sigma(t)\mathfrak{B}$ is a Banach space with norm $|f| = |\sigma^{-1}f|_{\mathfrak{B}}$.

As in Chapter VIII, by $\mathfrak{B}_n$ and $\mathfrak{B}_{n\times n}$ we denote respectively the set of all n-dimensional vectors and the set of all nth order matrices with elements from $\mathfrak{B}$. The set $\mathfrak{B}_n$ becomes a Banach space if the norm of an element $f = \{f_1, \cdots, f_n\} \in \mathfrak{B}_n$ is defined by, for example, the equality

$$|f| = \sum_1^n |f_j|_{\mathfrak{B}}.$$

The spaces $\sigma(t)\mathfrak{B}_n$ and $\sigma(t)\mathfrak{B}_{n\times n}$ are defined analogously.

Everywhere in the sequel h denotes some fixed positive number. If an element of $e^{-h|t|}\tilde{L}_{n\times n}$ or $e^{-h|t|}\tilde{L}_n$ is denoted by a small letter, its Fourier transform is denoted by the corresponding capital letter.

The definitions and notation introduced in § 11 of Chapter I are also utilized in the sequel.

2. It follows easily from Theorem 6.1 of Chapter VIII that if the matrix function $k(t)$ belongs to $e^{-h|t|}\tilde{L}_{n\times n}$, then the condition

$$\det(I - K(\lambda + ih)) \neq 0 \qquad (-\infty < \lambda < \infty) \tag{1.1}$$

is necessary and sufficient for the operator

$$(I - \mathscr{K})\varphi = \varphi(t) - \int_0^\infty k(t-s)\varphi(s)\,ds \qquad (0 < t < \infty)$$

to be a Φ-operator in each of the spaces $e^{ht}E_n$. If condition (1.1) is fulfilled, the index[1] of the operator $I - \mathscr{K}$ in each of the spaces $e^{ht}E_n$ is calculated from the formula

$$\kappa(I - \mathscr{K}) = -\operatorname{ind}\det(I - K(\lambda + ih)). \tag{1.2}$$

Let us now consider in the space $e^{h|t|}\tilde{E}_n$ the equation

$$\varphi(t) - \int_0^\infty k_{11}(t-s)\varphi(s)\,ds - \int_{-\infty}^0 k_{12}(t-s)\varphi(s)\,ds = g(t) \tag{1.3}$$
$$(0 < t < \infty),$$
$$\varphi(t) - \int_0^\infty k_{21}(t-s)\varphi(s)\,ds - \int_{-\infty}^0 k_{22}(t-s)\varphi(s)\,ds = g(t)$$
$$(-\infty < t < 0),$$

where $k_{jr}(t) \in e^{-h|t|}\tilde{L}_{n\times n}$.

The space $e^{h|t|}\tilde{E}_n$ is isomorphic to the space F consisting of all pairs $\hat{f}(t) = \{f_1(t), f_2(t)\}$ of vector functions belonging to $e^{ht}E_n$ ($f_j(t) \in e^{ht}E_n, j = 1, 2$) with norm $|\hat{f}| = |f_1| + |f_2|$.

Under this isomorphism a pair $\hat{f} = \{f_1, f_2\} \in F$ is put into correspondence with

---

[1] That is, dim Ker $(I - \mathscr{K})$ − dim Ker $(I - \mathscr{K}^*)$.

each vector $f(t) \in e^{h|t|}\tilde{E}_n$ in accordance with the rule $f_1(t) = f(t), f_2(t) = f(-t)$, $0 < t < \infty$. In this notation, equation (1.3) can be written in the form

$$\varphi_1(t) - \int_0^\infty k_{11}(t-s)\varphi_1(s)\,ds - \int_0^\infty k_{12}(t+s)\varphi_2(s)\,ds = g_1(t),$$

$$\varphi_2(t) - \int_0^\infty k_{21}(-t-s)\varphi_1(s)\,ds - \int_0^\infty k_{22}(s-t)\varphi_2(s)\,ds = g_2(t)$$

$(0 < t < \infty)$. Let us write the operator B defined by the left side of this system in the form of a matrix[2)]

$$B = \begin{Vmatrix} I - \mathscr{K}_{11} & -\hat{\mathscr{K}}_{12} \\ -\hat{\mathscr{K}}'_{21} & I - \mathscr{K}'_{22} \end{Vmatrix}.$$

For any matrix function $k(t)$ of $L_{n\times n}$ the operator

$$(\hat{\mathscr{K}}\varphi)(t) = \int_0^\infty k(t+s)\varphi(s)\,ds$$

is completely continuous in each of the spaces E_n. It is sufficient to show this for the case $n = 1$. If a function $k(t) \in L_1(0, \infty)$ has the form $k(t) = e^{-t}p(t)$, $0 < t < \infty$, where $p(t)$ is a polynomial, then, as is easy to see, the operator $\hat{\mathscr{K}}$ is finite-dimensional in E. Since the set of all functions of the form $e^{-t}p(t)$ is dense in $L_1(0, \infty)$, the operator $\hat{\mathscr{K}}$ is completely continuous, whatever the function $k(t)$ may be.

It is not difficult to obtain by means of this the fact that the operator

$$T = \begin{Vmatrix} 0 & -\hat{\mathscr{K}}_{12} \\ -\hat{\mathscr{K}}_{21} & 0 \end{Vmatrix}$$

is completely continuous in the space F. Consequently the operator

$$B = \begin{Vmatrix} I - \mathscr{K}_{11} & 0 \\ 0 & I - \mathscr{K}'_{22} \end{Vmatrix} + T$$

is a Φ-operator if and only if the operators $I - \mathscr{K}_{11}$ and $I - \mathscr{K}'_{22}$ are. The latter is equivalent to fulfilling the conditions

$$\det(I - K_{11}(\lambda + ih)) \neq 0, \qquad \det(I - K_{22}(\lambda - ih)) \neq 0 \qquad (1.4)$$
$$(-\infty < \lambda < \infty).$$

If conditions (1.4) are fulfilled, then $\kappa(B) = \kappa(I - \mathscr{K}_{11}) + \kappa(I - \mathscr{K}'_{22})$. But by virtue of (1.2)

$$\kappa(I - \mathscr{K}_{11}) = -\operatorname{ind}\det(I - K_{11}(\lambda + ih)),$$

$$\kappa(I - \mathscr{K}'_{22}) = -\operatorname{ind}\det\left(I - \int_{-\infty}^{\infty} k_{22}(-t)e^{-ht}e^{i\lambda t}dt\right)$$

$$= \operatorname{ind}\det(I - K_{22}(\lambda - ih)).$$

2) For a function $k(t)$, $-\infty < t < \infty$, the integral operators on the semiaxis with kernels $k(t-s)$, $k(t+s)$ and $k(s-t)$ are denoted by the symbols $\mathscr{K}$, $\hat{\mathscr{K}}$ and $\mathscr{K}'$.

Thus the index of B is calculated from the formula

$$\kappa(B) = \text{ind}\det(I - K_{22}(\lambda - ih)) - \text{ind}\det(I - K_{11}(\lambda + ih)).$$

Since under the aforementioned isomorphism between the spaces $e^{h|t|}\tilde{E}_n$ and F the operator B corresponds to the operator A defined by the left side of equation (1.3), the following lemma has been established.

LEMMA 1.1. *Let the matrix functions $k_{jr}(t)$ belong to $e^{-h|t|}\tilde{L}_{n\times n}$ $(j, r = 1,2)$. For an operator A defined by the left side of equation* (1.3) *to be a Φ-operator in each of the spaces $e^{h|t|}\tilde{E}_n$, it is necessary and sufficient that conditions* (1.4) *be fulfilled.*

Under fulfillment of these conditions the index of A in each of the spaces $e^{h|t|}\tilde{E}_n$ is calculated from the formula

$$\kappa(A) = \text{ind}\det(I - K_{22}(\lambda - ih)) - \text{ind}\det(I - K_{11}(\lambda + ih)). \tag{1.5}$$

REMARK. If the matrix functions $k_{jr}(t)$ of $e^{-h|t|}\tilde{L}_{n\times n}$ satisfy the conditions

$$\det(I - K_{rr}(\lambda \pm ih)) \neq 0 \qquad (-\infty < \lambda < \infty;\ r = 1, 2),$$

then

$$\kappa(A) = \kappa_{-h}(A) + m_1 = \kappa^h(A) + m_2, \tag{1.6}$$

where $\kappa_{\pm h}(A)$ is the index of A in the space $e^{\pm ht}\tilde{E}_n$, and m_r $(r = 1, 2)$ is the number of zeros, with regard for their multiplicities, of the function $\det(I - K_{rr}(\lambda))$ in the strip $|\text{Im}\,\lambda| < h$.

Indeed, formula (1.5) implies that

$$\kappa(A) = \text{ind}\,\frac{\det(I - K_{22}(\lambda - ih))}{\det(I - K_{11}(\lambda - ih))} + \text{ind}\,\frac{\det(I - K_{11}(\lambda - ih))}{\det(I - K_{11}(\lambda + ih))}.$$

It is not difficult to observe that the first addend of the right side of the latter equality equals $\kappa_{-h}(A)$, while the second, by virtue of Rouché's theorem, equals m_1 since the function $\det(I - K_{11}(\lambda))$ is analytic in the strip $|\text{Im}\,\lambda| < h$ and does not vanish on the boundary. The second equality of (1.6) is established analogously.

We need the following proposition in the sequel.

1°. *Let the Banach space $\mathfrak{B}_1$ be contained in the Banach space $\mathfrak{B}$ and be dense in it, and let A_1 and A be bounded Φ-operators with equal indices which operate in $\mathfrak{B}_1$ and $\mathfrak{B}$, respectively. If A is a continuation of A_1, then* $\text{Ker}\,A = \text{Ker}\,A_1$.

PROOF. The inequality $\alpha(A_1) \leqq \alpha(A)$[3] holds since $\mathfrak{B}_1 \subset \mathfrak{B}$. We obtain the relation $\beta(A_1) \leqq \beta(A)$ from the inclusion $\mathfrak{B}^* \subset \mathfrak{B}_1^*$, which is a consequence of the fact that $\mathfrak{B}_1$ is dense in $\mathfrak{B}$. The inequalities just obtained, together with the equality $\kappa(A) = \kappa(A_1)$, reduce to the equality $\alpha(A) = \alpha(A_1)$, from which the assertion follows.

[3] For the notation, cf. § 11 of Chapter I.

THEOREM 1.1. *Let the matrix functions* $k_{jr}(t)$ *belong to* $e^{-h|t|}\tilde{L}_{n\times n}$, $j, r = 1,2$, *and satisfy condition* (1.4). *Then equation* (1) *has the same solutions in each of the spaces* $e^{h|t|}\tilde{E}$.

PROOF. Proposition 1° and Lemma 1.1 imply immediately that equation (1) has the same solutions in the spaces $e^{h|t|}(\tilde{L}_p)_n$ $(p \geqq 1)$ and $e^{h|t|}(\tilde{C}_0)_n$, since the intersection of these spaces is dense in each of them. Now let $\varphi(t)$ be a solution of equation (1) which belongs to $e^{h|t|}\tilde{M}_n$. Then each coordinate of the vector function

$$\psi(t) = \int_{-\infty}^{0} k_{12}(t-s)\varphi(s)\,ds$$

is the sum of n functions of the form

$$\omega(t) = \int_{-\infty}^{0} k(t-s)\chi(s)\,ds,$$

where $k(t) \in e^{-h|t|}\tilde{L}$ and $\chi(t) \in e^{h|t|}\tilde{M}$. Considering the function $\omega(t)$ for $t > 0$, we obtain $\omega(t) \in e^{-ht}C_0$. As a matter of fact, by setting $k(t) = \hat{k}(t)e^{-ht}$ and $\chi(t) = \hat{\chi}(t)e^{-ht}$, we obtain

$$\begin{aligned} |e^{ht}\omega(t)| &\leqq \int_{-\infty}^{0} |\hat{k}(t-s)\hat{\chi}(s)|ds \\ &\leqq \sup_{-\infty<s<0} |\hat{\chi}(s)| \int_{t}^{\infty} |\hat{k}(s)|ds \to 0 \qquad (t \to \infty), \end{aligned}$$

and the function $\omega(t)$ is continuous (cf. Gel'fand, Raĭkov and Šilov [1], § 16, Lemma).

Thus the vector function $\psi(t)$ belongs to $e^{-ht}(C_0)_n$ $(t > 0)$, and a fortiori to the space $e^{ht}(C_0)_n$. Let us write the first equality of (1) in the form

$$\hat{\varphi}(t) - \int_{0}^{\infty} \hat{k}_{11}(t-s)\hat{\varphi}(s)\,ds = \hat{\psi}(t) \qquad (0 < t < \infty), \tag{1.7}$$

where $\hat{\psi}(t) = e^{-ht}\varphi(t)$, $\hat{k}_{11}(t) = e^{-ht}k_{11}(t)$ and $\hat{\chi}(t) = e^{-ht}\psi(t)$. Equation (1.7) is obviously solvable in the space M_n. Since $\hat{\psi}(t) \in (C_0)_n$, we obtain easily from Theorems 4.2 and 6.1, Chapter VIII, the fact that every solution of (1.7) which belongs to M_n also belongs to $(C_0)_n$. Therefore $\varphi(t) \in e^{ht}\,(C_0)_n$ $(t > 0)$.

The relation $\varphi(-t) \in e^{ht}(C_0)_n$ $(t > 0)$ is established analogously. Thus $\varphi(t) \in e^{h|t|}\,(\tilde{C}_0)_n$; consequently $\varphi(t)$ belongs to any space $e^{h|t|}\tilde{E}_n$.

§ 2. Asymptotic expansions of solutions of the general equation

1. Let the matrix functions $k_j(t)$ belong to $e^{-h|t|}\tilde{L}_{n\times n}$, $j = 1, 2$. Let us consider the equation

$$\varphi(t) - \int_0^\infty k_1(t-s)\varphi(s)\,ds - \int_{-\infty}^0 k_2(t-s)\varphi(s)\,ds \tag{2.1}$$

in the space $e^{h|t|}\tilde{L}_n$ $(-\infty < t < \infty)$.

Let us show that the equation (2.1) is equivalent to the boundary value problem

$$\left.\begin{aligned} [I - K_1(\lambda + ih)]\Phi_+(\lambda + ih) &= \Omega(\lambda + ih), \\ -[I - K_2(\lambda - ih)]\Phi_-(\lambda - ih) &= \Omega(\lambda - ih) \end{aligned}\right\} \quad (-\infty < \lambda < \infty), \tag{2.2}$$

where

$$\Phi_+(\lambda + ih) = \int_0^\infty \varphi(t)e^{-ht}e^{i\lambda t}\,dt, \qquad \Phi_-(\lambda - ih) = \int_{-\infty}^0 \varphi(t)e^{ht}e^{i\lambda t}\,dt, \tag{2.3}$$

and the vector function $\Omega(\lambda)$ is the Fourier transform of some vector function of $e^{-h|t|}\tilde{L}_n$.

Indeed, let us denote by P the projection defined in the space $e^{h|t|}\tilde{L}_n$ by the equality

$$(P\varphi)(t) = \begin{cases} \varphi(t), & 0 < t < \infty, \\ 0, & -\infty < t < 0, \end{cases}$$

and let us set $Q = I - P$. Let us rewrite equation (2.1) in the following form:

$$(P\varphi)(t) - \int_{-\infty}^\infty k_1(t-s)(P\varphi)(s)\,ds = -(Q\varphi)(t) + \int_{-\infty}^\infty k_2(t-s)(Q\varphi)(s)\,ds.$$

It is easy to verify that the left side of the latter equality is a vector function of $e^{ht}\tilde{L}_n$, and the right, of $e^{-ht}\tilde{L}_n$. Let us denote their common value by $\omega(t)$. The vector function $\omega(t)$ obviously belongs to $e^{-h|t|}\tilde{L}_n$. We obtain (2.2) by applying the Fourier transform to the equalities

$$e^{-ht}(P\varphi)(t) - e^{-ht}\int_{-\infty}^\infty k_1(t-s)(P\varphi)(s)\,ds = e^{-ht}\omega(t),$$

$$-e^{ht}(Q\varphi)(t) + e^{ht}\int_{-\infty}^\infty k_2(t-s)(Q\varphi)(s)\,ds = e^{ht}\omega(t).$$

The vector functions $\Phi_\pm(z)$ and $\Omega(z)$ admit continuation to vector functions which are holomorphic respectively in the regions $\operatorname{Im} z > h$, $\operatorname{Im} z < -h$ and $|\operatorname{Im} z| < h$ and continuous there, including the boundaries. Equalities (2.2) represent an n-dimensional homogeneous Hilbert problem with respect to the contour consisting of the pair of parallel lines $z = \lambda \pm ih$, $-\infty < \lambda < \infty$.

We shall henceforth write the vector functions $\{a_1(t), \cdots, a_n(t)\}$ concisely as $\{a_r(t)\}_1^n$.

THEOREM 2.1. *Let a matrix function $k(t) \in e^{-h|t|}\tilde{L}_{n\times n}$ possess the property*

$$\det (I - K(\lambda \pm ih)) \neq 0 \qquad (-\infty < \lambda < \infty) \tag{2.4}$$

and let the α_j, $j = 1,\cdots,m$, *be all of the distinct zeros of the function* $\det (I - K(\lambda))$ *in the strip* $|\operatorname{Im} \lambda| < h$, *and the* P_j, *their multiplicities. Then the vector functions*

$$\varphi_{jq}(t) = \{e^{-i\alpha_j t} P_{jqr}(t)\}_{r=1}^{n} \qquad (j = 1,\cdots, m;\ q = 1,\cdots,P_j),$$

where the $P_{jqr}(t)$ *are polynomials of degrees* $\leqq p_j - 1$, *form a basis of all solutions of the equation*

$$\varphi(t) - \int_{-\infty}^{\infty} k(t - s)\varphi(s)\, ds = 0 \qquad (-\infty < t < \infty) \tag{2.5}$$

in any space $e^{h|t|} \tilde{E}_n$.

PROOF. By virtue of Theorem 1.1, equation (2.5) has the same solutions in all of the spaces $e^{h|t|}\tilde{E}_n$. Therefore it is sufficient to establish the theorem for the space $e^{h|t|}\tilde{L}_n$. In this case equation (2.5) is equivalent to the boundary value problem

$$\left.\begin{aligned} [I - K(\lambda + ih)]\, \Phi_+(\lambda + ih) &= \Omega(\lambda + ih), \\ -[I - K(\lambda - ih)]\, \Phi_-(\lambda - ih) &= \Omega(\lambda - ih) \end{aligned}\right\} \qquad (-\infty < \lambda < \infty). \tag{2.6}$$

The vector functions $\Phi_\pm(\lambda)$ and $\Omega(\lambda)$ which are a solution of problem (2.6) are uniquely determined by each of them. Therefore, when speaking of a solution of problem (2.6), we shall have one of these vector functions in mind. If $\Omega(\lambda)$ is a solution of problem (2.6), it is not difficult to ascertain that the vector function $[I - K(\lambda)]^{-1}\Omega(\lambda)$ is holomorphic in the entire plane except, perhaps, for the points α_j, $j = 1,\cdots,m$, which are poles for it of order $\leqq p_j$ and tend to zero at infinity. Consequently it is rational and has the form

$$[I - K(\lambda)]^{-1}\Omega(\lambda) = \left\{ \sum_{j=1}^{m} \frac{Q_{jr}(\lambda)}{(\lambda - \alpha_j)^{p_j}} \right\}_{r=1}^{n},$$

where the $Q_{jr}(\lambda)$ are polynomials of degrees $\leqq p_j - 1$. Since $[I - K(\lambda)]^{-1}\Omega(\lambda)$ coincides with $\Phi_+(\lambda)$ for $\operatorname{Im} \lambda \geqq h$ and with $-\Phi_-(\lambda)$ for $\operatorname{Im} \lambda \leqq -h$, we obtain, by virtue of (2.3), the fact that every solution of equation (2.5) has the form

$$\varphi(t) = \left\{ \sum_{j=1}^{m} e^{-i\alpha_j t} P_{jr}(t) \right\}_{r=1}^{n} \qquad (-\infty < t < \infty), \tag{2.7}$$

where the $P_{jr}(t)$ are polynomials of degrees $\leqq p_j - 1$.

Let us establish that $p = p_j$ linearly independent solutions of equation (2.5) correspond to each zero $\alpha = \alpha_j$. To this end let us transform the matrix $I - K(\lambda)$.

A constant nonsingular matrix A_1 exists such that the point $\lambda = \alpha$ is a zero of multiplicity $q_1 \geqq 1$ of the first column of the matrix $(I - K(\lambda))A_1^{-1}$, which we represent in the form $(I - K(\lambda))A_1^{-1} = G_1 D_1$. Here

$$D_1 = \begin{Vmatrix} \left(\dfrac{\lambda-\alpha}{\lambda-\beta}\right)^{q_1} & 0\cdots\cdots 0 \\ 0 & 1\cdots\cdots 0 \\ \cdots\cdots & \cdots\cdots \\ 0 & 0\cdots\cdots 1 \end{Vmatrix},$$

where β is an arbitrary number with $|\operatorname{Im}\beta| > h$, while the first column of the matrix G_1 does not vanish at the point $\lambda = \alpha$, (i.e. at least one of its elements does not vanish).

If $q_1 < p$, the determinant of G_1 vanishes at the point $\lambda = \alpha$, and we represent G_1 analogously: $G_1 A_2^{-1} = G_2 D_2$. Continuing this process, we obtain the following representation of the matrix $I - K(\lambda)$:

$$I - K(\lambda) = G_r D_r A_r D_{r-1} A_{r-1} \cdots D_1 A_1,$$

where the determinant of G_r does not vanish at the point $\lambda = \alpha$, and $q_1 + \cdots + q_r = p$. The vector functions $F_+^{(lk)}(\lambda)$ defined by the equalities

$$A_k D_{k-1} A_{k-1} \cdots D_1 A_1 F_+^{(lk)} = \begin{pmatrix} (\lambda-\alpha)^{-l} \\ 0 \\ \vdots \\ 0 \end{pmatrix}$$

($k = 1,\cdots,r$; $l = 1,\cdots, q_k$) are linearly independent and are solutions of problem (2.6).

Thus we have found $p_1 + \cdots + p_m$ linearly independent solutions of problem (2.6):

$$\Phi_+^{(jq)}(\lambda) = \left\{ \frac{Q_{jqr}(\lambda)}{(\lambda-\alpha_j)^{p_j}} \right\}_{r=1}^n \qquad (j = 1,\cdots,m;\ q = 1,\cdots, p_j),$$

where the $Q_{jqr}(\lambda)$ are polynomials of degrees $\leqq p_j - 1$. Defining the vector functions $\varphi_{jq}(t)$ relative to them, we obtain $p_1 + \cdots + p_m$ linearly independent solutions of (2.5):

$$\varphi_{jq}(t) = \{e^{-i\alpha_j t} P_{jqr}(t)\}_{r=1}^n$$
$$(-\infty < t < \infty;\ j = 1,\cdots, m;\ q = 1,\cdots, p_j), \tag{2.8}$$

where the P_{jqr} are polynomials of degrees $\leqq p_j - 1$.

In order to complete the theorem's proof, it remains for us to show that all linearly independent solutions of equation (2.5) are exhausted by the vector functions (2.8).

Lemma 1.1 implies that the index of the operator

$$(I - \mathscr{K})\varphi = \varphi(t) - \int_{-\infty}^{\infty} k(t-s)\varphi(s)\,ds$$

in the space $e^{h|t|}\tilde{L}_n$ is defined by the equality

$$\kappa(I - \mathscr{K}) = \operatorname{ind}\det(I - K(\lambda - ih)) - \operatorname{ind}\det(I - K(\lambda + ih)).$$

By virtue of Rouché's theorem, the right side of the latter equality equals $p_1 + \cdots + p_m$. But $\beta(I - \mathscr{K}) = \alpha(I - \mathscr{K})^* = 0$. As a matter of fact, the operator $(I - \mathscr{K})^*$ operating in the space $e^{-h|t|}\tilde{M}_n$ is of the same type as the operator $I - \mathscr{K}$. The part of the theorem which has been proved implies that all solutions of the equation $(I - \mathscr{K})^*\varphi = 0$ in the space $e^{h|t|}\tilde{M}_n$ have form (2.7) but none of them get into the space $e^{-h|t|}\tilde{M}_n$. Thus the equation $(I - \mathscr{K})^*\varphi = 0$ has only the trivial solution in the space $e^{-h|t|}\tilde{M}_n$; consequently

$$\alpha(I - \mathscr{K}) = \kappa(I - \mathscr{K}) = p_1 + \cdots + p_m.$$

The theorem is proved.

The theorem just proved implies that in the case $n = 1$ the functions $\varphi_{jq}(t) = e^{-i\alpha_j t}t^{q-1}$ $(j = 1,\cdots,m; q = 1,\cdots,p_j)$ form a basis of all solutions of equation (2.5). This result is well known (cf. E. Titchmarsh [1], Theorem 146).

LEMMA 2.1. *Let the matrix functions $k_r(t) \in e^{-h|t|}\tilde{L}_{n\times n}$ $(r = 1,2)$ possess the property*

$$\det(I - K_r(\lambda \pm ih)) \neq 0 \qquad (-\infty < \lambda < \infty; r = 1,2), \tag{2.9}$$

and let B_1 and B_2 be bounded linear operators in the space $e^{h|t|}\tilde{E}_n$ which map it into $e^{-ht}\tilde{E}_n$ and $e^{ht}\tilde{E}_n$, respectively. Then every solution $\varphi(t) \in e^{h|t|}\tilde{E}_n$ of the equation

$$\left.\begin{aligned} &\varphi(t) - \int_0^\infty k_1(t-s)\varphi(s)\,ds - (B_1\varphi)(t) = 0 \qquad (0 < t < \infty),\\ &\varphi(t) - \int_{-\infty}^0 k_2(t-s)\varphi(s)\,ds - (B_2\varphi)(t) = 0 \qquad (-\infty < t < 0)\end{aligned}\right\} \tag{2.10}$$

admits the representations

$$\begin{aligned} \varphi(t) &= \left\{\sum_{j=1}^{m_1} e^{-i\alpha_{1j}t} P_{1jk}(t)\right\}_{k=1}^{n} + \varphi_1(t),\\ \varphi(t) &= \left\{\sum_{j=1}^{m_2} e^{-i\alpha_{2j}t} P_{2jk}(t)\right\}_{k=1}^{n} + \varphi_2(t),\end{aligned} \tag{2.11}$$

where the vector functions $\varphi_1(t)$ and $\varphi_2(t)$ belong to the spaces $e^{-ht}\tilde{E}_n$ and $e^{ht}\tilde{E}_n$ respectively, α_{rj} $(r = 1,2; j = 1,\cdots,m_r)$ are all of the distinct zeros of the function $\det(I - K_r(\lambda))$ *in the strip $|\operatorname{Im}\lambda| < h$, p_{rj} are their multiplicities, and $P_{rjk}(t)$ are polynomials of degrees $\leq p_{rj} - 1$.*

PROOF. Let us observe first of all that if the condition

$$\det(I - K(\lambda - ih)) \neq 0 \qquad (-\infty < \lambda < \infty)$$

is fulfilled for a matrix function $k(t) \in e^{-h|t|}\tilde{L}_{n\times n}$, the operator

$$(I - \mathscr{K})f = f(t) - \int_{-\infty}^{\infty} k(t-s)f(s)\,ds$$

is invertible in the space $e^{-ht}\tilde{E}_n$. Indeed, by virtue of Wiener's theorem[4)] a matrix function $q(t) \in L_{n\times n}$ exists such that $I - Q(\lambda) = [I - K(\lambda - ih)]^{-1}$. Now it is not difficult to verify that the operator inverse to $I - \mathscr{K}$ is defined by the equality

$$(I - \mathscr{K})^{-1}f = f(t) - \int_{-\infty}^{\infty} q(t-s)\,e^{h(s-t)}f(s)\,ds.$$

Now let the vector function $\varphi(t) \in e^{h|t|}\tilde{E}_n$ be a solution of equation (2.10). Let us write the first equality of (2.10) in the form

$$\varphi(t) - \int_{-\infty}^{\infty} k_1(t-s)\varphi(s)ds = (B_1\varphi)(t) - \int_{-\infty}^{0} k_1(t-s)\varphi(s)\,ds + \psi_-(t)$$

$(-\infty < t < \infty)$, where the vector function $\psi_-(t) = 0$ for $t > 0$. Thus $\varphi(t)$ satisfies the relation

$$\varphi(t) - \int_{-\infty}^{\infty} k_1(t-s)\varphi(s)ds = \psi(t) \qquad (-\infty < t < \infty),$$

where $\psi(t) \in e^{-ht}\tilde{E}_n$. The operator defined by the left side of the last equality (let us denote it by A) is invertible in the space $e^{-ht}\tilde{E}_n$. Therefore there exists a unique solution $\varphi_1(t) \in e^{-ht}\tilde{E}_n$ of the equation $A\varphi = \psi$. Thus $A(\varphi - \varphi_1) = 0$, and application of Theorem 2.1 leads to the first representation of (2.11). The second equality of (2.11) is established analogously.

THEOREM 2.2. *Let the matrix functions* $k_{jr}(t) \in e^{-h|t|}\tilde{L}_{n\times n}$ $(j, r = 1,2)$ *possess the property*

$$\det(I - K_{rr}(\lambda \pm ih)) \neq 0 \qquad (-\infty < \lambda < \infty;\ r = 1,2).$$

Then every solution $\varphi(t) \in e^{h|t|}\tilde{E}_n$ *of equation* (1) *admits the asymptotic expansions*

$$\left.\begin{aligned} \varphi(t) &= \left\{\sum_{j=1}^{m_1} e^{-i\alpha_{1j}t} P_{1jk}(t)\right\}_{k=1}^{n} + o(e^{-ht}) && (t \to \infty), \\ \varphi(t) &= \left\{\sum_{j=1}^{m_2} e^{-i\alpha_{2j}t} P_{2jk}(t)\right\}_{k=1}^{n} + o(e^{ht}) && (t \to -\infty), \end{aligned}\right\} \tag{2.12}$$

where α_{rj} $(r = 1,2;\ j = 1,\cdots,m_r)$ *are all of the distinct zeros of the function* $\det(I - K_{rr}(\lambda))$ *in the strip* $|\operatorname{Im}\lambda| < h$, p_{rj} *are their multiplicities, and* $P_{rjk}(t)$ *are polynomials of degrees* $\leqq p_{rj} - 1$.

PROOF. Lemma 2.1 is applicable to equation (1), since the operators

4) Cf. Gel'fand, Raĭkov and Šilov [1], § 17.

$$(\mathscr{K}_{12}f)(t) = \int_{-\infty}^{0} k_{12}(t-s)f(s)ds$$

and

$$(\mathscr{K}_{21}f)(t) = \int_{0}^{\infty} k_{21}(t-s)f(s)ds$$

map the spaces $e^{h|t|}\tilde{E}_n$ into the spaces $e^{-ht}\tilde{E}_n$ and $e^{ht}\tilde{E}_n$, respectively. Theorem 1.1 implies that the vector functions $\varphi_1(t)$ and $\varphi_2(t)$ in representations (2.11) belong to the spaces $e^{-ht}(\tilde{C}_0)_n$ and $e^{ht}(\tilde{C}_0)_n$, respectively. The latter is equivalent to the asymptotic expansions (2.12).

COROLLARY 2.1. *Let the matrix function* $k(t) \in e^{-h|t|}\tilde{L}_{n\times n}$ *satisfy conditions* (2.4). *Then every solution* $\varphi(t) \in e^{ht} E_n$ *of the system of Wiener-Hopf equations*

$$\varphi(t) - \int_{0}^{\infty} k(t-s)\varphi(s)\,ds = 0 \qquad (0 < t < \infty)$$

admits the asymptotic expansion

$$\varphi(t) = \left\{\sum_{j=1}^{m} e^{-i\alpha_j t} P_{jr}(t)\right\}_{r=1}^{n} + o(e^{-ht}) \qquad (t \to \infty), \tag{2.13}$$

where $\alpha_j, j = 1,\cdots, m$, *are all the distinct zeros of* $\det(I - K(\lambda))$ *in the strip* $|\operatorname{Im}\lambda| < h$, p_j *are their multiplicities, and* $P_{jr}(t)$ *are polynomials of degrees* $\leqq p_j - 1$.

COROLLARY 2.2. *Let the matrix functions* $k_j(t) \in e^{-h|t|}\tilde{L}_{n\times n}$, $j = 1,2$, *satisfy conditions* (2.9). *Then every solution* $\varphi(t) \in e^{h|t|}\tilde{E}_n$ *of the system of pair integral equations*

$$\varphi(t) - \int_{-\infty}^{\infty} k_1(t-s)\varphi(s)ds = 0 \qquad (0 < t < \infty),$$

$$\varphi(t) - \int_{-\infty}^{\infty} k_2(t-s)\varphi(s)ds = 0 \qquad (-\infty < t < 0)$$

and the transpose system

$$\varphi(t) - \int_{0}^{\infty} k_1(t-s)\varphi(s)ds - \int_{-\infty}^{0} k_2(t-s)\varphi(s)ds = 0$$
$$(-\infty < t < \infty)$$

admits the asymptotic expansions (2.12). *where* α_{rj} $(r = 1,2; j = 1,\cdots, m_r)$ *are all of the distinct zeros of the function* $\det(I - K_r(\lambda))$ *in the strip* $|\operatorname{Im}\lambda| < h$, p_{rj} *are their multiplicities, and* $P_{rjk}(t)$ *are polynomials of degrees* $\leqq p_{rj} - 1$.

2. As already noted, the results cited above remain valid for the discrete ana-

logues of the respective equations. Let us, for example, formulate the discrete analogue of Theorem 2.2. To this end let us agree on the following notation. Let F be one of the spaces $\tilde{l}_p$ $(p \geqq 1;\ \tilde{l}_1 = \tilde{l})$, $\tilde{m}$, $\tilde{c}$ or $\tilde{c}_0$ of sequences of complex numbers $\xi = \{\xi_m\}_{-\infty}^{\infty}$. By F_n we denote as usual the linear space of sequences of n-dimensional vectors whose coordinates with the same index k $(= 1, \cdots, n)$ form a sequence of F, and by $F_{n\times n}$ the set of sequences of square nth order matrices the corresponding elements of which form a sequence of F.

Let $\gamma = \{\gamma_m\}_{-\infty}^{\infty}$ be a numerical sequence such that $\gamma_m \neq 0$ for all m. We shall write $\xi \in \gamma F$ if $\{\gamma_m^{-1}\xi_m\}_{-\infty}^{\infty} \in F$. The spaces γF_n and $\gamma F_{n\times n}$ are defined analogously.

THEOREM 2.3. *Let the sequences of matrices* $\{A_m^{(jr)}\}_{m=-\infty}^{\infty}$ $(j, r = 1, 2)$ *belong to* $\{h^{-|m|}\}_{-\infty}^{\infty}\tilde{l}_{n\times n}$ *for some* $h > 1$, *and let the matrix functions*

$$U^{(rr)}(\zeta) = \sum_{j=-\infty}^{\infty} A_j^{(rr)}\zeta^j \qquad (r = 1, 2)$$

possess the property

$$\det U^{(rr)}(\zeta h^{\pm 1}) \neq 0 \qquad (\,|\zeta| = 1;\, r = 1, 2).$$

Then every solution $\xi = \{\xi_m\}_{-\infty}^{\infty}$ *of the infinite system of equations*

$$\sum_{j=0}^{\infty} A_{m-j}^{(11)}\xi_j + \sum_{j=-\infty}^{-1} A_{m-j}^{(12)}\xi_j = 0 \qquad (m = 0, 1, \cdots),$$

$$\sum_{j=0}^{\infty} A_{m-j}^{(21)}\xi_j + \sum_{j=-\infty}^{-1} A_{m-j}^{(22)}\xi_j = 0 \qquad (m = -1, -2, \cdots)$$

which is contained in $\{h^{|m|}\}_{-\infty}^{\infty}F_n$ *admits the asymptotic expansions*

$$\xi_m = \left\{\sum_{j=1}^{q_1} \alpha_{1j}^{-m}P_{1jk}(m)\right\}_{k=1}^{n} + o(h^{-m}) \qquad (m \to \infty),$$

$$\xi_m = \left\{\sum_{j=1}^{q_2} \alpha_{2j}^{-m}P_{2jk}(m)\right\}_{k=1}^{n} + o(h^{m}) \qquad (m \to -\infty),$$

where α_{rj} $(r = 1, 2;\ j = 1, \cdots, m_r)$ *are all of the distinct zeros of the function* $\det U^{(rr)}(\xi)$ *in the annulus* $h^{-1} < |\zeta| < h$, p_{rj} *are their multiplicities, and* $P_{rjk}(t)$ *are polynomials of degrees* $\leqq p_{rj} - 1$.

§ 3. Transpose equation of a pair

Let us denote by $\mathfrak{R}_h$ the algebra of all functions of the form

$$\mathscr{A}(z) = c + \begin{cases} \int_{-\infty}^{\infty} a_1(t)e^{izt}\,dt & (\operatorname{Im} z = h), \\ \int_{-\infty}^{\infty} a_2(t)e^{izt}\,dt & (\operatorname{Im} z = -h), \end{cases}$$

where $a_1(t) \in e^{ht}\tilde{L}$, $a_2(t) \in e^{-ht}\tilde{L}$ and c is an arbitrary complex number.

The functions $\mathscr{A}(z)$ of the algebra $\mathfrak{R}_h$ are defined on the contour consisting of the pair of straight lines Im $z = \pm h$. Let us denote by $\mathfrak{R}_h^+$ the subalgebra of $\mathfrak{R}_h$ which consists of all functions $\mathscr{A}(z)$ for which $a_1(t) = 0$ for $t < 0$ and $a_2(t) = 0$ for $t > 0$, and let us denote by $\mathfrak{R}_h^-$ the subalgebra of $\mathfrak{R}_h$ which consists of all functions $\mathscr{A}(z)$ for which $a_1(t) = a_2(t) \in e^{-h|t|}\tilde{L}$. Every function of $\mathfrak{R}_h^+$ admits continuation to a function holomorphic in the region $|\text{Im } z| > h$ (i. e. outside the strip $|\text{Im } z| \leqq h$) and continuous right up to and including the boundary, while every function of $\mathfrak{R}_h^-$ admits continuation to a function holomorphic inside the strip $|\text{Im } z| < h$ and continuous right up to and including the boundary.

Every function $\mathscr{A}(z) \in \mathfrak{R}_h$ is representable as the sum of two functions $\mathscr{A}_\pm(z) \in \mathfrak{R}_h^\pm$. As a matter of fact, it is sufficient to set

$$\mathscr{A}_+(z) = c + \begin{cases} \int_0^\infty [a_1(t) - a_2(t)]\, e^{izt}\, dt & (\text{Im } z = h), \\ \int_{-\infty}^0 [a_2(t) - a_1(t)]\, e^{izt}\, dt & (\text{Im } z = -h), \end{cases}$$

$$\mathscr{A}_-(z) = \int_{-\infty}^0 a_1(t) e^{izt}\, dt + \int_0^\infty a_2(t)\, e^{izt}\, dt \qquad (\text{Im } z = \pm h).$$

It is easy to see that this representation is unique to within a constant addend.

Let the conditions

$$\mathscr{A}(z) \neq 0 \quad (\text{Im } z = \pm h), \qquad \mathscr{A}(\infty) = 1 \tag{3.1}$$

be fulfilled for a function $\mathscr{A}(z) \in \mathfrak{R}_h$. Let us introduce the notation

$$\nu_1 = \text{ind } \mathscr{A}(\lambda + ih)\big|_{\lambda=-\infty}^{\infty}, \qquad \nu_2 = \text{ind } \mathscr{A}(\lambda - ih)\big|_{\lambda=-\infty}^{\infty},$$
$$\kappa = \nu_2 - \nu_1 \ (= \kappa(\mathscr{A})).$$

Let us consider an auxiliary function $\mathscr{A}_1(z)$ of $\mathfrak{R}_h$ which is defined by the equality

$$\mathscr{A}_1(z) = \begin{cases} \left(\dfrac{\lambda - a}{\lambda + a}\right)^\kappa & \text{for Im } z = h, \\ 1 & \text{for Im } z = -h, \end{cases}$$

where $\lambda = \text{Re } z$ and Im $a > 2h$. For it, $\kappa(\mathscr{A}_1) = -\kappa$; therefore $\kappa(\mathscr{A}_1, \mathscr{A}) = 0$, i.e. the function $\mathscr{A}_1(z)$ annihilates the index of the function $\mathscr{A}(z)$. From results of I.C. Gohberg **[1]** it follows that the function $\mathscr{A}_1(z)\mathscr{A}(z)$ admits the factorization

$$\mathscr{A}_1(z)\mathscr{A}(z) = G_+(z)G_-(z),$$

where $G_\pm(z) \in \mathfrak{R}_h^\pm$, $G_\pm^{-1}(z) \in \mathfrak{R}_h^\pm$ and $G_\pm(\infty) = 1$. The equalities

$$\left.\begin{aligned} \mathscr{A}(\lambda + ih) &= G_+(\lambda + ih)G_-(\lambda + ih)\left(\frac{\lambda + a}{\lambda - a}\right)^\kappa, \\ \mathscr{A}(\lambda - ih) &= G_+(\lambda - ih)G_-(\lambda - ih) \end{aligned}\right\} \quad (-\infty < \lambda < \infty) \tag{3.2}$$

follow from this factorization.

Now let us consider in the space $e^{h|t|}\bar{E}$ the homogeneous integral equation

$$\varphi(t) - \int_0^\infty k_1(t-s)\varphi(s)\,ds - \int_{-\infty}^0 k_2(t-s)\varphi(s)\,ds = 0 \qquad (3.3)$$
$$(-\infty < t < \infty),$$

which is the transpose of a pair, where the $k_j(t) \in e^{-h|t|}\bar{L}$ $(j = 1, 2)$ and satisfy the conditions

$$1 - K_j(\lambda \pm ih) \neq 0 \qquad (-\infty < \lambda < \infty; j = 1, 2). \qquad (3.4)$$

Since, by virtue of Theorem 1.1, equation (3.3) has the same solutions in all spaces $e^{h|t|}E$, it is sufficient to consider this equation in the space $e^{h|t|}L$.

By virtue of (2.2), equation (3.3) is equivalent to the boundary value problem

$$\left.\begin{aligned}[1 - K_1(\lambda + ih)]\Phi_+(\lambda + ih) &= \Omega(\lambda + ih),\\ -[1 - K_2(\lambda - ih)]\Phi_-(\lambda - ih) &= \Omega(\lambda - ih)\end{aligned}\right\} \quad (-\infty < \lambda < \infty), \qquad (3.5)$$

where the functions

$$\Phi(z) = \begin{cases}\Phi_+(\lambda + ih), & \text{for } \operatorname{Im} z = h,\\ \Phi_-(\lambda - ih), & \text{for } \operatorname{Im} z = -h\end{cases}$$

and $\Omega(z)$ belong to the algebras $\mathfrak{R}_h^+$ and $\mathfrak{R}_h^-$, respectively.

The function

$$\mathscr{A}(z) = \begin{cases}1 - K_1(\lambda + ih), & \operatorname{Im} z = h,\\ 1 - K_2(\lambda - ih), & \operatorname{Im} z = -h,\end{cases}$$

belongs to $\mathfrak{R}_h$ and satisfies conditions (3.1), and consequently admits factorization (3.2).

Factorization (3.2) implies that for $\kappa(\mathscr{A}) = \kappa = \nu_2 - \nu_1 > 0$ the general solution of problem (3.5) is given by the equalities

$$\left.\begin{aligned}\Omega(\lambda) &= \frac{P_{\kappa-1}(\lambda)}{(\lambda - a - ih)^\kappa} G_-(\lambda),\\ \Phi_+(\lambda + ih) &= \frac{P_{\kappa-1}(\lambda + ih)}{G_+(\lambda + ih)\,(\lambda + a)^\kappa},\\ \Phi_-(\lambda - ih) &= -\frac{P_{\kappa-1}(\lambda - ih)}{G_+(\lambda - ih)\,(\lambda - a - 2ih)^\kappa},\end{aligned}\right\} \qquad (3.6)$$

where $P_{\kappa-1}(z)$ is an arbitrary polynomial of degree $\leqq \kappa - 1$. If, however, $\kappa \leqq 0$, problem (3.5), and consequently also equation (3.3), has only the trivial solution.

LEMMA 3.1. *Let $a(t) \in e^{-ht}L_1(0, \infty)$, and let n be a nonnegative integer. Then for* $\operatorname{Im} \delta > -h$

$$\int_t^\infty (t-s)^n e^{-i\delta(t-s)} a(s)\,ds = o(e^{-ht}) \qquad (t \to \infty) \tag{3.7}$$

and for Im $\delta < -h$

$$\int_0^t (t-s)^n e^{-i\delta(t-s)} a(s)\,ds = o(e^{-ht}) \qquad (t \to \infty). \tag{3.8}$$

PROOF. Let us note that for any $r > 0$ there is a constant $c > 0$ such that $t^n \leqq ce^{rt}$ $(0 \leqq t < \infty)$. In particular, the inequality

$$(s-t)^n \leqq ce^{(h+\operatorname{Im}\delta)(s-t)} \qquad (s \geqq t)$$

holds for Im $\delta > -h$.

Relation (3.7) now follows from the following estimates:

$$\left| \int_t^\infty (t-s)^n e^{-i\delta(t-s)} a(s)\,ds \right| \leqq e^{\operatorname{Im}\delta t} \int_t^\infty (s-t)^n e^{-\operatorname{Im}\delta s} |a(s)|\,ds$$

$$\leqq e^{\operatorname{Im}\delta t} \int_t^\infty ce^{(h+\operatorname{Im}\delta)(s-t)} e^{-\operatorname{Im}\delta s} |a(s)|\,ds$$

$$= ce^{-ht} \int_t^\infty e^{hs} |a(s)|\,ds = o(e^{-ht}) \qquad (t \to \infty).$$

Turning to the proof of (3.8), we make use of the estimates

$$\left| \int_0^t (t-s)^n e^{-i\delta(t-s)} a(s)\,ds \right| \leqq e^{-ht} \int_0^t (t-s)^n e^{(h+\operatorname{Im}\delta)(t-s)} a^{hs} |a(s)|\,ds$$

$$= e^{-ht} \int_0^t s^n e^{-\varepsilon s} \rho(t-s)\,ds,$$

where $\rho(t) = e^{ht}|a(t)|$ and $\varepsilon = -(h + \operatorname{Im}\delta) > 0$.

Let us show that the last integral tends to zero as $t \to \infty$. Indeed, whatever the positive number $b < t$ may be, the relations

$$\int_0^t s^n e^{-\varepsilon s} \rho(t-s)\,ds = \int_0^b s^n e^{-\varepsilon s} \rho(t-s)\,ds + \int_b^t s^n e^{-\varepsilon s} \rho(t-s)\,ds$$

$$\leqq \sup_{0 \leqq s < \infty} s^n e^{-\varepsilon s} \int_{t-b}^t \rho(s)\,ds + \sup_{b \leqq s < \infty} s^n e^{-\varepsilon s} \int_b^t \rho(t-s)\,ds$$

are valid, which implies that

$$\lim_{t \to \infty} \int_0^t s^n e^{\varepsilon - s} \rho(t-s)\,ds = 0.$$

The lemma is proved.

As already noted, the solutions of equation (3.3) belong to each of the spaces

$e^{h|t|}\tilde{E}$. In order to formulate a theorem concerning the asymptotics of these solutions, let us introduce in addition the following notation: the α_j, $j = 1,\cdots, n$, are all of the distinct zeros of the function $1 - K_1(\lambda)$ in the strip $|\mathrm{Im}\lambda| < h$, and are so numbered that $\mathrm{Im}\,\alpha_1 \geqq \mathrm{Im}\,\alpha_2 \geqq \cdots \geqq \mathrm{Im}\,\alpha_n$; the p_j are their multiplicities; and $m = p_1 + \cdots + p_n$. Let κ be some positive integer. If $\kappa < m$, let us denote by τ the largest integer such that $p_1 + \cdots + p_\tau \leqq k$ and $\mathrm{Im}\,\alpha_\tau > \mathrm{Im}\,\alpha_{\tau+1}$.

Let conditions (3.4) be fulfilled for the functions $k_j(t)$ of $e^{-h|t|}\tilde{L}$ $(j = 1, 2)$. Let us set $\kappa = \nu_2 - \nu_1$, where

$$\nu_1 = \mathrm{ind}\,(I - K_1(\lambda + ih)), \qquad \nu_2 = \mathrm{ind}(1 - K_2(\lambda - ih)).$$

Let us set

$$G(\lambda) = 1 + \int_0^\infty \gamma(t) e^{i\lambda t}\, dt.$$

for a function $\gamma(t)$ prescribed on the semiaxis $0 < t < \infty$.

THEOREM 3.1 *Let $\kappa > 0$. Then the following assertions are true.*

a) *If $m \leqq \kappa$, a function $\gamma(t) \in e^{-ht}L_1(0, \infty)$ exists such that to each zero $\alpha = \alpha_j, j = 1,\cdots, n$, of the function $1 - K_1(\lambda)$ in the strip $|\mathrm{Im}\,\lambda| < h$ there correspond $p = p_j$ solutions of equation* (3.3) *with the asymptotic expansion*

$$\varphi_r(t) = e^{-i\alpha t} \sum_{j=0}^{r} \frac{(-i)^j t^{r-j}}{j!(r-j)!} G^{(j)}(\alpha) + o(e^{-ht}) \quad (r = 0, 1,\cdots, p-1) \qquad (3.9)$$

as $t \to \infty$. The remaining $\kappa - m$ linearly independent solutions belong to any of the spaces $e^{-ht}\tilde{E}$.

b) *If $m > \kappa$, a function $\gamma(t) \in e^{ct}L(0, \infty)$ $(\mathrm{Im}\,\alpha_{\tau+1} < c < \mathrm{Im}\,\alpha_\tau)$ exists such that to each "upper" zero $\alpha = \alpha_j$, $j = 1,\cdots, \tau$, there correspond $p = p_j$ solutions of equation* (3.3) *with the asymptotic expansion*

$$\varphi_r(t) = e^{-i\alpha t} \sum_{j=0}^{r} \frac{(-i)^j t^{r-j}}{j!(r-j)!} G^{(j)}(\alpha) + \sum_{j=\tau+1}^{n} P_{jr}(t)\, e^{-i\alpha_j t} + o(e^{-ht}) \qquad (3.10)$$

$(r = 0, 1,\cdots, p-1)$ as $t \to \infty$, where the $P_{jr}(t)$ are polynomials of degrees exactly $p_j - 1$ $(j = \tau + 1,\cdots, n,)$; the remaining $\kappa - p_1 - p_2 - \cdots - p_\tau$ linearly independent solutions admit asymptotic expansions of the form

$$\sum_{j=\tau+1}^{n} P_j(t) e^{-i\alpha_j t} + o(e^{-ht})$$

as $t \to \infty$, where the $P_j(t)$ are polynomials of degrees $\leqq p_j - 1$ $(j = \tau + 1,\cdots, n)$.

PROOF. Let us note that if the function $\Phi_+(\lambda + ih)$ is the Fourier transform of the function $e^{-ht}\varphi(t) \in L_1(0, \infty)$ and

$$\Phi_+(\lambda + ih) = -(-i)^r G(\lambda + ih)/(\lambda - \delta)^r,$$

where Im $\delta < 0$ and

$$G(\lambda) = 1 + \int_0^\infty \gamma(t)e^{i\lambda t}\, dt, \qquad \gamma(t) \in e^{ht}L_1(0, \infty),$$

then

$$\varphi(t) = \frac{t^{r-1}e^{-i(\delta+ih)t}}{(r-1)!} + \frac{1}{(r-1)!}\int_0^t (t-s)^{r-1}e^{-i(\delta+ih)(t-s)}\gamma(s)\, ds. \tag{3.11}$$

Equality (3.11) is obtained by applying the inverse Fourier transform to the function $\Phi_+(\lambda + ih)$.

The general solution of problem (3.5) is, by virtue of (3.6), defined by the equality

$$\Phi_+(\lambda + ih) = \frac{P_{\kappa-1}}{G_+(\lambda + ih)(\lambda + a)^\kappa}. \tag{3.12}$$

In the case $\kappa \geqq m$ let us rewrite the latter equality in the form

$$\Phi_+(\lambda + ih) = \frac{P_{\kappa-1}G(\lambda + ih)}{\prod_{j=1}^n (\lambda + ih - \alpha_j)^{p_j}(\lambda + a)^{\kappa-m}}, \tag{3.13}$$

where the function

$$G(\lambda) = \frac{\prod_{j=1}^n (\lambda - \alpha_j)^{p_j}}{G_+(\lambda)(\lambda + a - ih)^m},$$

which is holomorphic in the region Im $\lambda > h$, admits, by virtue of the first equality of (3.2), continuation to a function holomorphic in the region Im $\lambda > -h$. It is not difficult to infer from this that the function $G(\lambda)$ is representable in the form

$$G(\lambda) = 1 + \int_0^\infty \gamma(t)e^{i\lambda t}\, dt,$$

where $\gamma(t) \in e^{-ht}L_1(0, \infty)$.

By virtue of equality (3.13), the functions

$$\Phi_+^{(j,k)}(\lambda + ih) = \frac{-(-i)^k G(\lambda + ih)}{(\lambda + ih - \alpha_j)^k} \qquad (j = 1,\cdots,n;\ k = 1,\cdots, p_j) \tag{3.14}$$

and

$$\Phi_+^{(r)}(\lambda + ih) = \frac{-(-i)^r G(\lambda + ih)}{(\lambda + a)^r} \qquad (r = 1,\cdots, \kappa - m) \tag{3.15}$$

form a basis of all solutions of boundary value problem (3.5).

Equalities (3.14) and (3.11) imply that to each zero $\alpha = \alpha_j$, $j = 1,\cdots, n$, there correspond the $p = p_j$ solutions of equation (3.3)

$$\varphi_k(t) = \frac{t^k e^{-i\alpha t}}{k!} + \frac{1}{k!}\int_0^t (t-s)^k e^{-i\alpha(t-s)}\gamma(s)\, ds \qquad (t > 0;\ k = 0,\cdots, p-1).$$

They admit asymptotic expansions (3.9).

Indeed,

$$\varphi_k(t) = \frac{e^{-i\alpha t}}{k!}\left[t^k + \int_0^\infty (t-s)^k e^{i\alpha s}\gamma(s)\,ds\right] - \frac{1}{k!}\int_t^\infty (t-s)^k e^{-i\alpha(t-s)}\gamma(s)\,ds.$$

By virtue of (3.7), the last integral is $o(e^{-ht})$ as $t \to \infty$, and the expression in square brackets equals

$$\sum_{j=0}^{k} (-i)^j C_k^j t^{k-j} G^{(j)}(\alpha),$$

since

$$\int_0^\infty s^j e^{i\alpha s}\gamma(s)\,ds = (-i)^j G^{(j)}(\alpha).$$

Relative to the functions (3.15) we find $\kappa - m$ linearly independent solutions of equation (3.3), of the form

$$\varphi(t) = \frac{t^r e^{-i\beta t}}{r!} + \frac{1}{r!}\int_0^t (t-s)^r e^{-i\beta(t-s)}\gamma(s)\,ds,$$

where $\beta = -\alpha + ih$ ($\operatorname{Im}\beta < -h$). They all belong to any space $e^{-ht}\bar{E}$. Indeed, this is obvious for the first addend and follows from estimates (3.8) for the second.

Let us observe that the existence of $\kappa - m$ linearly independent solutions of equation (3.3) which belong to any space $e^{-ht}\bar{E}$ follows also from the equality $\kappa = \kappa_{-h} + m$, where κ_{-h} is the index of equation (3.3) in the space $e^{-ht}\bar{E}$ (cf. (1.6)), and from the fact that in the case $\kappa_{-h} \geqq 0$ equation (3.3) has precisely κ_{-h} linearly independent solutions in the space $e^{-ht}\bar{E}$.

Turning to consideration of the case $\kappa < m$, let us rewrite equality (3.12), which gives the general solution of problem (3.5), in the form

$$\Phi_+(\lambda + ih) = \frac{P_{\kappa-1}G(\lambda + ih)}{\prod_{j=1}^{\tau}(\lambda + ih - \alpha_j)^{p_j}(\lambda + a)^{\kappa-q}}, \tag{3.16}$$

where $q = p_1 + \cdots + p_\tau$, and the function

$$G(\lambda) = \frac{\prod_{j=1}^{\tau}(\lambda - \alpha_j)^{p_j}}{G_+(\lambda)(\lambda + a - ih)^q},$$

which is holomorphic in the region $\operatorname{Im}\lambda > h$, admits, by virtue of the first equality of (3.2), continuation to a function holomorphic in the region $\operatorname{Im}\lambda > -h$, with the exception of the points α_j, $j = \tau + 1, \cdots, n$, which are poles for it of order p_j. By detaching the sum of the principal parts of the function $G(\lambda)$ relative to its poles α_j, $j = \tau + 1, \cdots, n$, it is easy to obtain the fact that

$$G(\lambda) - \sum_{j=\tau+1}^{n} \frac{d_j(\lambda)}{(\lambda - \alpha_j)^{p_j}} = 1 + \int_0^\infty \rho(t) e^{i\lambda t}\,dt,$$

where $\rho(t) \in e^{-ht}L_1(0, \infty)$ and $d_j(\lambda)$, $j = \tau + 1, \cdots, n$, are polynomials of degrees $\leqq p_j - 1$. For $\operatorname{Im} \lambda > \operatorname{Im} \alpha_{\tau+1}$ the sum of the principal parts of the function $G(\lambda)$ admits the representation

$$\sum_{j=\tau+1}^{n} \frac{d_j(\lambda)}{(\lambda - \alpha_j)^{p_j}} = \int_0^\infty \omega(t) e^{i\lambda t}\, dt \qquad (\operatorname{Im} \lambda > \operatorname{Im} \alpha_{\tau+1}),$$

where

$$\omega(t) = \sum_{j=\tau+1}^{n} e^{-i\alpha_j t} Q_j(t),$$

and the $Q_j(t)$ are polynomials of degrees exactly $p_j - 1$.

Thus

$$G(\lambda) = 1 + \int_0^\infty \gamma(t) e^{i\lambda t}\, dt \qquad (\operatorname{Im} \lambda > \operatorname{Im} \alpha_{\tau+1}),$$

where $\gamma(t) = \rho(t) + \omega(t) \in e^{ct}L_1(0, \infty)$ $(\operatorname{Im} \alpha_{\tau+1} < c < \operatorname{Im} \alpha_\tau)$.

By virtue of (3.16), the functions

$$\Phi_+^{(j,k)}(\lambda + ih) = \frac{-(-i)^k G(\lambda + ih)}{(\lambda + ih - \alpha_j)_k} \qquad (j = 1, \cdots, \tau;\ k = 1, \cdots, p_j) \tag{3.17}$$

and

$$\Phi_+^{(r)}(\lambda + ih) = \frac{-(-i)^r G(\lambda + ih)}{(\lambda + a)^r} \qquad (r = 1, \cdots, \kappa - q) \tag{3.18}$$

form a basis of all solutions of problem (3.5).

Applying (3.11) to the functions (3.17), we obtain the fact that to each upper zero $\alpha = \alpha_j$, $j = 1, \cdots, \tau$, there correspond the $p = p_j$ solutions of (3.3):

$$\begin{aligned}
\varphi_k(t) &= \frac{t_k e^{-i\alpha t}}{k!} + \frac{1}{k!} \int_0^t (t - s)^k e^{-i\alpha(t-s)} \gamma(s)\, ds \\
&= \left[\frac{t_k e^{-i\alpha t}}{k!} + \frac{1}{k!} \int_0^\infty (t - s)^k e^{-i\alpha(t-s)} \gamma(s)\, ds \right] \\
&\quad - \frac{1}{k!} \int_t^\infty (t - s)^k e^{-i\alpha(t-s)} \omega(s)\, ds - \frac{1}{k!} \int_t^\infty (t - s)^k e^{-i\alpha(t-s)} \rho(s)\, ds \\
&\qquad (t > 0;\ k = 1, \cdots, p - 1).
\end{aligned}$$

As was shown above, the expression in square brackets equals

$$e^{-i\alpha t} \sum_{j=0}^{k} \frac{(-i)^j t^{k-j}}{j!\,(k-j)!} G^{(j)}(\alpha);$$

the last integral in the expression for $\varphi_k(t)$ is, by virtue of (3.7), $o(e^{-ht})$ as $t \to \infty$; and the penultimate integral, as is easy to see, represents a function of the same type as $\omega(t)$. Thus we obtain expansion (3.10).

To the functions (3.18) there correspond, by virtue of (3.11), $\kappa - q$ linearly independent solutions of equation (3.3) of the form

$$\varphi(t) = \frac{t^r e^{-i\beta t}}{r!} + \frac{1}{r!}\int_0^t (t-s)^r e^{-i\beta(t-s)}\rho(s)\,ds$$
$$+\frac{1}{r!}\int_0^t (t-s)^r e^{-i\beta(t-s)}\omega(s)\,ds \qquad (t>0),$$

where $\beta = -a + ih$ ($\operatorname{Im}\beta < -h$). We have already evaluated the sum of the first two terms. It is $o(e^{-ht})$ ($t \to \infty$); the last integral, as is easy to see, equals

$$\sum_{j=\tau+1}^{n} P_j(t)\,e^{-i\alpha_j t} + o(e^{-ht}) \qquad (t \to \infty),$$

where the $P_j(t)$, $j = \tau + 1, \cdots, n$, are polynomials of degrees $\leqq p_j - 1$.

The theorem is proved.

Analogous asymptotic expansions hold as $t \to -\infty$. They are determined by the zeros of the function $1 - K_2(\lambda)$.

§ 4. Systems of Wiener-Hopf equations

The results of the present section represent refinements of the asymptotic expansions (2.13) of the solutions of the homogeneous system of Wiener-Hopf equations

$$\varphi(t) - \int_0^\infty k(t-s)\varphi(s)\,ds = 0 \qquad (0 < t < \infty), \tag{4.1}$$

where the matrix function $k(t)$ belongs to $e^{-h|t|}\tilde{L}_{n\times n}$ and its Fourier transform $K(\lambda)$ satisfies the conditions

$$\det\,(I - K(\lambda \pm ih)) \neq 0 \qquad (-\infty < \lambda < \infty). \tag{4.2}$$

Equation (4.1) is considered in the space $e^{ht}E_n$.

1. It follows easily from Theorems 2.4 and 6.1 of Chapter VIII that if conditions (4.2) are fulfilled, the matrix function $I - K(\lambda + ih)$ admits the factorization

$$I - K(\lambda + ih) = G_-(\lambda)\left\| \left(\frac{\lambda - a}{\lambda + a}\right)^{\kappa_j}\delta_{jk}\right\|_1^n G_+(\lambda) \tag{4.3}$$
$$(-\infty < \lambda < \infty),$$

where $\operatorname{Im} a > 2h$, $\kappa_1 \geqq \kappa_2 \geqq \cdots \geqq \kappa_n$ are integers called the *right indices* of the matrix function $I - K(\lambda + ih)$, and $G_\pm(\lambda)$ are matrix functions respectively holomorphic in the upper and lower half-plane and continuous there right up to and including the boundary. Their determinants do not vanish in the respective half-planes. Moreover,

$$G_\pm(\lambda) = I + \int_0^\infty g_\pm(\pm t)e^{\pm i\lambda t}\,dt,$$

where the matrix functions $g_{\pm}(t)$ belong to $L_{n\times n}$ and in all of the spaces $e^{ht}E_n$ equation (4.1) has the same solutions, which form the linear manifold $\sigma = -(\kappa_\beta + \kappa_{\beta+1} + \cdots + \kappa_n)$ of finite dimension, where the κ_j are all of the negative right indices.

Let us introduce the following notation: the α_j, $j = 1,\cdots,r$, are all of the distinct zeros of the function $\det(I - K(\lambda))$ in the strip $|\mathrm{Im}\lambda| < h$, so numbered that $\mathrm{Im}\,\alpha_1 \geqq \mathrm{Im}\,\alpha_2 \geqq \cdots \geqq \mathrm{Im}\,\alpha_r$; the p_j are their multiplicities; and $m = p_1 + \cdots + p_r$. If the inequality $m > -\kappa_j$ is fulfilled for some negative right index κ_j (< 0), let us denote by τ_j the largest integer such that $p_1 + \cdots + p_\tau \leqq -\kappa_j$ and $\mathrm{Im}\ \alpha_{\tau_j} > \mathrm{Im}\ \alpha_{\tau_{j+1}}$.

THEOREM 4.1. *Let conditions* (4.2) *be fulfilled for a matrix function $k(t)$ of $e^{-h|t|}\bar{L}_{n\times n}$, and let $\kappa_\beta,\cdots,\kappa_n$ be all of the negative right indices of the matrix function $I - K(\lambda + ih)$. Then the subspace of all solutions $\varphi(t) \in e^{ht}E_n$ of equation* (4.1) *can be decomposed into the direct sum of the subspaces R_j, $j = \beta,\cdots,n$, with bases*

$$\varphi_1^{(j)}(t),\cdots,\varphi_{-\kappa_j}^{(j)}(t) \tag{4.4}$$

of the following structure:

a) *If $-\kappa_j \geqq m$, the first m vector functions of the basis* (4.4) *admit the following asymptotic expansions as $t \to \infty$*:

$$\{e^{-i\alpha_\nu t}P_{\nu\mu l}^{(j)}\}_{l=1}^n + o(e^{-ht}) \qquad (\nu = 1,\cdots, r;\ \mu = 1,\cdots,p_\nu), \tag{4.5}$$

where the $P_{\nu\mu l}^{(j)}(t)$ are polynomials of degrees $\leqq \mu - 1$, and the vector functions $\varphi_{m+1}^{(j)}(t),\cdots, \varphi_{-\kappa_j}^{(j)}(t)$ belong to any space $e^{-ht}E_n$.

b) *If $-\kappa_j < m$, the first $q_j = p_1 + \cdots + p_{\tau_j}$ vector functions of the basis* (4.4) *admit the following asymptotic expansions as $t \to \infty$*:

$$\Big\{e^{-i\alpha_\nu t}P_{\nu\mu l}^{(j)}(t) + \sum_{s=\tau_j+1}^{r} e^{-i\alpha_s t}\, P_{\nu\mu s l}^{(j)}(t)\Big\}_{l=1}^n + o(e^{-ht})$$
$$(\nu = 1,\cdots,\tau_j;\ \mu = 1,\cdots, p_\nu), \tag{4.6}$$

where the $P_{\nu\mu l}^{(j)}(t)$ are some polynomials of degrees $\leqq \mu - 1$, the $P_{\nu\mu s l}^{(j)}(t)$ are polynomials of degrees $\leqq p_s - 1$ $(s = \tau_j + 1,\cdots, r)$, and as $t \to \infty$ the vector functions $\varphi_{q_j+1}^{(j)}(t),\cdots, \varphi_{-\kappa_j}^{(j)}(t)$ admit asymptotic expansions of the form

$$\Big\{\sum_{s=\tau_j+1}^{r} e^{-i\alpha_s t}P_{sl}^{(j)}(t)\Big\}_{l=1}^n + o(e^{-ht}), \tag{4.7}$$

where the $P_{sl}^{(j)}(t)$ are polynomials of degrees $\leqq p_s - 1$ $(s = \tau_j + 1,\cdots, r)$.

PROOF. Equation (4.1) considered in the space $e^{ht}L_n$ is equivalent to Hilbert's boundary value problem

$$[I - K(\lambda + ih)]\Phi_+(\lambda + ih) = B_-(\lambda + ih) \tag{4.8}$$
$$(-\infty < \lambda < \infty),$$

where

$$\left.\begin{aligned} \Phi_+(\lambda + ih) &= \int_0^\infty \varphi(t)e^{-ht}e^{i\lambda t}\,dt, \\ B_-(\lambda + ih) &= \int_{-\infty}^0 b(t)e^{-ht}e^{i\lambda t}\,dt, \\ b(t) = - \int_0^\infty k\,(t - s)\varphi(s)ds &\qquad (-\infty < t < 0). \end{aligned}\right\} \tag{4.9}$$

The vector functions $\Phi_+(\lambda)$ and $B_-(\lambda)$ admit continuation to vector functions which are respectively holomorphic in the half-planes $\operatorname{Im}\lambda > h$ and $\operatorname{Im}\lambda < h$ and continuous there, including the boundary.

The general solution of problem (4.8) is, by virtue of factorization (4.3), defined by the equality

$$\Phi_+(\lambda + ih) = G_+^{-1}(\lambda)\left\{0,\cdots, \frac{P_{-\kappa_\beta-1}(\lambda)}{(\lambda + a)^{-\kappa_\beta}},\cdots, \frac{P_{-\kappa_n-1}(\lambda)}{(\lambda + a)^{-\kappa_n}}\right\},$$

where $P_{-\kappa_j-1}(\lambda)$ is an arbitrary polynomial of degree $\leqq -k_j - 1$ ($j = \beta, \beta + 1,\cdots, n$), there being in correspondence with each right index $\kappa_j < 0$ the $-\kappa_j$ linearly independent solutions of problem (4.8) which are defined by the equality

$$\Phi_+^{(j)}(\lambda + ih) = G_+^{-1}(\lambda)\left\{0,\cdots, \frac{P_{-\kappa_j-1}(\lambda)}{(\lambda + a)^{-\kappa_j}}, 0, \cdots, 0\right\}. \tag{4.10}$$

In the case $-\kappa_j \geqq m$ let us rewrite the latter equality as

$$\Phi_+^{(j)}(\lambda + ih) = G(\lambda + ih)\left\{0,\cdots, \frac{P_{-\kappa_j-1}(\lambda)}{\prod_{s=1}^r(\lambda + ih - \alpha_s)^{p_s}(\lambda + a)^{-\kappa_j-m}}, 0, \cdots, 0\right\}, \tag{4.11}$$

where the matrix function

$$G(\lambda) = G_+^{-1}(\lambda - ih)\,\frac{\prod_{s=1}^r(\lambda - \alpha_s)^{p_s}}{(\lambda - ih + a)^m},$$

which is holomorphic in the region $\operatorname{Im}\lambda > h$, admits, by virtue of (4.3), continuation to a matrix function which is holomorphic in the region $\operatorname{Im}\lambda > -h$. Hence it is not difficult to infer that it is representable in the form

$$G(\lambda) = I + \int_0^\infty \gamma(t)e^{i\lambda t}\,dt,$$

where $\gamma(t) \in e^{-ht}L_{n\times n}$.

By virtue of (4.11) the vector functions

$$\left.\begin{aligned}\Phi_+^{(j\nu\mu)}(\lambda+ih) &= G(\lambda+ih)\left\{0,\cdots,\frac{1}{(\lambda+ih-\alpha_\nu)^\mu},\ 0,\cdots,0\right\}\\ &(\nu=1,2,\cdots,r;\ \mu=1,2,\cdots,p_\nu),\\ \Phi_+^{(j\nu)}(\lambda+ih) &= G(\lambda+ih)\left\{0,\cdots,\frac{1}{(\lambda+a)^\nu},\ 0,\cdots,0\right\}\\ &(\nu=1,2,\cdots,-\kappa_j-m)\end{aligned}\right\}\qquad(4.12)$$

form a basis of the subspace of those solutions of problem (4.8) which correspond to the right index κ_j.

Let us define relative to the vector functions $\Phi_+^{(j\nu\mu)}(\lambda+ih)$, and with the help of the first equality of (4.9), vector functions $\varphi_{\nu\mu}^{(j)}(t)$ which are solutions of equation (4.1). Each coordinate of one of these vector functions has the form

$$f(t) = ct^{\mu-1}e^{-i\alpha_\nu t} + \int_0^t (t-s)^{\mu-1}e^{-i\alpha_\nu(t-s)}a(s)\,ds,$$

where c is a number and $a(t)\in e^{-ht}L$ is a function depending on j, ν, μ and the subscript of the coordinate. It is easy to see that

$$f(t) = e^{-i\alpha_\nu t}P(t) + \int_t^\infty (t-s)^{\mu-1}e^{-i\alpha_\nu(t-s)}a(s)\,ds,$$

where $P(t)$ is a polynomial of degree $\leqq \mu-1$. By virtue of equality (3.7), the latter integral is $o(e^{-ht})$ as $t\to\infty$.

Thus the vector functions $\varphi_{\nu\mu}^{(j)}(t)$ admit asymptotic expansions (4.5).

Relative to the vector functions $\Phi_+^{(j\nu)}$ we find $-\kappa_j-m$ linearly independent solutions of equation (4.1). By developing an analogous argument it is not difficult to obtain the fact that these solutions belong to any space $e^{-ht}E_n$.

Turning to consideration of the case $-\kappa_j<m$, let us rewrite equality (4.10) in the form

$$\Phi_+^{(j)}(\lambda+ih) = G_j(\lambda+ih)\left\{0,\cdots,\frac{P_{-\kappa_j-1}(\lambda)}{\prod_{s=1}^{\tau_j}(\lambda+ih-\alpha_s)^{p_s}(\lambda+a)^{-\kappa_j-q_j}},\ 0,\cdots,0\right\}\quad(4.13)$$

where the matrix function

$$G_j(\lambda) = G_+^{-1}(\lambda-ih)\,\frac{\prod_{s=1}^{\tau_j}(\lambda-\alpha_s)^{p_s}}{(\lambda-ih+a)^{q_j}},$$

which is holomorphic in the region $\operatorname{Im}\lambda>h$, admits, by virtue of (4.3), continuation to a matrix function which is holomorphic in the region $\operatorname{Im}(\lambda)>-h$, with the exception of the points α_s, $s=\tau_j+1,\cdots r$, which are poles for it of order $\leqq p_s$. By detaching the sum of the principal parts of the matrix function $G_j(\lambda)$ relative to its poles α_s, $s=\tau_j+1,\cdots,r$, it is easy to obtain the fact that for $\operatorname{Im}\lambda>\operatorname{Im}\alpha_{\tau_j+1}$ it is representable in the form

$$G_j(\lambda) = I + \int_0^\infty \gamma_j(t) e^{i\lambda t}\, dt \qquad (\operatorname{Im} \lambda > \operatorname{Im} \alpha_{\tau_j+1}),$$

where $\gamma_j(t) = \rho_j(t) + \omega_j(t)$, $\rho_j(t) \in e^{-ht} L_{n\times n}$, and $\omega_j(t)$ is a matrix function whose elements have the form

$$\sum_{s=\tau_j+1}^{r} e^{-i\alpha_s t} Q_s(t),$$

where the $Q_s(t)$ are polynomials of degrees $\leqq p_s - 1$.

By virtue of equality (4.13), the vector functions

$$\Phi_+^{(j\nu\mu)}(\lambda + ih) = G_j(\lambda + ih) \left\{0, \cdots, \frac{1}{(\lambda + ih - \alpha_\nu)^\mu}, 0, \cdots, 0\right\}$$
$$(\nu = 1,2,\cdots,\tau_j,\ \mu = 1,2,\cdots, p_\nu)$$

and

$$\Phi_+^{(j\nu)}(\lambda + ih) = G_j(\lambda + ih) \left\{0, \cdots, \frac{1}{(\lambda + a)^\nu}, 0, \cdots, 0\right\}$$
$$(\nu = 1,2,\cdots, -\kappa_j - q_j)$$

form a basis of the subspace of those solutions of boundary value problem (4.8) which correspond to the right index κ_j.

By making use of the estimates obtained in the preceding section it is not difficult to establish the fact that the solutions of equation (4.1) which are defined by means of the vector functions $\Phi_+^{(j\nu\mu)}(\lambda + ih)$ and $\Phi_+^{(j\nu)}(\lambda + ih)$ admit asymptotic expansions (4.6) and (4.7).

In order to complete the proof of the theorem, let us observe that the solutions of equation (4.1) which correspond to distinct right indices are linearly independent.

Remarks on the Literature

Chapter I

Basically the results of I. C. Gohberg **[1, 2, 3]** are presented in this chapter. The presentation is developed in the setting of his article **[3]**. The abstract results are cited in somewhat less space than in the article cited above; however they are presented in more detail and are supplemented with several new ones.

§ 1. Propositions 5°–8° were established jointly with F. V. Širokov.

§ 4. The results of this section are published here for the first time.

§ 5. Theorem 5.1 in its sufficiency part generalizes a well-known theorem of M. G. Kreĭn **[4]** on the factorization of functions expanding into absolutely convergent Fourier series. The proof carried out here is simpler than in the cited article.

P. Masani **[1]** was the first to prove a lemma analogous to Lemma 5.1, by a method different from the one advanced here (cf. M. S. Budjanu and I.C. Gohberg **[1]**). The latter proof was presented earlier by H. Cartan **[1]** in a specific case. There are propositions similar in content and proof to Lemma 5.1 in G. Baxter **[1]**, G. F. Mandžavidze **[1]**, G. F. Mandžavidze and B. V. Hvedelidze, **[1]**, I. B. Simonenko **[2]**, F. B. Atkinson **[1]**, Gohberg and Kreĭn (**[5]**, Chapter IV, § 4), and Budjanu and Gohberg **[1, 3]**. In the proof we basically follow the book of Gohberg and Kreĭn **[5]**, where this lemma is put forth in a more general setting.

§ 7. The results of this section are generalizations of M. G. Kreĭn's results **[4]** established for functions expanding into absolutely convergent Fourier series.

The concept of the symbol of a singular integral operator was introduced for the first time by S. G. Mihlin **[2]**. Concerning the relationship between the symbol introduced in this section and the symbol of a singular integral operator, cf. the Remarks on § 6 of Chapter V.

§ 8. Theorems 8.1 and 8.3 were established by M. G. Kreĭn **[4]**. The formula for the resolvent $\gamma(t, a)$ at the end of this section is also due to Kreĭn **[8]**. Theorem 8.2 was obtained in a somewhat different form by I. S. Čebotaru **[1]**.

Equations with vanishing symbol are not considered in this book. A whole series of articles is devoted to these questions (for example, F. D. Gahov and V. I. Smagina **[1]**, V. B. Dybin **[1]**, V. B. Dybin and N. K. Karapetjanc **[1]**, Z. Presdorf **[1, 2]**, M. I. Haĭkin **[1, 2]**, and G. N. Čebotarev **[1, 2]**).

Chapter II

§ 1. Theorem 1.1 was communicated to the authors by A. S. Markus.

§ 2. Theorem 2.1 was established by N. I. Pol'skiĭ [1].

§ 3. Theorem 3.1 was proved by S. G. Mihlin [1] for the case when the perturbed operator is positive definite and the projections $P_\tau = Q_\tau$ are orthogonal.

L. E. Lerer [1] recently proved a theorem for the Hilbert space case which in a certain sense is a converse of Theorem 3.1.

§ 4. Theorem 4.1 is due to A. S. Markus [1].

§ 5. All the results of this section were established by A. S. Markus [1]. Theorem 5.1 represents a refinement of a theorem of G. M. Vaĭnikko [1], in which it is proved that the operator A has form (5.2) if $A \in \Pi\{P_n, P_n\}$, where P_n $(n = 1, 2, \cdots)$ is an arbitrary sequence of finite-dimensional orthoprojections converging strongly to the identity operator. Vaĭnikko's method is utilized in the proof of Theorem 5.1.

§ 6. Theorem 6.2 was established by A. S. Markus [1]. Its proof is based on Theorem 6.1, which was obtained by A. Brown and C. Pearcy [1] in connection with a problem on commutators.

Chapter III

§ 1. The results of this section were obtained by the authors [1, 4].

Condition 2) in Theorem 1.1 is essential. This can be ascertained from the following example. Let an isometric operator V, a left inverse of it $V^{(-1)}$, and the projections P_n $(n = 1, 2, \cdots)$ be defined in l_2 by the equalities

$$V\{\xi_j\}_1^\infty = \{0, \xi_1, 0, \xi_2, \cdots\}, \qquad V^{(-1)}\{\xi_j\}_1^\infty = \{\xi_2, \xi_4, \xi_6, \cdots\},$$
$$P_n\{\xi_j\}_1^\infty = \{\xi_1, \cdots, \xi_n, 0, 0, \cdots\}.$$

Let us denote by A an operator in $\mathfrak{R}(V)$, by $A(\zeta)$ $(|\zeta| = 1)$ its symbol, and by a_j $(j = 0, \pm 1, \cdots)$ the Fourier coefficients of $A(\zeta)$. The following proposition holds.

For the projection method (P_n, P_n) to be applicable to an operator A of $\mathfrak{R}(V)$, it is necessary and sufficient that $A(\zeta) \neq 0$ $(|\zeta| = 1)$, that ind $A(\zeta) = 0$, and that all the matrices $\|a_{j-k}\|_{j,k=1}^n$ $(n = 1, 2, \cdots)$ be invertible.

This result was established by A. S. Markus and I. A. Fel'dman.

§ 2. The second assertion of Theorem 2.1 was established for the first time by G. Baxter [2] in the space l_1 under the condition that the function $a(\zeta)$ expand into an absolutely convergent Fourier series. His proof, which is different from ours, is cited in this section.

Theorem 2.2 was proved by G. Szegö for the case when the function $a(\zeta)$ is nonnegative and the functions $a(\zeta)$ and $\log a(\zeta)$ are absolutely integrable. It was proved by G. Baxter [2] on the assumption that the functions $a(\zeta)$ and $\log a(\zeta)$ belong to a Wiener algebra with weight.

Theorems concerning projection methods for solving Wiener-Hopf equations with vanishing symbol have been proved by Gohberg and Levčenko [1, 2, 3].

§ 3. The second assertion of Theorem 3.1 was proved in the space L_1 by I. S. Čebotaru **[1]**.

The second assertion of Theorem 3.2, as well as the remark concerning the Galerkin method for the nonzero index case, was established by V. V. Ivanov and E. A. Karagodova **[1]** (cf. also V. V. Ivanov **[2]**).

§ 4. Further development of the theory of multivariate Wiener-Hopf equations is contained in I. B. Simonenko's articles **[4, 5]**.

§ 5. The method cited for calculating the index is due to V. L. Zaguskin and A. V. Haritonov **[1]**.

The Schur-Kohn direct method of determining the index of a polynomial relative to the circle should also be noted. This method is presented in N. N. Meĭman's article **[1]** (cf. also M. G. Kreĭn and M. A. Naĭmark **[1]**).

§ 6. Theorem 6.1, in a different form, under the additional assumption that the polynomials $x(\zeta) = \sum_0^n x_j\zeta^j$ and $y(\zeta) = \sum_0^n y_{-j}\zeta^j$ do not vanish on the unit disk $|\zeta| \leqq 1$, was established by G. Baxter and I. Hirschman **[1]** (cf. also M. Shinbrot's article **[1]**). A. A. Semencul **[1, 2]** established formula (6.3) under the same assumption. A clearer proof of Baxter and Hirschman's theorem also belongs to this author. This proof, which makes use of a transition to infinite-dimensional discrete Wiener-Hopf equations, was cited in the Russian edition of the book.

The results of this section were established in Gohberg and Semencul's article **[2]**.

The determinant of the matrix $\mathscr{P}$ in the formulation of Theorem 6.3 is the *resultant* of the polynomials $y(\zeta) = y_0\zeta^n + y_{-1}\zeta^{n-1} + \cdots + y_{-n}$ and $x(\zeta) = x_n\zeta^n + x_{n-1}\zeta^{n-1} + \cdots + x_0$; thus the condition of the nonsingularity of the matrix $\mathscr{P}$ can be replaced by the following: the polynomials $x(\zeta)$ and $y(\zeta)$ do not have common roots.

For a hermitian Toeplitz matrix $\mathscr{A}_n$ the latter condition is equivalent to the fact that the polynomial $y(\zeta)$ does not have roots on the unit circle and that among its roots there is no pair of roots specularly situated relative to that circle. M. G. Kreĭn **[5]** proves Theorem 6.3, so formulated, by a different method in the case when a Toeplitz matrix $\mathscr{A}_n$ is hermitian.

§ 7. The results of this section were established by I. C. Gohberg and N. Ja. Krupnik **[8]**.

Comparing the results of §§ 6 and 7, we naturally arrive at the following question. Do there exist for every nonsingular Toeplitz matrix $\mathscr{A}_n = \|a_{j-k}\|_{j,k=0}^n$ numbers ν and μ $(0 \leqq \nu < \mu \leqq n)$ such that the matrix $\mathscr{A}_n^{-1}$ is uniquely determined by the solutions $s_0,\cdots, s_n$ and $t_0,\cdots, t_n$ of the equations

$$\sum_{k=0}^n a_{j-k}\, x_k = \delta_{j\nu}, \qquad \sum_{k=0}^n a_{j-k}\, z_k = \delta_{j\mu}, \qquad j = 0, 1,\cdots, n?$$

An affirmative answer to this question would mean that the numbers a_j $(j = 0, \pm 1,\cdots, \pm n)$ are uniquely determined by the equations

$$\sum_{k=0}^{n} a_{j-k} s_k = \delta_{j\nu}, \qquad \sum_{k=0}^{n} a_{j-k} t_k = \delta_{j\mu}, \qquad j = 0, 1, \cdots, n. \tag{$*$}$$

From the example of the matrix

$$\mathscr{A}_3 = \begin{Vmatrix} 0 & 0 & 1 & 1 \\ 0 & 0 & 0 & 1 \\ 1 & 0 & 0 & 0 \\ 1 & 1 & 0 & 0 \end{Vmatrix}$$

it can be ascertained that the latter does not hold; i. e., whatever the numbers ν and μ $(0 \leqq \nu \leqq \mu \leqq 3)$, the numbers a_j $(j = 0, \pm 1, \pm 2, \pm 3)$ are not determined uniquely by the equations $(*)$.

§ 8. It is possible to derive Theorem 8.1 in a different form, under some additional assumptions, from the abstract results of I. Hirschman **[1, 2]**. A. A. Semencul **[1]** established formula (8.4) under these additional assumptions. His proof of this theorem, which is based on the transition to Wiener-Hopf equations, was cited in the Russian edition of the book. The results of § 8 were established by Gohberg and Semencul **[1]**. Theorem 8.2, in a different form for hermitian kernels, was established by M. G. Kreĭn.

The results of §§ 6 and 8 have some points of contiguity with the fundamental research of M. G. Kreĭn **[1 – 3]** on the theory of inverse problems of the spectral analysis of differential equations (cf. also Gohberg and Kreĭn **[5]**). Unfortunately, this connection has remained beyond the limits of the book.

Questions on approximating the spectrum for the operators considered here are studied in L. E. Lerer's paper **[2]**.

Chapter IV

The basic results of this chapter with their detailed proofs are being published here for the first time.

§ 1. Lemma 1.2 is a special case of one of I. B. Simonenko's propositions **[2]**.

§ 2. The results of this section are formulated in I. C. Gohberg's article **[4]**. Part of them were obtained earlier by H. Widom **[1]**. The generalization of the results of this section for h_p spaces (consisting of the sequences of the Fourier coefficients of the functions in H_p) was obtained by Gohberg and Krupnik **[7]**. Recently P. V. Dudučava **[1]** generalized the results of this section to operators defined in l_p $(1 < p < \infty;\ p \neq 2)$ by Toeplitz matrices $\|a_{j-k}\|_{j,k=0}^{\infty}$ consisting of the Fourier coefficients of sectionally Wiener functions.

§ 4. The results of this section were established by Gohberg **[4]**. They have not yet been generalized to the spaces h_p and l_p.

§ 5. Theorem 5.1 is a special case of a theorem of I.B. Simonenko **[2]**. Theorems 5.2 and 5.3 were established by Gohberg **[4]**.

The class of functions M_R was introduced in Simonenko's article [2]. Another equivalent definition of this class is given there. Let us cite it.

A measurable function $f(\zeta)$ belongs to the class M_R if it satisfies the following two conditions:

a) $\operatorname{ess\,inf}_{|\zeta|=1} |f(\zeta)| > 0$ and $\operatorname{ess\,sup}_{|\zeta|=1} |f(\zeta)| < \infty$.

b) A covering of the circle $|\zeta| = 1$ by open arcs exists such that the values of the function $f(\zeta)$ on each of them lie inside some angle less than π, with its vertex at the point $\zeta = 0$.

Chapter V

The presentation in this chapter is conducted in the setting of I. C. Gohberg's article [3]. The general theorems for several classes of pair operators are first established in the abstract case, and then the basic statements concerning various concrete pair systems and their transposes are derived from them.

In contrast to Gohberg's article, here the simple relationship between pair equations and corresponding equations of the first chapter is exploited. This relationship noticeably simplifies the proofs of the necessity of the conditions in the basic theorems.

§ 1. The results of this section are being published here for the first time.

§ 2. Theorem 2.1 was established by Gohberg [3].

§ 3. Theorem 3.1 was obtained by Gohberg [3]. It is a generalization of earlier results of Gohberg and Kreĭn [3].

§ 4. Theorem 4.1 was established by Gohberg and Kreĭn [3].

§ 5. Theorem 5.2 was obtained by Gohberg and Krupnik [1]. There is a generalization of this theorem in articles [2, 3] of the same authors.

§ 6. Well-known propositions concerning singular integral equations (cf. N. I. Mushelišvili [1] and S. G. Mihlin [2]) can be obtained from the results of § 2. Theorem 6.1 was established by I. C. Gohberg [3], I. B. Simonenko [1], and B. V. Hvedelidze [1, 2]. Theorem 6.2 was established by Gohberg and Krupnik [1].

The definition of symbol cited here is due to S. G. Mihlin [2]. Natural considerations lead to the fact that the function $c(\zeta)$ should be regarded as the symbol of the operator $Pc(\zeta)P\,|\,H_p$. This agrees with the definition of the symbol of a Wiener-Hopf operator.

Chapter VI

The basic results of this chapter were established by the authors [1, 3].

§ 2. The last assertions in Theorems 2.1 and 2.4, as well as Theorems 2.2 and 2.3, were established by I. C. Gohberg and V. G. Čeban [1] under certain additional restrictions. Let us note that the conditions of applicability of the projection method analyzed in the case when the functions $a(\zeta)$ and $b(\zeta)$ are piecewise continuous still have not been determined.

§ 4. The sufficiency part of Theorem 4.1 under certain additional restrictions was established for the first time by V. V. Ivanov [**1, 2**].

Some of the results of this chapter have been generalized to the case of singular integral equations along an arbitrary closed contour (cf. Gohberg and Špigel′ [**1, 2**] and Špigel′ [**1, 2**]).

Chapter VII

The authors' results [**5, 6**] are presented in this chapter.

Part of the results of this chapter were obtained by R. G. Douglas and L. A. Coburn simultaneously with the authors. Further development of this chapter's results can be found in Semencul [**3**] and Gohberg and Semencul [**1**].

§ 6. The question concerning the conditions of the applicability of the projection method to pair integral difference equations and their transposes remains open.

Chapter VIII

§ 1. The results of this section generalize certain results of Gohberg and Kreĭn [**2**].

§ 2. Theorem 2.1 was established by Gohberg [**2, 3**]. (A generalization of it can be found in Budjanu and Gohberg [**3, 4**].) This theorem is a generalization of Theorems 2.2 and 2.4 established earlier by Gohberg and Kreĭn [**2**].

§ 3. Theorem 3.1 was established in another form by B. V. Hvedelidze and G. F. Mandžavidze [**1**], and by I. B. Simonenko [**1, 3**].

§ 4. Theorems 4.1 and 4.2 were established by Gohberg, and Theorem 4.3, by the authors [**2, 4**]. Lemma 4.1 is due to N. Ja. Krupnik [**1**].

§ 5. Theorem 5.1 was established by Gohberg [**3**]. It is a generalization of a result of Gohberg and Kreĭn [**2**]. Theorems 5.2, 5.3 and 5.4 were obtained by the authors [**2, 3, 4**]. The second assertion of Theorem 5.2 was proved independently of the authors for the case of the algebra $W_{n \times n}$ by I. I. Hirschman [**3**] (cf. also N. Bowers [**1**]). The generalization to the matrix case of the results of §§ 6– 8 of Chapter III is contained in the same article [**3**] of I. I. Hirschman. There is a generalization of Theorem 5.1 to the case of a piecewise continuous matrix function in Gohberg and Krupnik's article [**4**]. The remaining theorems of this section have not yet found generalizations to the latter case.

A certain modification of the projection method for systems of Wiener-Hopf equations with nonzero partial indices is considered in I. S. Čebotaru's articles [**2, 3**]. The generalization of Theorems 5.1 and 5.3 to the case of systems of discrete Wiener-Hopf equations with operator coefficients was obtained by M. S. Budjanu [**1**].

§ 6. Theorem 6.1 was established by Gohberg and Kreĭn [**2**]. We note that this theorem has not yet found a generalization to the case of systems of integral-difference equations, i.e. equations of the form (1), Chapter VII. Theorems 6.2 and 6.3 were obtained by the authors **[2, 4]**.

I. A. Fel′dman **[1]** has generalized the results of this section to the case of Wiener-Hopf equations with operator kernels. He also considered applications to the radiative transfer equation.

§ 7. The results of this section are being published here for the first time.

§ 9. Results of the authors **[2, 4]** are presented in this section. All of the theorems in this section (except Theorem 9.3) are of a sufficiency character. The question of necessity of the conditions of these theorems remains open.

§ 10. The results of this section were established by Gohberg and Krupnik **[6]**.

Appendix

The results of the Appendix are due to I. A. Fel′dman **[1–3]**.

1. The complete continuity of an operator with kernel depending on the sum of the arguments was established by Gohberg and Kreĭn **[2]**.

2. The equivalence of equation (2.1) and boundary value problem (2.2) is established here analogously to the way it was done by F. D. Gahov and Ju. I. Čerskiĭ **[1]** in the one-dimensional case.

The asymptotic expansions obtained in §§ 3 and 4 are generalizations of M. G. Kreĭn's results **[4]** concerning the asymptotics of the solutions of a homogeneous Wiener-Hopf equation.

The theorems of this Appendix have been generalized by Fel′dman **[4, 5, 6]** to Wiener-Hopf integral equations with operator-valued kernels. Applications of these results to the radiative transfer equation are also obtained in those papers.

Bibliography

F. V. Atkinson

1. *Some aspects of Baxter's functional equation,* J. Math. Anal. Appl. **7** (1963), 1–30. MR **27** #5135.
2. *The normal solubility of linear equations in normed spaces,* Mat. Sb. **28 (70)** (1951), 3–14. (Russian) MR **13**,46.

N. K. Bari

1. *Trigonometric series,* Fizmatgiz, Moscow, 1961; English transl., Macmillan, New York; Pergamon Press, Oxford, 1964. MR **23** #A3411; **30** #1347.

M. A. Barkar' and I. C. Gohberg

1. *On factorization of operators relative to a discrete chain of projections in Banach space,* Mat. Issled. **1** (1966), no. 1, 32–54; English transl., Amer. Math. Soc. Transl. (2) **90** (1970), 81–103. MR **34** #6529.

G. Baxter

1. *An operator identity,* Pacific J. Math **8** (1958), 649–663. MR **21** #2298.
2. *A norm inequality for a "finite-section" Wiener-Hopf equation,* Illinois J. Math. **7** (1963), 97–103. MR **26** #2818.

G. Baxter and I. I. Hirschman, Jr.

1. *An explicit inversion formula for finite-section Wiener-Hopf operators,* Bull. Amer. Math. Soc. **70** (1964), 820–823. MR **30** #414.

N. Bowers, Jr.

1. *Toeplitz forms associated with matrix valued functions,* Thesis, University of Minnesota, Minneapolis, Minn., 1965.

A. Brown and Pearcy

1. *Structure of commutators of operators,* Ann. of Math. (2) **82** (1965), 112–127. MR **31** #2612.

M. S. Budjanu

1. *Solution of certain classes of Wiener-Hopf equations with operator coefficients,* Bul. Akad. Štiince RSS Moldoven. **1966**, no. 4, 18–31. (Russian) MR **34** #8225.

M. S. Budjanu and I. C. Gohberg

1. *The factorization problem in abstract Banach algebras.* I. *Splitting algebras,* Mat. Issled. **2** (1967), no. 2, 25–51; English transl., Amer. Math. Soc. transl. (2) (to appear). MR **37** #5697.
2. *The factorization problem in abstract Banach algebras.* II. *Irreducible algebras,* Mat. Issled. **2** (1967), no. 3, 3–19; English transl., Amer. Math. Soc. Transl. (2) (to appear). MR **37** #5698.
3. *General theorems on the factorization of matrix valued functions.* I. *The fundamental theorem,* Mat. Issled. **3** (1968), no. 2 (8), 87–103; English transl., Amer. Math. Soc. Transl. (2) **102** (1973), 1–14. MR **41** #4246a.
4. *General theorems on the factorization of matrix-valued functions.* II. *Some tests and their consequences,* Mat. Issled. **3** (1968), no. 3 (9), 3–18; English transl., Amer. Math. Soc. Transl. (2) **102** (1973), 15–26. MR 41 #4246b.

H. Cartan

1. *Sur les matrices holomorphes de n variables complexes,* J. Math. Pures Appl. **19** (1940), 1–26. MR **1**, 312.

G. N. Čebotarev
1. *An equation of convolution type of the first kind,* Izv. Vysš. Učebn. Zaved. Matematika **1967**, no. 2 (57), 80–92. (Russian) MR **35** #680.
2. *Normal solvability of the Wiener-Hopf equations in certain singular cases,* Izv. Vysš. Učebn. Zaved. Matematika **1968**, no. 3 (70), 113–118. (Russian) MR **37** #3374.

I. S. Čebotaru
1. *An approximate method of solution of Wiener-Hopf type equations,* Studies in Algebra and Math. Anal., Izdat. "Karta Moldovenjaske", Kishinev, 1965, pp. 79–96. (Russian) MR **34** #6463.
2. *The reduction of systems of Wiener-Hopf equations to systems with vanishing indices,* Bul. Akad. Štiince RSS Moldoven. **1967**, no. 8, 54–66. (Russian) MR **37** #3375.
3. *The projection method of solution of systems of discrete Wiener-Hopf equations,* Mat. Issled. **3** (1968), no. 1 (7), 159–183. (Russian) MR **41** #9018.

L. A. Coburn and R. G. Douglas
1. *Translation operators on the half-line,* Proc. Nat. Acad. Sci. U.S.A. **62** (1969), 1010–1013. MR **43** #985.

R. V. Dudučava
1. *Discrete Wiener-Hopf equations in spaces with a weight,* Soobšč. Akad. Nauk Gruzin. SSSR **67** (1972), no. 1, 17–20. (Russian)

V. B. Dybin
1. *An exceptional case of an integral pair equation of the convolution type,* Dokl. Akad. Nauk SSSR **176** (1967), 251–254 = Soviet Math. Dokl. **8** (1967), 1073–1077. MR **37** # 1917.

V. B. Dybin and N. K. Karapetjanc
1. *Application of the normalization method to a class of infinite systems of linear algebraic equations,* Izv. Vysš. Učebn. Zaved. Matematika **1967**, no. 10 (65), 39–49. (Russian) MR **39** #4643.

I. A. Fel'dman
1. *Asymptotic behavior of solutions of some systems of integral equations,* Dokl. Akad. Nauk SSSR **154** (1964), 57–60 = Soviet Math. Dokl. **5** (1964), 52–56. MR **28** #443.
2. *The asymptotics of solutions of systems of integral equations of Wiener-Hopf type,* Sibirsk. Mat. Ž. **6** (1965), 596–615. (Russian) MR **31** #3808.
3. *Asymptotic behavior of solutions of a system of Wiener-Hopf equations,* Studies in Algebra and Math. Anal., Izdat. "Karta Moldovenjaske", Kishinev, 1965, pp. 147–152. (Russian) MR **34** #1814.
4. *An equation of radiation energy transfer and Wiener-Hopf operator equations,* Funkcional. Anal. i Priložen. **5** (1971), no. 3, 106–108 = Functional Anal. Appl. **5** (1971), 262–264. MR **44** #858.
5. *Wiener-Hopf operator equations and their application to transport equations,* Mat. Issled. **6** (1971), no. 3, 115–132. (Russian)
6. *On an iteration method for the equation of radiant energy transfer,* Dokl. Akad. Nauk SSSR **199** (1971), 36–39 = Soviet Math. Dokl. **12** (1971), 1034–1038.
7. *On some projection methods for the solution of the equations of radiative transfer* Mat. Issled. **7** (1972), no. 4, 228–236. (Russian)

F. D. Gahov and Ju. I. Čerskiĭ
1. *Singular integral equations of convolution type,* Izv. Akad. Nauk SSSR Ser. Mat. **20** (1956), 33–52. (Russian) MR **18**,134.

F. D. Gahov and V. I. Smagina
1. *Exceptional cases of integral equations of convolution type and equations of first kind,* Izv. Akad. Nauk SSSR Ser. Mat. **26** (1962), 361–390. (Russian) MR **30** #1374.

F. R. Gantmaher
1. *Matrix theory,* 2nd ed., "Nauka", Moscow, 1966; English transl. of 1st ed., Chelsea, New York, 1959. MR **21** #6372c; **34** #2585.

I. M. Gel'fand, D. A. Raĭkov and G. E. Šilov
1. *Commutative normed rings,* Fizmatgiz, Moscow, 1960; English transl., Chelsea, New York, 1964. MR **23** #A1242; **34** #4940.

I. C. GOHBERG

1. *Tests for one-sided invertibility of elements in normed rings, and their applications,* Dokl. Akad. Nauk SSSR **145** (1962), 971–974 = Soviet Math. Dokl. **3** (1962), 1119–1123. MR **27** #6147.
2. *A general theorem concerning the factorization of matrix-functions in normed rings,* Dokl. Akad. Nauk SSSR **146** (1962), 284–287 = Soviet Math. Dokl. **3** (1962), 1281–1284. MR **25** #4376.
3. *The factorization problem in normed rings, functions of isometric and symmetric operators, and singular integral equations,* Uspehi Mat. Nauk **19** (1964), no. 1 (115), 71–124 = Russian Math. Surveys **19** (1964), no. 1, 63–114. MR **29** #487.
4. *Toeplitz matrices composed of the Fourier coefficients of piecewise continuous functions,* Funkcional. Anal. i Priložen. **1** (1967), no. 2, 91–92 = Functional Anal. Appl. **1** (1967), 166–167. MR **35** #4763.

I. C. GOHBERG AND V. G. ČEBAN

1. *On a reduction method for discrete analogues of equations of Wiener-Hopf type,* Ukrain. Mat. Ž. **16** (1964), 822–829; English transl., Amer. Math. Soc. Transl. (2) **65** (1967), 41–49. MR **30** #2244.

I. C. GOHBERG AND I. A. FEL'DMAN

1. *Approximate solutions of some classes of linear equations,* Dokl. Akad. Nauk SSSR **160** (1965), 750–753 = Soviet Math. Dokl. **6** (1965), 174–177. MR **34** #6572.
2. *Reduction method for systems of equations of Wiener-Hopf type,* Dokl. Akad. Nauk SSSR **165** (1965), 268–271 = Soviet Math. Dokl. **6** (1965), 1433–1436. MR **32** #8085.
3. *On truncated Wiener-Hopf equations,* Abstracts of Short Scientific Reports, Internat. Congress of Math. (Moscow, 1966), Section 5, pp. 44–45. (Russian)
4. *Projection methods for solving Wiener-Hopf equations,* Acad. Sci. Moldovian SSR. Inst. Math. with Computing Center, Akad. Nauk Moldov. SSR, Kishinev, 1967. (Russian) MR **37** #1915.
5. *On Wiener-Hopf integral-difference equations,* Dokl. Akad. Nauk SSSR **183** (1968), 25–28 = Soviet Math. Dokl. **9** (1968), 1312–1316.
6. *Integro-difference Wiener-Hopf equations,* Acta Sci. Math. (Szeged) **30** (1969), 199–224. (Russian) MR **40** #7880.

I. C. GOHBERG AND M. G. KREĬN

1. *The basic propositions on defect numbers, root numbers and indices of linear operators,* Uspehi Mat. Nauk **12** (1957), no. 2 (74), 43–118; English transl., Amer. Math. Soc. Transl. (2) **13** (1960), 185–264. MR **20** # 3459; **22** #3984.
2. *Systems of integral equations on a half line with kernels depending on the difference of arguments,* Uspehi Mat. Nauk **13** (1958), no. 2 (80), 3–72; English transl., Amer. Math. Soc. transl. (2) **14** (1960), 217–287. MR **21** #1506; **22** #3954.
3. *On a dual integral equation and its transpose.* I, Teoret. Prikl. Mat. Vyp. 1 (1958), 58–81. (Russian) MR **35** #5877.
4. *Introduction to the theory of linear nonselfadjoint operators in Hilbert space,* "Nauka", Moscow, 1965; English transl., Transl. Math. Monographs, vol. 18, Amer. Math. Soc., Providence, R.I., 1969. MR **36** #3137.
5. *Theory and applications of Volterra operators in Hilbert space,* "Nauka", Moscow, 1967; English transl., Transl. Math. Monographs, vol. 24, Amer. Math. Soc., Providence, R.I., 1970. MR **36** #2007.

I. C. GOHBERG AND JA. KRUPNIK

1. *The spectrum of one-dimensional singular integral operators with piece-wise continuous coefficients,* Mat. Issled. **3** (1968) no. 1 (7), 16–30; English transl., Amer. Math. Soc. Transl. (2) 103 (1973), 181–193. MR **41** #2469.
2. *The spectrum of singular integral operators in L_p spaces,* Studia Math. **31** (1968), 347–362. (Russian) MR **38** #5068.
3. *On the spectrum of singular integral operators in L_p spaces with weight,* Dokl. Akad. Nauk SSSR **185** (1969), 745–748 = Soviet Math. Dokl. **10** (1969), 406–410. MR **40** #1817.
4. *Systems of singular integral equations in L_p spaces with a weight,* Dokl. Akad. Nauk SSSR

186 (1969), 998–1001 = Soviet Math. Dokl. **10** (1969), 688–691. MR **40** #1818.

5. *The algebra generated by the one-dimensional singular integral operators with piecewise continuous coefficients*, Funkcional. Anal. i Priložen. **4** (1970), no. 3, 26–36 = Functional Anal. Appl. **4** (1970), 193–201. MR **42** #5057.
6. *On complex linear singular integral equations*, Mat. Issled. **4** (1969), no. 4, 20–32; English transl., Amer. Math. Soc. Transl. (2) MR **43** #996.
7. *On the algebra generated by Toeplitz matrices in h_p spaces*, Mat. Issled **4** (1969), no. 3, 54–62. (Russian)
8. *A formula for the inversion of finite Toeplitz matrices*, Mat. Issled. 7 (1972), no. 2, 274–283. (Russian)

I. C. Gohberg and V. I. Levčenko

1. *On the convergence of a projection method of solution of a degenerate discrete Wiener-Hopf equation*, Mat. Issled. **6** (1971), no. 4, 20–36. (Russian)
2. *Projection methods for the solution of degenerate Wiener-Hopf equations*, Funkcional. Anal. i Priložen. **5** (1971), no. 4, 69–70. (Russian)
3. *On a projection method for a degenerate discrete Wiener-Hopf equation*, Mat. Issled. 7 (1972), no. 3, 238–253. (Russion)

I. C. Gohberg and A. A. Semencul

1. *Toeplitz matrices composed of the Fourier coefficients of functions with discontinuities of almost-periodic type*, Mat. Issled. **5** (1970), no. 4, 63–83. (Russian)
2. *On the inversion of finite Toeplitz matrices and their continuous analogs*, Mat. Issled. 7 (1972), no. 2, 201–223. (Russian)

I. C. Gohberg and E. M. Špigel′

1. *A projection method for the solution of singular integral equations*, Dokl. Akad. Nauk SSSR **196** (1971), 1002–1005 = Soviet Math. Dokl. **12** (1971), 289–293. MR **43** #3755.
2. *On the projection method of solution of singular integral equations with polynomial coefficients*, Mat. Issled. **6** (1971), no. 3, 45–61. (Russian)

I. C. Gohberg, A. S. Markus and I. A. Fel′dman

1. *Normally solvable operators and ideals associated with them*, Bul. Akad. Štiince RSS Moldoven. **1960**, no. 10 (76), 51–70; English transl., Amer. Math. Soc. Transl. (2) **61** (1967), 63–84. MR **36** #2004.

L. S. Gol′denšteĭn

1. *Tests for one-sided inverses of functions of several isometric operators and their applications*, Dokl. Akad. Nauk SSSR **155** (1964), 28–31 = Soviet Math. Dokl. **5** (1964), 330–334. MR **28** #4403.
2. *A discrete analog of the multidimensional Wiener-Hopf integral equation*, Mat. Issled. **2** (1967), no. 3, 52–63. (Russian) MR **37** #3376.

L. S. Gol′denšteĭn and I. C. Gohberg

1. *On a multidimensional integral equation on a half-space whose kernel is a function of the difference of the arguments, and on a discrete analogue of this equation*, Dokl. Akad. Nauk SSSR **131** (1960), 9–12 = Soviet Math. Dokl. **1** (1960), 173–176. MR **22** #8298.

M. I. Haĭkin

1. *An integral equation of convolution type of the first kind*, Izv. Vysš. Učebn. Zaved. Matematika **1967**, no. 3 (58), 105–116. (Russian) MR **35** #2098.
2. *On the regularization of operators with open range*, Izv. Vysš. Učebn. Zaved. Matematika **1970**, no. 8, 118–123. (Russian)

E. Hille and R. S. Phillips

1. *Functional analysis and semi-groups*, rev. ed., Amer. Math. Soc. Colloq. Publ., vol. 31, Amer. Math. Soc., Providence, R.I., 1957; Russian transl., IL, Moscow, 1962. MR **19**, 664.

I. I. Hirschman, Jr.

1. *Finite section Wiener-Hopf equations on a compact group with ordered dual*, Bull. Amer. Math. Soc. **70** (1964), 508–510. MR **29** #490.
2. *Szegö functions on a locally compact Abelian group with ordered dual*, Trans. Amer. Math. Soc. **121** (1966), 133–159. MR **32** #8042.
3. *Matrix-valued Toeplitz operators*, Duke Math. J. **34** (1967), 403–415. MR **36** #3071.

K. Hoffman

1. *Banach spaces of analytic functions*, Prentice-Hall Series in Modern Analysis, Prentice-Hall, Englewood Cliffs, N.J., 1962; Russian transl., IL, Moscow, 1963. MR **24** #A2844.

B. V. Hvedelidze

1. *Linear discontinuous boundary problems in the theory of functions, singular integral equations and some of their applications*, Akad. Nauk Gruzin. SSR. Trudy Tbiliss. Mat. Inst. Razmadze **23** (1956), 3–158. (Russian) MR **21** #5873.
2. *A remark on my work "Linear discontinuous boundary problems in the theory of functions, singular integral equations and some of their applications"*, Soobšč. Akad. Nauk Gruzin. SSR **21** (1958), 129–130. (Russian) MR **21** #5874.

V. V. Ivanov

1. *The use of the method of moments and the "mixed" method for an approximate solution of singular integral equations*, Dokl. Akad. Nauk SSSR **114** (1957), 945–948. (Russian) MR **21** #2883.
2. *The theory of approximation methods and its application to the numerical solution of singular integral equations*, "Naukova Dumka", Kiev, 1968 (Russian)

V. V. Ivanov and E. A. Karagodova

1. *Approximate solution of singular integral equations of convolution type by Galerkin's method*, Ukrain. Mat. Ž. **13** (1961), no. 1, 28–38. (Russian) MR **25** #1415.

L. V. Kantorovič and G. P. Akilov

1. *Functional analysis in normed spaces*, Fizmatgiz, Moscow, 1959; English transl., Internat. Series of Monographs in Pure and Appl. Math., vol. 46, Macmillan, New York, 1964. MR **22** #9837; **35** #4699.

S. Karlin

1. *Bases in Banach spaces*, Duke Math. J. **15** (1948), 971–985. MR **10**, 548.

M. G. Kreĭn

1. *On integral equations generating differential equations of 2nd order*, Dokl. Akad. Nauk SSSR **97** (1954), 21–24. (Russian) MR **16**, 372; **17**, 1436.
2. *Continuous analogues of propositions on polynomials orthogonal on the unit circle*, Dokl. Akad. Nauk SSSR **105** (1955), 637–640. (Russian) MR **18**, 291.
3. *On the theory of accelerants and S-matrices of canonical differential systems*, Dokl. Akad. Nauk SSSR **111** (1956), 1167–1170. (Russian) MR **19**, 277.
4. *Integral equations on a half-line with kernel depending upon the difference of the arguments*, Uspehi Mat. Nauk **13** (1958), no. 5 (83), 3–120; English transl., Amer. Math. Soc. Transl. (2) **22** (1962), 163–288. MR **21** #1507.
5. *Distribution of roots of polynomials orthogonal on the unit circle with respect to a sign-alternating weight*, Teor. Funkciĭ Funkcional. Anal. i Priložen. Vyp. 2 (1966), 131–137. (Russian) MR **34** #1584.

M. G. Kreĭn and M. A. Naĭmark

1. *The method of symmetric and Hermitian forms in the theory of isolated solutions of integral equations*, ONTI, DNVTU, Kharkov, 1936. (Russian)

N. Ja. Krupnik

1. *On the question of normal solvability and the index of singular integral equations*, Kišinev. Gos. Univ. Učen. Zap. **82** (1965), 3–7. (Russian) MR **34** #4950.

L. E. Lerer

1. *On a class of perturbations for operators admitting reductions*, Mat. Issled. **6** (1971), no. 1, 168–173. (Russian)
2. *On the asymptotic distribution of the spectrum*, Mat. Issled. **7** (1972), no. 4, 141–146. (Russian)

B. M. Levitan

1. *Almost-periodic functions*, GITTL, Moscow, 1953. (Russian) MR **15**, 700.

G. F. Mandžavidze

1. *Approximate solution of boundary problems of the theory of analytic functions*, Fizmatgiz, Moscow, 1960. (Russian) MR **22** #6890.

G. F. Mandžavidze and B. V. Hvedelidze

1. *On the Riemann-Privalov problem with continuous coefficients*, Dokl. Akad. Nauk SSSR **123** (1958), 791–794. (Russian) MR **22** #3806.

A. S. Markus

1. *The reduction method for operators in Hilbert space*, Mat. Issled. **4** (1969), no. 1 (11), 71–79; English transl., Amer. Math. Soc. Transl. (2) **103** (1973), 194–200. MR 40 #7782.

P. Masani

1. *The Laurent factorization of operator-valued functions*, Proc. London Math. Soc. (3) **6** (1956), 59–69. MR **18**, 138.

N. N. Meĭman

1. *Some problems on the distribution of the zeros of polynomials*, Uspehi Mat. Nauk **4** (1949), no. 6 (34), 154–188. (Russian) MR **11**, 661.

S. G. Mihlin

1. *Variational methods in mathematical physics*, 2nd ed., "Nauka", Moscow, 1971; English transl. of 1st ed., Macmillan, New York, 1964. MR **30** #2712.
2. *Singular integral equations*, Uspehi Mat. Nauk 3 (1948), no. 3 (25), 29–112; English transl., Amer. Math. Soc. Transl. (1) **10** (1962), 84–198. MR **10**, 305.

N. I. Mushelišvili

1. *Singular integral equations. Boundary problems of function theory and their application to mathematical physics*, 3rd ed., "Nauka", Moscow, 1967; English transl. of 1st ed., Noordhoff, Groningen, 1953. MR **15**, 434.

N. I. Pol'skiĭ

1. *Projection methods for the solution of linear problems*, Uspehi Mat. Nauk **18** (1963), no. 2 (110), 179–180. (Russian)

Z. Presdorf

1. *Singular integral equations with symbol vanishing at a finite number of points*, Mat. Issled. **7** (1972), no. 1, 116–132. (Russian)
2. *On systems of singular integral equations with vanishing symbol*, Mat. Issled. **7** (1972), no. 2, 129–142. (Russian)

F. Riesz and B. Sz.-Nagy

1. *Leçons d'analyse fonctionnelle*, 2nd ed., Akad. Kiadō, Budapest, 1953; Russian transl., IL, Moscow, 1954; English transl., Ungar, New York, 1955. MR **15**, 132; **17**, 175.

A. A. Semencul

1. *Inversion of finite Toeplitz matrices and their continuous analogs*, Appendix II to Gohberg and Fel'dman [4], pp. 140–156. (Russian)
2. *The inversion of finite sections of paired operators and their transposes*, Mat. Issled. **3** (1968), no. 1 (7), 100–107. (Russian) MR **41** #9025.
3. *On singular integral equations whose coefficients have discontinuities of almost-periodic type*, Mat. Issled. **6** (1971), no. 3, 92–114. (Russian)

M. Shinbrot

1. *A class of difference kernels*, Proc. Amer. Math. Soc. **13** (1962), 399–406. MR **25** #1414.

G. E. Šilov

1. *Letter to the editor: On locally analytic functions*, Uspehi Mat. Nauk **21** (1966), no. 6 (132) 177–182. (Russian) MR **36** #1696.

I. B. Simonenko

1. *The Riemann boundary value problem for n pairs of functions with continuous coefficients*, Izv. Vysš. Učebn. Zaved. Matematika **1961**, no. 1 (20), 140–145. (Russian) MR **24** #A838.
2. *The Riemann boundary value problem for n pairs of functions with measurable coefficients and its application to the study of singular integrals in L_p spaces with weights*, Izv. Akad. Nauk SSSR Ser. Mat. **28** (1964), 277–306. (Russian) MR **29** #253.
3. *Some general questions on the theory of the Riemann boundary problem*, Izv. Akad. Nauk SSSR Ser. Mat. **32** (1968), 1138–1146 = Math. USSR Izv. **2** (1968), 1091–1100. MR **38** #3447.

4. *Operators of convolution type in cones,* Mat. Sb. **74 (116)** (1967), 298–313 = Math. USSR Sb. **3** (1967), 279–294. MR **36** #5773.
5. *Multidimensional discrete convolutions,* Mat. Issled. **3** (1968), no. 1 (7), 108–122. (Russian) MR **41** #2412.

E. M. Špigel'
1. *On a projection method for solution of singular integral equations with rational coefficients,* Mat. Issled. **7** (1972), no. 1, 163–185. (Russian)
2. *A projection method for solving singular integral equations with rational coefficients along a multiply-connected contour,* Mat. Issled. **7** (1972), no. 2, 181–200. (Russian)

E. C. Titchmarsh
1. *Introduction to the theory of Fourier integrals,* Clarendon Press, Oxford, 1937; Russian transl., GITTL, Moscow, 1948.

G. M. Vaĭnikko
1. *On the question of convergence of Galerkin's method,* Tartu Riikl. Ül. Toimetised Vih. 177 (1965), 148–153. (Russian) MR **36** #1094.

H. Widom
1. *On the spectrum of a Toeplitz operator,* Pacific J. Math. **14** (1964), 365–375. MR **29** #476.

N. Wiener and E. Hopf
1. *Über eine Klasse singulärer Integralgleichungen,* S.-B. Akad. Wiss. Berlin **1931,** 696–706.

B. Yood
1. *Properties of linear transformations preserved under addition of a completely continuous transformation,* Duke Math. J. **18** (1951), 599–612. MR **13,** 355.

V. L. Zaguskin and A. V. Haritonov
1. *An iteration method for solving a stability problem,* Ž. Vyčisl. Mat. i Mat. Fiz. **3** (1963), 361–364 = USSR Comput. Math. and Math. Phys. **3** (1963), 474–479. MR **27** #5361.